Proceedings of the Steklov Institute of Mathematics

edited by

S. M. Nikol'skiĭ

Number 119 (1973)

Problems in the theory of point explosion in gases

by

V. P. Korobeĭnikov

Providence, Rhode Island

AMERICAN MATHEMATICAL SOCIETY

1976

Академия Наук
Союза Советских Социалистических Республик

ТРУДЫ
ордена Ленина
МАТЕМАТИЧЕСКОГО ИНСТИТУТА
имени В. А. СТЕКЛОВА
CXIX

Ответственный редактор
академик С. М. НИКОЛЬСКИЙ

ЗАДАЧИ ТЕОРИИ ТОЧЕЧНОГО ВЗРЫВА В ГАЗАХ

В. П. КОРОБЕЙНИКОВ

Издательство "Наука"
Москва 1973

Translated by J. Schulenberger

AMS (MOS) subject classifications (1970). Primary 76L05; Secondary 76W05.

ABSTRACT. The theory of point explosion arose from the necessity of describing process of propagation of expolosions from concentrated charges in continuous media.

The monograph is devoted to the development of the theory of point explosion in gases. It contains formulations of new problems; theoretical models of the motion of the medium are constructed which take account of various physical phenomena and properties of the medium; methods of solving the new problems which arise are developed; a detailed study of problems of the theory which have previously been formulated is presented. Certain applications to physical problems are considered.

Library of Congress Cataloging in Publication Data

Korobeĭnikov, Viktor Pavlovich.
Problems in the theory of point explosion in gases.

(Proceedings of the Steklov Institute of Mathematics ; no. 119)
Translation of Zadachi teorii tochechnogo vzryva v gazakh.
Bibliography: p.
1. Gas dynamics. 2. Exploding wire phenomena.
I. Title. II. Series: Akademii͡a nauk SSSR. Matematicheskiĭ institut. Proceedings ; no. 119.
QA1.A413 no. 119 [QC168] 510'.8s [533'.2] 75-45104
ISBN 0-8218-3019-8

Translation authorized by Meždunarodnaja Kniga, Moscow

**PROCEEDINGS OF THE STEKLOV INSTITUTE OF MATHEMATICS
IN THE ACADEMY OF SCIENCES OF THE USSR**

(Труды математического института В. А. Стеклова, т. CXIX, 1973)

TABLE OF CONTENTS

INTRODUCTION

The present investigation is devoted to the study of questions in the theory of point explosion in gases which arose in connection with the necessity of describing phenomena taking place in continuous media on explosion of charges of small volume and weight but having large specific energy. The results of this theory make it possible to obtain with sufficient accuracy many necessary data on the character of the transient motion and the propagation of shock waves occurring during explosions. The theory of point explosions in gases has found application to such phenomena as nuclear explosions in the atmosphere, processes of explosion in cosmic space, the motion of slender, blunt bodies (including meteorites) in the earth's atmosphere, the propagation of shock waves caused by powerful laser radiation and electric discharges in a gas, and electric explosion of conductors. This theory was developed about 25 years ago in the work of L. I. Sedov and also G. I. Taylor, K. P. Stanjukovič, and J. von Neumann. Many problems in this theory have been solved in the work of a number of Soviet and foreign scientists.

We shall discuss briefly the essential features of a concentrated explosion. Suppose that in a volume which is sufficiently small as compared with the volume of the surrounding medium there is a concentrated a small mass of explosive substance and that at some instant of time an explosion occurs which is accompanied by a rapid release of energy, where the density of the energy released (the amount of energy per unit volume) is much greater than the energy density of the surrounding medium. Here an almost instantaneous increase in the pressure and temperature of the medium occurs in a neighborhood of the explosion, and a strong shock wave arises which is damped out as it propagates away from the site of the explosion. Consideration of all the factors affecting the propagation of the blast wave under real conditions makes the problem of theoretical description of the explosion phenomenon very complicated. Here, as in a number of other complex phenomena, for successful solution of the problem a certain idealization is required which consists in considering only those factors which are dominant in the development of the total phenomenon. In questions related to explosions the basic idealization is the assumption that the process of energy release takes place instantaneously and the volume occupied by the explosive substance and the mass of the charge are equal to zero; in the case of spherical symmetry the explosion occurs at a point; for cylindrical symmetry, along a line; and for plane symmetry, over a plane. This idealized process of energy release is called a point explosion. In addition, we construct various idealized schemes of the evolution of the explosion process and various models for the motion of the medium.

In this work we study known models for the motion of the gas during point explosion, and also new models which take various physical processes into account; the mathematical problems for these models are formulated, and methods for their solution are studied. The questions considered here are mainly those which have been treated by the author or with his participation. The book consists of eight chapters and an appendix. The bibliography is compiled separately for each chapter. It contains, as a rule, only works which are closely related to our investigations or results which we use in studying various questions.

Chapter 1 is mainly of an auxiliary nature. Here the well-known basic equations of hydrodynamics and magnetohydrodynamics are presented, as are also conditions on the shock waves which follow from the conservation laws; conditions on the magnetohydrodynamic waves are included (§§1–4). Certain properties of these equations are indicated. In §§5 and 6 the necessary background on the theory of dimensions and group theory is given, and the concepts of self-similar and invariant solutions of the hydrodynamic and magnetohydrodynamic equations are introduced. A derivation of the integrals of adiabatic motions for self-similar solutions in one-dimensional hydrodynamics and magnetohydrodynamics is also given. These integrals are used in the sequel for the study of specific problems.

§7 is devoted to formulations and examples of solutions to problems for certain equations of mathematical physics; these problems are related to determining perturbations caused by an instantaneous local release of energy. Here the concept of a point explosion is introduced, and exact solutions of the point explosion problem are given for the heat equation, the wave equation, and the equations for the spherically symmetric motions of an incompressible fluid. The problem for the equations of gas dynamics is formulated here, and certain properties of the solutions and the behavior of the moving gas are discussed.

In Chapter 2 a strong explosion in a perfect gas is studied, certain questions of accounting for real physical processes in gases at high temperatures are considered, and the main applications of the theory of a strong explosion to physical phenomena are briefly discussed.

In §1 L. I. Sedov's well-known solution of the self-similar problem of strong explosion in a gas with variable initial density $\rho_1 = Ar^{-\omega}$ is presented. Certain analytic properties of this solution are investigated in §2 of this chapter. We mainly consider the case of constant density $(\omega = 0)$. Here justification is given for the representation of the solution in a neighborhood of the center in the form of special power series in the dimensionless coordinate, and a theorem on the continuous dependence of the solution on the adiabatic exponent γ is proved. The actual variation of the solution for γ varying in the range $1.1 \leqslant \gamma \leqslant 7$ is clarified by computations. In particular, the energy parameter α, which is of importance for applications and characterizes the dimensionless total energy of the moving gas, is computed for this range of γ. The $\alpha(\gamma)$ dependence is approximated by simple analytic formulas with relative error not exceeding 2%.

For $\omega \neq 0$ the behavior of the solutions is studied in a neighborhood of singular values of ω where the analytic expression for the solution is indeterminate.

In §3 of this chapter the concept of homothermic motions is introduced, and the solution of the problem of strong explosion is considered for the model of homothermic motions for spherical as well as for cylindrical and plane shock waves.

In the remaining sections of this chapter explosion problems are formulated which take account of equilibrium ionization and dissociation as well as thermal conductivity. Applications of the theory to phenomena of nuclear explosion, discharges in gases, explosion of conductors under the action of a current, and focusing of a laser beam are considered.

In Chapter 3 we consider a method of linearization by means of which it is possible to take into account corrections to the self-similar solution stipulated by various physical parameters of the medium. This method of linearization consists essentially in the following.

Suppose that f_{i0} are functions of the self-similar solution. We seek another solution in the form $f_i = f_{i0} + \sigma f_{i1}$, where the index 1 denotes the new unknown functions (corrections to the base solution or perturbations) and σ is some small parameter. If we then substitute the solution f_i and the boundary conditions into the system of equations of motion and observe that f_{i0} is a solution of these equations, then as $\sigma \to 0$ we obtain linear differential equations (partial differential equations, in general) for the functions f_{i1}, i.e. a linearized system. The applicability of this method is completely justified when the functions f_{i1} and their derivatives are finite. Here the following one-dimensional problems are solved: the problem of constant counterpressure in the medium with variable and constant initial density, the case of an initial velocity for the gas motion, and the problem involving variable counterpressure for an isothermic, spherically symmetric atmosphere. The investigation of solutions of these problems is very useful in studying questions of the dependence of a solution of the problem of point explosions on constants and variable parameters entering in the formulation of the problem. Thus, for example, a detailed investigation of the point explosion problem taking constant counterpressure into account initiated by N. S. Mel′nikova and A. Sakurai showed that the solution for various γ depends continuously on the parameter p_1, the constant initial pressure. Computations were carried out for this problem and special tables of functions characterizing the effect of the counterpressure p_1 on the gas flow have been published.

The method of linearization makes it possible to consider at the same time two or more parameters which destroy self-similarity. Thus it is possible to consider the effect of the velocity of motion in the initial state of the gas and the effect of counterpressure at the same time.

We point out also that in this chapter a new point explosion problem in a moving gas is formulated. Applications of this problem are considered in §4 and also in Chapter 8.

Chapter 4 contains a qualitative and quantitative study of questions of the propagation of shock waves and gas motion following explosion at a point, along a line, or over a plane in a homogeneous gas with constant counterpressure p_1. Here a problem which is important in both its theoretical and its applied aspects is considered. The first two sections are devoted to an investigation of the solution near the center of symmetry and to the asymptotic damping laws for shock waves at large distances from the explosion site. In §3 a numerical solution of the point explosion problem involving counterpressure for plane, cylindrical, and spherical shock waves and various values of γ is given. This solution was obtained jointly with P. I. Čuškin. The method of solution is based on applying the scheme of integral relations used earlier in other hydrodynamical problems by A. A. Dorodnicyn, O. M. Belocerkovskiĭ, P. I. Čuškin, and others. The system of equations for transient one-dimensional motions is taken in divergence form and is reduced to dimensionless form by using the independent variables $\xi = (r/r_n)^\nu$, $q = a_\infty^2/D^2$, where r is the linear coordinate, a_∞ is the speed of sound in the unperturbed gas, r_n and D give the coordinate and velocity of the shock wave, and ν is the symmetry parameter. Dimensionless quantities related to the velocity, density, and pressure are taken as the basic unknown functions.

Near the center of the explosion ($\xi = 0$) an interval is distinguished which is bounded by a line $\xi_0(q)$ corresponding to a fixed Lagrange coordinate. Asymptotic formulas for the functions are used inside this central interval. The differential equations of gas dynamics are replaced by a system of integral relations. The integrands of these relations in the region between ξ_0 and $\xi_n = 1$ are represented in terms of interpolation polynomials in ξ. Here n is the number of regions between ξ_0 and ξ_n for which the integral relations are written. The coefficients of these polynomials are determined by means of numerical integration of an approximating system of ordinary differential equations in q obtained from the integral relations of the problem. The integration of the approximating system of equations in the variable q, which is related to time, is taken from a certain small value $q = q_0$, where a linearized solution taking into account the counterpressure is chosen, to values of q close to unity. A method is indicated for computing all the unknown functions including the pressure impulses and the velocity head for a sequence of values of the spatial coordinate. The computations are carried out for various values of n, γ and ν. The main computations are made for the values $n = 4$ (the region $\xi_0 \leqslant \xi \leqslant 1$ is partitioned into four strips) and $n = 8$ (the region of integration is partitioned into eight strips) with various adiabatic exponents γ. The accuracy of the numerical solution is controlled by satisfying the integral conservation laws of mass and energy, determining the behavior of the solution for various n, comparison of the values of the unknown functions for the same ξ and q obtained from different systems of initial integral relations, and comparison with the results of other authors. On the basis of the results of the computations for the approximation $n = 8$, tables are made of the fields of the basic functions $(0.05 \leqslant q \leqslant 0.9)$ for plane, cylindrical, and spherical explosions and various γ.

In the following section the parameters of the front of the shock waves for $\gamma = 1.4$ are considered separately, and some conclusions regarding practical application of the solution are given.

Effective approximate formulas for computing the parameters of the wave front are given in §5 of Chapter 4. These formulas have a rather high accuracy and are convenient for applications. The derivation of these formulas is based on the approximation of the dependence of the speed of gas particles behind the front on the coordinate of the wave by using the analytic relations for the strong stage of the explosion and the asymptotic damping laws as the shock wave degenerates into a sound wave. §6 is devoted to questions of converting the solution when applying different systems of coordinates and similarity laws.

The next section of this chapter (§7) is devoted to the analogy between an explosion and hypersonic flow around slender, blunt bodies. Here data are given comparing results of determining the parameters of stationary flow found by direct computation of the problem and by conversion on the basis of the explosion analogy. A conclusion is drawn regarding the possibility of applying the reverse analogy to determine the parameters of cylindrical and plane blast waves on the basis of results of computing the flow around blunt plates and cylinders.

In the next two sections we consider questions of using the solutions obtained for the calculation of an explosion in a stationary, homogeneous gas flow and with reflection from an absolutely hard wall. Formulas are given for determining the flow parameters in the initial stage of regular reflection of shock waves of a point explosion from a stationary plane.

Chapter 5 is devoted to questions of a concentrated explosion in inhomogeneous media and explosion with nonsymmetric energy release. In §1 of this chapter we give an example of an exact solution of a nonlinear point explosion problem in a medium with variable density, taking the counterpressure into account. Formulas are presented which describe the field of gas flow and the variation of the parameters of the wave front.

In §2 we study questions of the energy distribution between half-spaces occupied by gases with different properties for an explosion on the boundary separating the two media. We consider special features of the formulations of these problems.

§3 is devoted mainly to an approximate method of determining the shape and parameters of the front of shock waves in an inhomogeneous atmosphere following a point explosion. The method considered here is essentially a generalization of our method for constructing approximate formulas for shock waves presented in §5 of

Chapter 4. Here a dimensional analysis is given and a dimensionless parameter $h = H/r^0$, $r^0 = (E_0/p_0)^\nu$, is introduced, where H is a characteristic dimension of the inhomogeneity (the height of the standard atmosphere in the isothermic case) and p_0 is the pressure at the point of explosion. The velocity behind the wave front, $v_2(r_2, \theta)$, is approximated; here r_2, θ are the spherical coordinates of the front. The form of the shock wave $r_2(t, \theta)$ is found from the solution of a Cauchy problem for a non-linear partial differential equation of first order defining the function $r_2(t, \theta)$. The initial data for $r_2(t, \theta)$ are obtained from the solution of the linearized problem due to V. P. Karlikov.

In the last section of the chapter some ideas are presented for using the principle of plane sections in the computation of a concentrated explosion along slightly curved lines and surfaces, and also for the case in which the specific energy varies over the line or plane of the explosion.

In recent years phenomena of a concentrated explosion in media where the propagation of the shock wave induces chemical reactions with energy release have come to be studied. An investigation of the problems arising here is the topic of Chapter 6. Here there are given formulations of problems on plane, cylindrical, and spherical point explosions in a combustible mixture of gases for various models of the course of chemical reactions. The general qualitative picture of the motion of the gas is briefly described, and those characteristic values of the coordinates of the shock wave are determined starting from which the energy of combustion of the combustible mixture affects the process of gas motion. Methods are worked out for solving the mathematical problems which arise in studying the phenomena in question. The following flow models are considered: 1) the model of a detonation wave; 2) the model of including an exothermic reaction directly behind the front of the shock wave with simple kinetics in which the rate of combustion of the mixture is given by the simple law of Arrhenius; 3) the model of two fronts (the shock wave and the flame front); 4) the model of including exothermic reactions behind the front of the shock wave after a time lag for ignition, while in the kinetics of the model both the direct and reverse course of the reactions are considered. For the model of a detonation wave (§2) a similarity law is formulated, a solution is given of the linearized problem taking into account the evolution of heat on the wave front, a method is developed for the numerical solution of the non-self-similar problem on a strong detonation wave analogous to the method applied in §3 of Chapter 4, and results are presented of the numerical solution of the point explosion problem for plane, cylindrical and spherical waves. Since the detonation wave formed as a consequence of the explosion is overcompressed, the question of its passage to the Chapman-Jouguet regime is considered, and an estimate is given of the distances for which the parameters of the overcompressed wave will become close to the parameters of the Chapman-Jouguet wave.

The last section of this chapter is devoted to a discussion of questions of the effect of the kinetics of chemical reactions on the picture of the gas flow. On the basis of dimensional analysis, conditions are obtained for self-similarity of solutions to the problems, taking into account the kinetics of chemical reactions. For model 2 the self-similar problem is considered for the case of an explosion in a medium with constant density. Using the flow model with an induction zone, one can draw a conclusion regarding possible damping of the combustion process near the front of the shock wave as it becomes weaker with increasing time. For the case of a strong explosion it is shown that at the initial instants of time the front of combustion reactions practically coincides with the shock wave, and there is essentially an overcompressed detonation wave. Computations made for the problem indicate that with increase in time the detonation wave decomposes into an ordinary shock wave with a flame front following it. Thus the uncoupling of a detonation wave is theoretically justified (this effect was observed in experiments of R. I. Solouhin, L. G. Gvosdeva, and others). It is also shown that the reverse reaction has an essential effect on the quantity of total heat evolution in the gas flow in the zone of the chemical reactions.

In connection with the development of magnetohydrodynamics and its applications, the question arose of studying the propagation of magnetohydrodynamic shock waves. The theory of point explosion in an electrically conducting gas with applied magnetic and electric fields has been developed in a number of our papers.(1) Chapter 7 of the present book is devoted to developing the foundations of this theory. Here formulations of point explosion problems are given for the equations of magnetohydrodynamics with various configurations of the magnetic fields both for infinite and for finite conductivity of the gas. Assuming an infinite conductivity (§2), a numerical solution is given of the problem on cylindrical and plane explosions with an applied magnetic field parallel to the axis or plane of the explosion. It is found that pressure gradients exist near the center of the explosion, and the velocities of the shock waves increase as compared with the case where the field is absent. A self-similar problem on a strong cylindrical explosion with an annular initial field is investigated, and its generalization to the case of the equations of magnetohydrodynamics for a rarefied plasma is given. In §3 approximate relations are found for the parameters of plane shock waves with an initial field inclined to the front of the plane wave.

In §§4 and 5 of this chapter two-dimensional and spatial transient problems are considered. For the case of an explosion at a point with a constant initial field a method of linearization is developed which considers the effect of the magnetic fields on the motion of the gas and on the shape of the wave. It is found that the shock wave will be elongated in directions perpendicular to the initial field. The magnetic fields drop to zero in a narrow layer of gas behind the front of a strong shock wave, and there is

---

(1) A survey of the basic results is given in my paper, *Gas dynamics of explosions*, Ann. Rev. Fluid Mech. **3** (1971), 317–346.

a peculiar current layer. An exact solution of the problem is given for arbitrary weak fields.

Problems for the case of finite conductivity are investigated in §6. Here we consider separately the case of a weak magnetic field and arbitrary magnetic Reynolds numbers R_m and the case of arbitrary fields but small numbers R_m. The structure of the magnetic field on passage of the front of a strong shock wave is determined for weak fields. Perturbation of the magnetic field in advance of the wave front is considered. For small numbers R_m a class of self-similar solutions is found and investigated.

In the last sections of Chapter 7 we describe problems related to determining electromagnetic waves excited by magnetohydrodynamic shock waves when there is a conductivity jump. For plane waves it is possible to construct exact solutions which take into account the radiated electromagnetic waves. Some solutions are considered for the spatial case as well.

The last chapter of the book is devoted to the application of the foregoing theory to the problem of the propagation of disturbances from solar flares. At the present time it is commonly assumed that the disturbance from a flare propagates as a shock wave. We proceed from the assumption that for a number of flares it is possible to use the hydrodynamic model of the flow caused by a point charge in a gas or by a local supply of energy and mass with variable initial density, pressure, and velocity. Theoretical and experimental data are analyzed regarding the law of motion of the wave and the retardation of its velocity. It is found that certain theoretical results agree well with observed data.

The author is deeply grateful to L. I. Sedov for the formulation of a number of problems, many discussions of the results, and his constant attention and support.

Thanks are due also to colleagues in the Mechanics Division of the V. A. Steklov Institute of Mathematics in the Academy of Sciences of the USSR, especially N. S. Mel′nikova and A. G. Kulikovskiĭ for their daily assistance and support in the work.

The author also wishes to thank his colleagues and collaborators, especially V. P. Karlikov, P. I. Čuškin, E. V. Rjazanov, and V. A. Levin, for their participation in the development of the theory presented here.

CHAPTER I

BASIC EQUATIONS AND FORMULATIONS OF PROBLEMS

§1. The equations of gas dynamics

1.1. *The equations in integral form.* In the presentation of known results concerning the basic equations of gas dynamics we shall follow [1]–[4]. In the investigation of questions of the motion of a compressible fluid or gas we shall consider the gas as a continuous medium, and we shall use hydrodynamic models to describe the motion of the material medium under study.

In studying the motion of a compressible continuous medium it is necessary to introduce scalar, vector, and tensor quantities: the temperature T, the velocity v, the stress tensor $\hat{P}$, etc. This motion can be considered in various coordinate systems. If our attention is concentrated on the study of a phenomenon relative to the reference system of the observer, we adopt the Euler point of view for studying the motion of the continuous medium. Thus, from the Euler point of view we are interested in the variation in velocity, density, temperature, and other quantities at a fixed site in space. If we wish to follow the variation of the parameters of individual points of the medium, we adopt the Lagrange point of view. We introduce a rectangular Cartesian coordinate system x^1, x^2, x^3 of the observer which is fixed in space. We shall identify the coordinate triple x^1, x^2, x^3 with a point of space and denote the vector by $\boldsymbol{r}$. The mathematical concept corresponding to the physical idea of the motion of a fluid is that of a continuous transformation of three-dimensional Euclidean space into itself. The parameter t describing the transformation is identified with time. The value $t = 0$ can be taken as the initial moment, and t can range over the entire real axis.

We introduce the Lagrange (or accompanying) coordinate system ξ^1, ξ^2, ξ^3. We denote the Lagrange point by ξ. Suppose that at the moment $t = 0$ a particle of the moving fluid is at the point $\xi = (\xi^1, \xi^2, \xi^3)$, and at the point $\boldsymbol{r}(x^1, x^2, x^3)$ at the moment t. We can then suppose that $\boldsymbol{r}$ is defined as a function of ξ and t, and from the kinematic point of view the motion of the medium can be considered as a transformation.

$$x^i = \varphi^i(\xi^i, t). \tag{1.1}$$

If ξ is fixed while t varies, then (1.1) defines the trajectory of the particle M which is initially located at the point ξ. On the other hand, for fixed t (1.1) defines a transformation of the region occupied by the fluid at the initial moment into the region occupied by the fluid at the moment t.

If points initially distinct remain distinct over the entire time of motion, one can introduce the inverse transformation

$$\xi^j = \Phi^j(r, t). \tag{1.2}$$

We shall assume that the functions φ^i and Φ^j have continuous derivatives to third order in all the variables except on certain singular surfaces, curves, or points.

We shall mainly use the Euler variables, although a number of questions will be studied by means of the Lagrange variables.

The equations of gas dynamics express the basic physical laws of the mechanics of a continuous medium. If nonequilibrium effects, thermal conductivity, and viscosity are neglected, then the integral form of these equations for some volume of fluid Ω^* has the form

$$\frac{d}{dt}\int\limits_{\Omega^*(t)} \rho\, d\Omega = 0 \tag{1.3}$$

—the law of conservation of mass,

$$\frac{d}{dt}\int\limits_{\Omega^*(t)} \rho \boldsymbol{v}\, d\Omega = \int\limits_{\Omega^*(t)} \boldsymbol{F}\rho\, d\Omega - \oint\limits_{\Sigma(t)} p\boldsymbol{n}\, d\Sigma \tag{1.4}$$

—the momentum equation, and

$$\frac{d}{dt}\int\limits_{\Omega^*(t)} \rho\left(\varepsilon + \frac{v^2}{2}\right) d\Omega = -\oint\limits_{\Sigma(t)} p\boldsymbol{v}\boldsymbol{n}\, d\Sigma + \int \rho(\boldsymbol{F}\boldsymbol{v})\, d\Omega \tag{1.5}$$

—the equation of conservation of energy.

Here $\Omega^*(t)$ denotes the volume of fluid bounded by the surface Σ, p is the pressure, ϵ is the internal energy, $\boldsymbol{F}$ is the inertial force (the field vector of external forces per unit mass), and $\boldsymbol{n}$ is the outer normal to the surface Σ.

If we consider the thermal conductivity with vector of heat flow q given by Fourier's law,

$$q = -\kappa \operatorname{grad} T,$$

where κ is the coefficient of thermal conductivity, then to the right side of (1.5) we must add a term of the form

$$-\oint\limits_{\Sigma(t)} (\kappa \operatorname{grad} T)\boldsymbol{n}\, d\Sigma. \tag{1.6}$$

Equations (1.3)–(1.5) can be applied to arbitrary volumes and in particular in the case where in the interior of Ω^* there is a surface Σ_0 on which the hydrodynamic functions are discontinuous.

We shall characterize a gas as a continuous medium in the Euler variables by the following basic quantities: the velocity vector $\boldsymbol{v}(\boldsymbol{r}, t)$, the density $\rho(\boldsymbol{r}, t)$, and the pressure $p(\boldsymbol{r}, t)$. As a rule, we shall assume that the internal energy is given by a thermodynamic function $\epsilon = \epsilon(p, \rho)$ or $\epsilon = \epsilon(T, \rho)$.

1.2. *The differential equations of gas dynamics and their properties.* We introduce the four-dimensional Euclidean space of coordinates and time E_4.

DEFINITION. The motion of the gas is said to be *continuous* in the region $\Omega \subset E_4$ if the functions v, ρ and p are continuous together with their first derivatives everywhere in Ω.

Continuous motions satisfy a system of differential equations which is equivalent to the system of equations in integral form (1.3)–(1.5). For the motion of a gas in which inertial forces and dissipative processes are neglected this system of equations can be written in the form

$$\frac{\partial\rho}{\partial t}+\operatorname{div}\rho\boldsymbol{v}=0, \tag{1.7}$$

$$\frac{\partial\rho\boldsymbol{v}}{\partial t}+\operatorname{div}\hat{P}=0, \tag{1.8}$$

$$\frac{\partial}{\partial t}\left(\rho\varepsilon+\frac{\rho v^2}{2}\right)+\operatorname{div}\boldsymbol{v}\left(\rho\varepsilon+p+\frac{\rho v^2}{2}\right)=0, \tag{1.9}$$

where $\epsilon = \epsilon(p, \rho)$, $\hat{P}$ is the tensor with components $P_{ik} = p\delta_{ik} + \rho v^i v^k$, and δ_{ik} is the unit tensor. The systems of equations (1.3)–(1.5) and (1.7)–(1.9) will form the basis for subsequent investigations. The form of the function $\epsilon(p, \rho)$ depends on the properties of the gas. For continuous motions we assume that it is a continuously differentiable function of the arguments p and ρ. A simple and important example of an ideal compressible medium is a perfect gas with a constant ratio of the specific heat capacities $\gamma = c_p/c_v$.

For a perfect gas the function $\epsilon(p, \rho)$ has the form

$$\varepsilon=\frac{1}{\gamma-1}\frac{p}{\rho}+\text{const}, \tag{1.10}$$

and the relation between the density, pressure, and temperature is given by the thermal equation of state

$$p=R\rho T, \tag{1.11}$$

where R is the gas constant.

For adiabatic, reversible motions of the gas the last equation of the system (1.9) can be transformed by means of (1.7) and (1.8) to the form

$$\frac{d\varepsilon}{dt}-\frac{p}{\rho^2}\frac{d\rho}{dt}=0, \tag{1.12}$$

where $d/dt = \partial/\partial t + v \operatorname{grad}$ is the material or total derivative. Introducing the entropy S and using the first law of thermodynamics, we have

$$TdS=d\varepsilon+pd\,\frac{1}{\rho}. \tag{1.13}$$

Using (1.12), we obtain

$$\frac{dS}{dt}=0. \tag{1.14}$$

Thus, for adiabatic, continuous motions of a nonviscous medium the energy equation can be taken in the form (1.14). For a perfect gas we have

$$S = c_v \ln \frac{p}{\rho^\gamma} + \text{const}, \tag{1.15}$$

where c_v is the heat capacity of the gas at constant volume. If $S = \text{const}$ everywhere in the region of motion of the gas, such motion is called *isentropic*.

We shall mention several important cases of gas motion which are characterized by a special geometric form of the trajectories in E_3 and also by the geometric form of the level surfaces of the basic quantities.

DEFINITION. A motion is called *two-dimensional* if the basic quantities can be expressed in terms of functions which depend, in addition to the time t, only on two spatial variables.

One of the forms of two-dimensional motions is plane-parallel or planar flow. The motion (or flow) is called plane-parallel if the trajectories of the particles in E_3 are plane curves the planes of which are parallel to a fixed plane Π while in directions perpendicular to Π the basic quantities do not vary and can depend only on t. If a Cartesian coordinate system is chosen such that the plane Π is the coordinate plane Ox^1x^2, we find that in plane-parallel motion the velocity vector $\mathbf{v}$ has two nonzero components $\mathbf{v}(v^1, v^2, 0)$, and the quantities v^1 and v^2 are functions only of the variables x^1, x^2 and t. The differential equations of plane-parallel motion in scalar notation are easily obtained from the system (1.7)–(1.9).

Another important example of two-dimensional motions are the axially symmetric flows. The motion (or flow) is called axially symmetric if the trajectories of the particles in E_3 are plane curves the planes of which pass through some fixed line—the axis of symmetry—while the basic unknown functions depend only on t on any circle in a plane perpendicular to the symmetry axis with center on this axis.

For the analysis of axially symmetry motion it is convenient to use cylindrical or spherical (polar) coordinates. If the system of cylindrical coordinates is chosen such that its axis coincides with the symmetry axis (suppose that this is the axis Ox^1), the basic quantities may be assumed to depend only on the coordinate x and the radius r. We denote by u the projection of the velocity on the axis Ox^1 and by v the modulus of the projection of the velocity on a plane perpendicular to the symmetry axis. For the Cartesian components of the velocity, density ρ, and pressure p we have $v^1 = u(x, r, t)$, $v^2 = v(x, r, t) \cos \varphi$, $v^3 = v(x, r, t) \sin \varphi$, $\rho = \rho(x, r, t)$, and $p = p(x, r, t)$, where $x = x^1$ and φ is the polar angle. For the chosen system of coordinates the differential equations of axially symmetric motion differ from the equations for plane-parallel motion in that the components v^1 and v^2 are replaced by u and v, the variable x^2 is replaced by the variable r, and for terms arising from the divergence there is an additional term containing the factor $1/r$, for

$$\operatorname{div} \psi \boldsymbol{v} = \frac{\partial \psi u}{\partial x} + \frac{\partial \psi v}{\partial r} + \frac{\psi v}{r}, \tag{1.16}$$

where ψ is some scalar function.

In some cases it is convenient to study axially symmetric motions in a polar spherical coordinate system r, θ, φ with center on the symmetry axis. The angle between the symmetry axis and the radius vector to a point of space is taken as the angle θ. If we introduce as unknown functions the projection v_r of the velocity on the radius vector of the point in question and the projection v_θ on the direction tangent to the circle with center at the origin of the coordinate system and passing through the symmetry axis and the given point of space, then the basic quantities will not depend on the angle φ. The equations of gas dynamics in this case can be written as follows:

$$\begin{gathered} r^2 \frac{\partial \rho}{\partial t} + \frac{\partial}{\partial r}(r^2 \rho v_r) + \frac{r}{\cos\theta}\frac{\partial}{\partial \theta}(\rho v_\theta \cos\theta) = 0, \\ \frac{dv_r}{dt} - \frac{v_\theta^2}{r} + \frac{1}{\rho}\frac{\partial p}{\partial r} = 0, \quad \frac{dS}{dt} = 0, \\ \frac{dv_\theta}{dt} - \frac{v_r v_\theta}{r} + \frac{1}{r\rho}\frac{\partial p}{\partial \theta} = 0, \\ \frac{d}{dt} = \frac{\partial}{\partial t} + v_r \frac{\partial}{\partial r} + \frac{v_\theta}{r}\frac{\partial}{\partial \theta}. \end{gathered} \tag{1.17}$$

DEFINITION. A motion is called *one-dimensional* if the basic quantities can be expressed in terms of functions which depend only on one spatial variable in addition to the time t.

We shall distinguish one-dimensional motions with plane, cylindrical, and spherical symmetry.

A one-dimensional motion with plane symmetry (or with plane waves) is a motion in which the trajectories of the particles in E_3 form a family of straight lines perpendicular to some fixed plane Π while along any plane parallel to Π the basic quantities can depend only on the time t and do not depend on the points of the plane. If we choose the axis Ox^1, or in another notation Or, of the Cartesian coordinate system perpendicular to Π, then the velocity vector v will have only one nonzero component $v^1 = v$, while $v = v(r, t)$, $p = p(r, t)$, and $\rho = \rho(r, t)$. The differential equations for this case are obtained very simply from the basic system (1.17)–(1.9).

A one-dimensional motion with cylindrical symmetry is an axially symmetric motion in which the trajectories of the particles in E_3 form a family of lines which intersect the symmetry axis orthogonally, while over surfaces of circular cylinders whose axes coincide with the symmetry axis the basic quantities do not vary and depend only on time.

Here it is convenient to use cylindrical coordinates x, r, φ and to take for the basic quantities the radial component of the velocity $v_r = v$, the density ρ, and the pressure p; we then have

$$v^1 = 0, \qquad v^2 = v(r, t)\cos\varphi, \qquad v^3 = v(r, t)\sin\varphi, \qquad p = p(r, t), \qquad \rho = \rho(r, t).$$

The system of differential equations can be obtained from the basic system (1.7)–(1.9) by making use of the remark made above for the axially symmetric case.

A one-dimensional motion with spherical symmetry (or with spherical waves) is a motion in which the trajectories of the particles in E_3 form a family of lines directed along rays issuing from a fixed point O, while the basic quantities do not vary over any sphere with center at O and can depend only on the time t. For the analytic description of such flows it is convenient to use a spherical coordinate system r, θ, φ with origin at O. If we denote by v the radial component of the velocity $\mathbf{v}$ ($v_r = v$), then for this flow we have $v = v(r, t), p = p(r, t)$, and $\rho = \rho(r, t)$.

The system of differential equations here can easily be obtained from (1.17) by setting $v_\theta = 0, v_r = v$ and assuming all functions to be independent of θ. In the sequel it will often be convenient for us to consider all three types of one-dimensional motions together. With this in mind we introduce a symmetry parameter ν such that $\nu = 1$ for plane symmetry, $\nu = 2$ for cylindrical symmetry, and $\nu = 3$ for spherical symmetry. The parameter ν indicates to us the dimension of that subspace of E_3 which it suffices to use in order to describe the dependence of the hydrodynamical functions on the coordinates.

Using the notation just introduced, we can write down a unified system of the differential equations of gas dynamics for one-dimensional flows as follows:

$$\frac{\partial}{\partial t}(r^{\nu-1}\rho) + \frac{\partial}{\partial r}(r^{\nu-1}v\rho) = 0, \tag{1.18}$$

$$\frac{\partial}{\partial t}(\rho v) + \frac{\partial}{\partial r}(p + \rho v^2) + \frac{(\nu - 1)\rho v^2}{r} = 0, \tag{1.19}$$

$$\frac{\partial}{\partial t}\rho r^{\nu-1}\left(\varepsilon + \frac{v^2}{2}\right) + \frac{\partial}{\partial r}\left[r^{\nu-1}\rho v\left(\varepsilon + \frac{p}{\rho} + \frac{v^2}{2}\right)\right] = 0. \tag{1.20}$$

The one-dimensional motions considered with plane, cylindrical and spherical symmetry are examples of motions admitted exactly by the system of equations (1.7)–(1.9).

In applications another form of one-dimensional motions is sometimes used which corresponds to quasi-one-dimensional flows of a compressible medium. An example of such flows are flows in pipes of variable cross-section described in the one-dimensional hydraulic approximation. We shall not consider this approximate model of one-dimensional motions in detail here. A derivation of the differential equations for the one-dimensional motions of the hydraulic approximation is given, for example, in [2]. The differential equations considered have been written in the Euler variables.

In the Lagrange variables the equations of gas dynamics have the form

$$\rho J = \rho_0, \tag{1.21}$$

$$\nabla \boldsymbol{r} \cdot \frac{d^2 \boldsymbol{r}}{dt^2} + \frac{1}{\rho} \nabla p = 0, \tag{1.22}$$

$$\frac{dS}{dt} = 0, \tag{1.23}$$

where $\rho_0 = \rho_0(\boldsymbol{r})$ is the initial density distribution, J is the Jacobian of the transformation (1.1),

$$J = \frac{\partial (x^1, x^2, x^3)}{\partial (\xi^1, \xi^2, \xi^3)} = \det\left(\frac{\partial x^i}{\partial \xi^j}\right),$$

and the gradient is taken in the space of the Lagrange coordinates ξ^j. In coordinate form (1.22) has the form

$$\frac{\partial x^i}{\partial \xi^j} \frac{d^2 x^i}{dt^2} = -\frac{1}{\rho} \frac{\partial p}{\partial \xi^j},$$

where summation on i is understood.

For the case of one-dimensional motions the system (1.21)–(1.23) can be written in the following form:

$$\begin{aligned} &\rho \frac{\partial r}{\partial \xi} = \rho_0 \frac{\xi^{\nu-1}}{r^{\nu-1}}, \qquad && v - \frac{\partial r}{\partial t} = 0, \\ &\frac{\partial p}{\partial \xi} = -\rho_0 \frac{\xi^{\nu-1}}{r^{\nu-1}} \frac{\partial v}{\partial t}, && \frac{\partial S}{\partial t} = 0. \end{aligned} \tag{1.24}$$

Here ξ is the initial coordinate of the particle, i.e. the distance the particle was away from the null plane, the symmetry axis, or the coordinate origin in the plane, cylindrical, and spherical cases respectively.

We shall mention certain properties of the equations of gas dynamics. Let φ, ψ, u, f, $\boldsymbol{v}$, and Φ be continuously differentiable functions. An equation of the form

$$\frac{\partial \varphi (u, f, \boldsymbol{r}, t)}{\partial t} + \operatorname{div} \psi (u, f, \boldsymbol{r}, t)\, \boldsymbol{v} = \Phi (u, \boldsymbol{v}, \boldsymbol{r}, t)$$

we shall call *an equation in the form of a conservation law* or *an equation in divergence form.*

PROPERTY 1. The systems of equations of gas dynamics (1.7)–(1.9) and (1.18)–(1.20) are written in divergence form. The system of equations (1.24) in the Lagrange variables can be brought to divergence form.

We mention that the equations (1.18)–(1.20) in divergence form were presented in [4].

PROPERTY 2. The systems (1.7)–(1.9), (1.18)–(1.20), (1.21)–(1.22), and (1.24) are hyperbolic systems of partial differential equations in Euclidean spaces of coordinate-time.

PROPERTY 3. The system of equations (1.7)–(1.9) is invariant under Newton-Galilean transformations $p_* = p$, $\rho_* = \rho$, $t_* = t + \tau$, $\boldsymbol{r}_* = \boldsymbol{r} + \boldsymbol{a}t + \boldsymbol{r}_0$, $\boldsymbol{v}_* = \boldsymbol{v} + \boldsymbol{a}$, where $\boldsymbol{a}$ and $\boldsymbol{r}_0$ are constant vectors, τ is a constant scalar, and quantities with an asterisk correspond to the state of motion after the transformation.

In considering many problems of gas dynamics the velocity $\boldsymbol{v}$, the pressure p, and the density ρ are taken as the basic quantities or the basic gas-dynamics functions. In a number of cases (especially for numerical solution) the basic quantities are taken to be ρ, $\boldsymbol{v}$, and the internal energy ϵ; the pressure is found from the caloric equation of state

$$p = p(\rho, \varepsilon). \tag{1.25}$$

It is sometimes convenient to use as the basic functions the velocity $\boldsymbol{v}$, the density ρ, and the total energy per unit volume E which for the case of gas dynamics is given by the expression

$$E = \rho\left(\varepsilon + \frac{v^2}{2}\right). \tag{1.26}$$

The system (1.7)–(1.9) in these variables assumes the form

$$\begin{gathered} \frac{\partial \rho}{\partial t} + \operatorname{div} \rho\boldsymbol{v} = 0, \qquad \frac{\partial \rho \boldsymbol{v}}{\partial t} + \operatorname{div}(\rho\boldsymbol{v}\cdot\boldsymbol{v} + \delta_{ik}p) = 0, \\ \frac{\partial E}{\partial t} + \operatorname{div} E\boldsymbol{v} + \operatorname{div}(p\boldsymbol{v}) = 0, \end{gathered} \tag{1.27}$$

where p is a given function of E, ρ and v^2 in correspondence with (1.25) and (1.26).

We mention also the following simple property of the system (1.7)–(1.9) or (1.27).

PROPERTY 4. The system of equations of gas dynamics (1.7)–(1.9) has a simplest solution

$$\boldsymbol{v} = 0, \qquad \rho = \rho_0(\boldsymbol{r}), \qquad p = p_0 = \text{const}. \tag{1.28}$$

The solution (1.28) is called the *equilibrium state* or the *state of rest.*

1.3. *Several simplifications and particular forms of the equations of motion of a fluid.* We shall consider certain instances of simplification of the equations of gas dynamics which will give approximate solutions of problems on the motion of the gas. One of the general methods of obtaining simplified equations is linearization. The meaning of linearization is as follows. Suppose that some base motion is known, i.e. an exact solution of the nonlinear equations of gas dynamics. If the equations of gas dynamics are taken in the form (1.7), (1.8), (1.14), then this solution has the form

$$\boldsymbol{v}_0(\boldsymbol{r}, t), \qquad \rho_0(\boldsymbol{r}, t), \qquad p_0(\boldsymbol{r}, t), \qquad S_0(\boldsymbol{r}, t). \tag{1.29}$$

We shall seek another solution of the form $\boldsymbol{v} = \boldsymbol{v}_0 + \sigma\boldsymbol{v}'$, $\rho = \rho_0 + \sigma\rho'$, $p = p_0 + \sigma p'$, $S = S_0 + \sigma S'$, where the primes denote new unknown functions of $\boldsymbol{r}$ and

t, i.e. increments to the base solution or perturbations; σ is some small parameter.

If this solution is substituted into the system (1.7), (1.8), (1.14), then on dividing by σ and recalling that the quantities v_0, ρ_0, p_0 and S_0 form a solution of this system, we obtain

$$\frac{\partial \rho'}{\partial t} + \operatorname{div}(\rho_0 v' + v_0 \rho' + \sigma v' \rho') = 0,$$

$$\frac{\partial}{\partial t}(\rho_0 v' + v_0 \rho' + \sigma v' \rho') + \operatorname{div}(\rho_0 v_0 v' + \rho' v_0 v_0 + \rho_0 v' v_0 \tag{1.30}$$

$$+ \sigma \rho' v_0 v' + \sigma \rho' v' v_0 + \sigma \rho_0 v' v' + \sigma^2 v' v' \rho') + \operatorname{grad} p' = 0,$$

$$\frac{\partial S'}{\partial t} + (v_0 \nabla) S' + (v' \nabla) S_0 + \sigma (v' \nabla) S' = 0,$$

$$p' = \frac{1}{\sigma}[p(\rho_0 + \sigma \rho', S + \sigma S') - p_0].$$

From this system it follows that the new unknown functions v', ρ', etc. depend not only on r and t but also on the parameter σ. Suppose that the functions v', ρ', S' and p', as solutions of the exact equations (1.30), and also their derivatives entering in these equations have a finite limit for $\sigma \to 0$. Then, passing to the limit in (1.30), we obtain the linearized system of the original equations

$$\frac{\partial \rho'}{\partial t} + \operatorname{div}(\rho_0 v' + v_0 \rho') = 0,$$

$$\frac{\partial}{\partial t}(\rho_0 v' + v_0 \rho') + \operatorname{div}(v' \rho_0 v_0 + \rho' v_0 v_0 + v_0 \rho_0 v') + \operatorname{grad} p' = 0, \tag{1.31}$$

$$\frac{\partial S'}{\partial t} + (v_0 \nabla) S' + (v' \nabla) S_0 = 0, \qquad p' = a_0^2 \rho' + \left(\frac{\partial p}{\partial S}\right)_0 S',$$

where $a_0^2 = (dp/d\rho)_s$ is the adiabatic speed of sound.

The process of obtaining equations of the type (1.31) is called linearization about the known solution v_0, ρ_0, p_0. The system of equations (1.31) is the linearized system of equations of gas dynamics. In what follows we shall make extensive use of the method of linearization.

If the base solution has the form

$$v_0 = 0, \ \rho_0 = \text{const}, \ S_0 = \text{const}, \ p_0 = \text{const},$$

then from (1.31) we obtain the well-known equations of acoustics:

$$\frac{\partial \rho'}{\partial t} + \rho_0 \operatorname{div} v' = 0, \qquad \frac{\partial}{\partial t}(\rho_0 v') + \nabla p' = 0, \qquad \frac{\partial S'}{\partial t} = 0, \qquad p' = a_0^2 \rho'. \tag{1.32}$$

Let us assume that $S' = \text{const}$. From (1.32) we then obtain the classical wave equation for the pressure

$$\frac{\partial^2 p'}{\partial t^2} - a_0^2 \Delta p' = 0. \tag{1.33}$$

In the theory of the propagation of sound the energy density of the perturbation is introduced [5]:

$$E' = \frac{\rho_0 v'^2}{2} - \frac{a_0^2}{2\rho_0}\rho'^2. \tag{1.34}$$

This energy corresponds to the difference between the total energy per unit volume of the moving fluid and the sum of the initial energy of the gas at rest with a part arising from the variation in the amount of matter, namely

$$E' = E - \rho_0\varepsilon_0 - \left(\varepsilon_0 + \frac{p_0}{\rho_0}\right)\rho' + o\,(\sigma^2\rho'^2).$$

An equation for the variation of E' can be obtained from the system (1.27):

$$\frac{\partial E'}{\partial t} + \operatorname{div} p'\boldsymbol{v}' = 0, \tag{1.35}$$

which expresses the law of conservation of energy of the perturbation.

For one-dimensional motions the system of acoustic equations takes the form

$$\frac{\partial \rho'}{\partial t} = -\rho_0 \frac{1}{r^{\nu-1}}\frac{\partial}{\partial r}(r^{\nu-1}v'), \quad p' = a_0^2\rho', \qquad \frac{\partial v'}{\partial t} = -\frac{1}{\rho_0}\frac{\partial p'}{\partial r}.$$

The wave equation (1.33) for these cases assumes the form

$$\frac{\partial^2 p'}{\partial t^2} = a_0^2 \frac{1}{r^{\nu-1}}\frac{\partial}{\partial r}\left(r^{\nu-1}\frac{\partial p'}{\partial r}\right). \tag{1.36}$$

For the plane case equation (1.36) has the solution

$$p' = A_1 r + A_2 t + f_1\left(t - \frac{r}{a_0}\right) + f_2\left(t + \frac{r}{a_0}\right). \tag{1.37}$$

In the case of spherical symmetry we obtain

$$p' = \frac{f_1\left(t - \frac{r}{a_0}\right) + f_2\left(t + \frac{r}{a_0}\right)}{r}. \tag{1.38}$$

For the case of a cylindrical wave the solution can be represented in integral form:

$$p' = \int_r^\infty \frac{f_1\left(t - \frac{R}{a_0}\right)}{\sqrt{R^2 - r^2}}\,dR + \int_r^\infty \frac{f_2\left(t + \frac{R}{a_0}\right)}{\sqrt{R^2 - r^2}}\,dR. \tag{1.39}$$

In the solutions (1.37)–(1.39) f_1 and f_2 are arbitrary functions of their arguments, $R = r^2 + z^2$, where z is the coordinate along the axis of cylindrical symmetry, and A_1 and A_2 are arbitrary constants.

We now consider another case of simplified equations of gas dynamics where it is assumed that the density ρ is constant (the case of an incompressible fluid). Then the form of (1.8) does not change, while (1.7) assumes the form

$$\operatorname{div} \boldsymbol{v} = 0. \tag{1.40}$$

The energy equation with consideration of the thermal conductivity [5] has the form

$$\frac{\partial}{\partial t}(\rho\varepsilon) + \operatorname{div}(\rho\varepsilon)\boldsymbol{v} - \operatorname{div}\kappa\nabla T + \frac{\partial}{\partial t}\left(\frac{\rho v^2}{2}\right) + \operatorname{div}\left[\boldsymbol{v}\left(p + \frac{\rho v^2}{2}\right)\right] = 0.$$

If we now use the equation of continuity and the equation for the amount of motion, we obtain

$$\frac{\partial}{\partial t}(\rho\varepsilon) + \boldsymbol{v}\nabla(\rho\varepsilon) - \operatorname{div}\kappa\nabla T = 0. \tag{1.41}$$

For $\kappa = \text{const}$, $\rho = \text{const}$, $v = 0$, and $\epsilon = c_v T + \text{const}$ the classical heat equation is obtained from (1.41)

$$\frac{\partial T}{\partial t} = \chi \Delta T, \tag{1.42}$$

where $\chi = \kappa / c_v \rho$, κ being the coefficient of thermal conductivity. Suppose now that $v = 0$ and $\kappa = \kappa(\epsilon)$. From (1.41) we obtain

$$\frac{\partial \varepsilon}{\partial t} = \operatorname{div} \frac{\kappa}{\rho}\left(\frac{dT}{d\varepsilon}\right)\nabla\varepsilon. \tag{1.43}$$

Using the fact that $c_v = d\epsilon/dT$, we can rewrite (1.43) as follows:

$$\frac{\partial \varepsilon}{\partial t} = \operatorname{div} \chi \nabla \varepsilon.$$

Equation (1.43) provides an example of a nonlinear heat equation. For cases of a one-dimensional process of heat flow this equation takes the form

$$\frac{\partial \varepsilon}{\partial t} = \frac{1}{r^{\nu-1}} \frac{1}{\partial r}\left(r^{\nu-1} \chi \frac{\partial \varepsilon}{\partial r}\right). \tag{1.44}$$

In addition to the equations of gas dynamics and certain simplified forms of these equations which have been mentioned, in studying problems on the motion of a gas it is often necessary to use more complex equations and to consider chemical reactions in the gas which are accompanied by the evolution or absorption of heat. We shall not write down the more complex equations for gas motion for these cases here, but we shall rather indicate them in the sections devoted to the investigation of the corresponding problems.

§2. The conditions on strong discontinuities

2.1. *Derivation of the relations for strong discontinuities.* In the sequel we shall consider more general motions of the gas which are not continuous motions. If the gas-dynamical functions are not continuously smooth but rather only piecewise smooth or piecewise continuous, they cannot satisfy the system of differential equations of gas dynamics. However, the equations of gas dynamics in integral form are valid for such motions. In this sense piecewise continuous solutions are sometimes called discontinuous or generalized solutions or even generalized motions. We shall consider discontinuous solutions for motions of a gas without taking viscosity into account.

A detailed analysis of various types of discontinuities is given in [6].

If in the domain of a generalized motion there exists a hypersurface $s(\boldsymbol{r}, t)$ on which the quantities $\boldsymbol{v}, \rho, p$ and ϵ have a discontinuity of first kind and away from which this motion is continuous, such a motion is called a motion with a strong discontinuity. The surface s for fixed t is called a surface of strong discontinuity. If the gas-dynamical functions themselves are continuous while only certain of their derivatives with respect to the coordinates and time are discontinuous, such a motion is said to have a weak discontinuity, and the corresponding surface is called a surface of weak discontinuity.

We now introduce the concept of the displacement velocity of a surface of discontinuity in the space E_3. On the surface $s(t)$ at the moment t we choose some point M and assume that at M there is a well-defined normal to s. We find further the point of intersection N of the normal to s at M with the surface $s(t + \Delta t)$. Let $H(\Delta t)$ be the length of the vector MN taken with a "plus" sign if the direction of MN and that of the unit vector normal $\boldsymbol{n}$ to s at M coincide and with a "minus" sign if these directions differ.

DEFINITION. The *displacement velocity* in the space E_3 of the surface s at the point M is the vector $\boldsymbol{D}$ normal to s defined as the limit

$$\boldsymbol{D} = \boldsymbol{n} \lim_{\Delta t \to 0} \frac{H(\Delta t)}{\Delta t}.$$

If the equation of the surface $s(t)$ is given in the form $s(\boldsymbol{r}, t) = 0$ and this surface is smooth, then for the computation of $\boldsymbol{D}$ we have

$$\boldsymbol{D} = -\frac{\partial s}{\partial t} \boldsymbol{n} / |\operatorname{grad} s|. \tag{1.45}$$

For coordinate transformations with passage to differently moving systems the velocity vector $\boldsymbol{D}$ depends on the choice of coordinate system.

By virtue of the conservation laws of hydrodynamics on a surface of strong discontinuity special relations exist between $\boldsymbol{v}, p, \rho$ and ϵ. These conditions are easily derived either by applying the conservation laws directly to elements of the moving or quiescent medium through which the shock wave is propagated or by using the equations of gas dynamics in integral form (1.3)–(1.5). We shall give a brief derivation of the conditions on surfaces of strong discontinuity using (1.3)–(1.5). As already mentioned, these equations can be applied to arbitrary volumes Ω and, in particular, in the case where inside a particular volume Ω^* there is a surface Σ_0 on which the hydrodynamical quantities are discontinuous. Together with the moving Lagrange volume Ω^* we introduce a moving volume Ω which coincides with Ω^* at the moment t in question but which moves with velocity D on the boundary (D is the normal velocity of the motion of the surface Σ bounding Ω). In general D is not equal to the normal velocity v_n of the motion of the gas particles.

From mathematical analysis (cf., for example, [7], [8]) there is the formula for differentiation of an integral with respect to a parameter:

$$\frac{d}{dt}\int_{\Omega} \Phi d\Omega = \int_{\Omega} \frac{\partial \Phi}{\partial t} d\Omega + \oint_{\Sigma} \Phi v_n d\Sigma, \tag{1.46}$$

where $\Phi(\boldsymbol{r}, t)$ is a function which is bounded, integrable with respect to the coordinates, and differentiable with respect to time, v_n is the normal component of the velocity of particles of the surface Σ, and $\Omega = \Omega(t)$ is a variable volume. If formula (1.46) is now written for the moving volumes Ω^* and Ω introduced earlier, on subtracting the corresponding sides of the equations we have

$$\frac{d}{dt}\int_{\Omega} \Phi d\Omega - \frac{d}{dt}\int_{\Omega^*} \Phi d\Omega = \int_{\Omega} \frac{\partial \Phi}{\partial t} d\Omega - \int_{\Omega^*} \frac{\partial \Phi}{\partial t} d\Omega + \oint_{\Sigma} \Phi (D - v_n) d\Sigma. \tag{1.47}$$

At time t by choice of $\Omega^*(t)$ we have $\Omega(t) = \Omega^*(t)$; hence

$$\int_{\Omega} \frac{\partial \Phi}{\partial t} d\Omega = \int_{\Omega^*} \frac{\partial \Phi}{\partial t} d\Omega.$$

This implies that (1.47) goes over into the equation

$$\frac{d}{dt}\int_{\Omega} \Phi d\Omega = \frac{d}{dt}\int_{\Omega^*} \Phi d\Omega + \oint_{\Sigma} \Phi (D - v_n) d\Sigma. \tag{1.48}$$

It is obvious that (1.48) remains true for a vector function $\boldsymbol{\Phi}$.

By means of (1.48) from the system of equations (1.3)–(1.5) we obtain the integral form of the equations of gas dynamics for an arbitrary moving volume $\Omega(t)$:

$$\frac{d}{dt}\int_{\Omega} \rho d\Omega = \oint_{\Sigma} \rho (D - v_n) d\Sigma, \tag{1.49}$$

$$\frac{d}{dt}\int_{\Omega} \rho \boldsymbol{v} d\Omega = \int_{\Omega} \rho \boldsymbol{F} d\Omega - \oint_{\Sigma} p \boldsymbol{n} d\Sigma + \oint_{\Sigma} \rho \boldsymbol{v} (D - v_n) d\Sigma, \tag{1.50}$$

$$\frac{d}{dt}\int_{\Omega} \rho \left(\frac{v^2}{2} + \varepsilon\right) d\Omega = -\oint_{\Sigma} p \boldsymbol{v}\boldsymbol{n} d\Sigma + \int_{\Omega} \rho (\boldsymbol{v}\boldsymbol{F}) d\Omega$$
$$+ \oint_{\Sigma} \rho \left(\frac{v^2}{2} + \varepsilon\right)(D - v_n) d\Sigma - \oint_{\Sigma} \boldsymbol{q}\boldsymbol{n} d\Sigma. \tag{1.51}$$

In (1.51) q is the vector of heat flow. For the case of one-dimensional motions we obtain the equations

$$\frac{d}{dt}\sigma_\nu \int_{r_1}^{r_2} \rho r^{\nu-1} dr = \left[\rho r^{\nu-1}\sigma_\nu\left(\frac{dr}{dt} - v\right)\right]\Bigg|_{r_1}^{r_2},$$

$$\sigma_\nu = 2(\nu - 1)\pi + (\nu - 2)(\nu - 3),$$

$$\frac{d}{dt}\sigma_\nu \int_{r_1}^{r_2} \rho v r^{\nu-1} dr = \sigma_\nu \int_{r_1}^{r_2} \rho F r^{\nu-1} dr + \sigma_\nu r^{\nu-1}\left[-p + \rho v\left(\frac{dr}{dt} - v\right)\right]\Bigg|_{r_1}^{r_2},$$

$$\frac{d}{dt}\sigma_\nu \int_{r_1}^{r_2} \rho\left(\frac{v^2}{2} + \varepsilon\right) r^{\nu-1} dr = \sigma_\nu \int_{r_1}^{r_2} \rho (vF) r^{\nu-1} dr$$

$$+ \sigma_\nu\left[-q - pv + \rho\left(\frac{v^2}{2} + \varepsilon\right)\left(\frac{dr}{dt} - v\right) r^{\nu-1}\right]\Bigg|_{r_1}^{r_2}. \qquad (1.52)$$

We now suppose that there exists an isolated surface of strong discontinuity and that some arbitrary part of it Σ_0 partitions the volume $\Omega(t)$ into two parts $\Omega_1(t)$ and $\Omega_2(t)$ and the surface $\Sigma(t)$ into the surfaces $\Sigma_1(t)$ and $\Sigma_2(t)$ respectively. Suppose that Ω_1 is bounded by the surface $s_1 = \Sigma_{01} + \Sigma_1$ and Ω_2 by $s_2 = \Sigma_{02} + \Sigma_2$, where Σ_{01} and Σ_{02} are the outer sides of the surface Σ_0 with respect to Ω_1 and Ω_2. We denote by the indices 1 and 2 the limits of the gas-dynamical functions as a spatial point tends to Σ_0 from the sides of Ω_1 and Ω_2 respectively.

Then from (1.49), on subtracting the analogous equations written for Ω_1 and Ω_2, we obtain

$$\int_\Sigma \rho(D - v_n)\, d\Sigma - \int_{s_1} \rho(D - v_n)\, d\Sigma - \int_{s_2} \rho(D - v_n)\, d\Sigma = 0.$$

Noting that the normals on the sides Σ_{01} and Σ_{02} have opposite directions, we thus have

$$\int_{\Sigma_0} [\rho_2(\boldsymbol{D} - \boldsymbol{v}_2) - \rho_1(\boldsymbol{D} - \boldsymbol{v}_1)]\, \boldsymbol{n}\, d\Sigma = 0,$$

where $\boldsymbol{n}$ is the unit outer normal vector to Σ_0.

By virtue of the arbitrary choice of the discontinuity surface Σ_0 we obtain the law of mass conservation on passage through a surface of discontinuity:

$$\rho_2(D - v_{n2}) = \rho_1(D - v_{n1}). \qquad (1.53)$$

Let us suppose further for generality that in a neighborhood of the discontinuity surface Σ_0 energy Q, computed per unit mass, is supplied or removed from the side of Ω_2. The energy Q may be thought of as the energy released instantaneously during the burning of a combustible mixture of gases or the energy absorbed during the vaporization of fluid particles in the gas flow. Then for the total energy per unit volume E on the side Σ_{02} of Σ_0 we can take

$$E_2 = \rho_2\left(\frac{v_2^2}{2} + \varepsilon_2 \pm Q\right),$$

where the "plus" sign indicates removal of energy and the "minus" sign supply of energy to the discontinuity surface and ϵ_2 is the internal energy of the medium in state 2.

If we now apply the argument used above to obtain the law of mass conservation to the momentum equation (1.50) and the energy equation (1.51), we obtain the remaining conditions on surfaces of strong discontinuity:

$$\rho_2 \boldsymbol{v}_2 (D - v_{n2}) - \rho_1 \boldsymbol{v}_1 (D - v_{n1}) = (p_2 - p_1)\,\boldsymbol{n},$$

$$\rho_2 (D - v_{n2})\left(\frac{v^2}{2} + \varepsilon_2 \mp Q\right) - q_{n2} - p_2 v_2$$

$$= \rho_1 (D - v_{n1})\left(\frac{v_1^2}{2} + \varepsilon_1\right) - q_{n1} - p_1 v_1. \qquad (1.54)$$

2.2. *Types of discontinuity surfaces and their properties for* $Q = 0$. If the differences of jumps of the quantities are denoted by square brackets, for example $[\rho] = \rho_2 - \rho_1$, then the relations (1.53)–(1.54) have the form

$$[\rho(D - v_n)] = 0, \qquad [\rho \boldsymbol{v}(D - v_n) - p\boldsymbol{n}] = 0,$$
$$\left[\rho(D - v_n)\left(\frac{v^2}{2} + \varepsilon\right) - q_n - pv\right] = 0. \qquad (1.55)$$

The velocities $D - v_{n1}$ and $D - v_{n2}$ may be considered as the speeds of the discontinuity surface with respect to particles on different sides of the discontinuity.

If $D - v_{n1} = 0$ and $D - v_{n2} = 0$, then the particles of the medium do not pass from one side of the discontinuity to the other, and $v_{n1} = v_{n2}$. In this case a discontinuity of the tangential component of the velocity on different sides of the discontinuity and an arbitrary discontinuity of the density are possible. Such a discontinuity is called a tangential or contant discontinuity. For a tangential discontinuity

$$[p] = 0, \qquad [q_n] = 0, \qquad [\rho] \neq 0, \qquad [v_n] = 0. \qquad (1.56)$$

If $[v_n] \neq 0$, then fluid particles pass from one side of the surface Σ_0 to the other changing their characteristics of state in a jumplike manner.

We now suppose that particles pass from side 1 to side 2 of the discontinuity surface, so that $[v_n] > 0$, and we make the following

DEFINITION. Surfaces of strong discontinuity for which $[v_n] > 0$, $[p] > 0$ and $[\rho] > 0$ are called *shock waves* or *compression jumps.*

If the particles pass from side 1 to side 2 of the surface Σ_0, but $[v_n] < 0$, $[p] < 0$ and $[\rho] < 0$, then such discontinuities are called *rarefaction jumps.*

In the propagation of a strong discontinuity through a quiescent medium ($v_{n1} = 0$)

for a shock wave the gas behind the jump Σ_0 invades the quiescent medium and compression takes place. If $v_{n2} < 0$, then the fluid velocity behind the jump is directed to the side opposite the velocity of propagation of the discontinuity in the stationary medium.

Suppose now that $[\rho] = 0$ and $D - v_{n1} \neq 0$. It then follows from (1.55) that

$$[p]=0, \qquad [v_n]=0, \qquad D-v_{n1}=\frac{[q_n]}{\rho_1 [\varepsilon]}. \tag{1.57}$$

Discontinuities for which $[\rho] = 0, D - v_{n1} \neq 0$, and condition (1.57) is satisfied are called fronts of phase changes or phase fronts. Phase fronts occur when $q_n \neq 0$. If $q_n = 0$, it follows from (1.57) that $[\epsilon] = 0$, and none of the quantities experience a jump on passing through Σ_0, i.e. there is no strong discontinuity.

We shall study flows with strong discontinuities of the type of ordinary shock waves and flows with the release of energy on the discontinuity surface $(Q \neq 0)$.

We now consider in more detail the case of discontinuity surfaces with no evolution of heat or supply of heat by conduction $(Q = 0, q = 0)$. In accordance with the increase in entropy $([S] > 0)$ only compression jumps or shock waves are possible in the gas. We shall formulate this property in more detail below.

That side of the surface Σ_0 from which the gas flows into it will be called the leading side or the side before the front; the opposite side will be called the back side or the side behind the front of the shock wave, while by the front of a shock wave we shall mean the discontinuity surface Σ_0.

In what follows we shall assume that the normal is directed to the side before the front of the shock wave. The gas parameters before the shock wave we shall denote by the index 1 or ∞ (sometimes 0) and the parameters of the gas directly behind the front of the shock wave by the index 2 or n (sometimes 1).

The choice of indices is a matter of convenience when studying specific problems. If we eliminate the velocities $v, D - v_{n1}$ and $D - v_{n2}$ from (1.55), we arrive at a relation of the form

$$\varepsilon_2-\varepsilon_1=\frac{1}{2}(p_2+p_1)\left(\frac{1}{\rho_1}-\frac{1}{\rho_2}\right). \tag{1.58}$$

The relation (1.58) gives the relation between the thermodynamic parameters ρ_1 and p_1 before the shock wave and p_2 and ρ_2 behind the shock wave, and is called the Hugoniot adiabatic or the shock adiabatic. For a perfect gas with constant heat capacities from (1.58) we have the following equation for the shock adiabatic:

$$\frac{p_2}{p_1}=\frac{(\gamma+1)V_1-(\gamma-1)V_2}{(\gamma+1)V_2-(\gamma-1)V_1}, \qquad V=\frac{1}{\rho}. \tag{1.59}$$

We shall mention the basic properties of shock waves. We call the quantity $[p]$

(or $[\rho]$) the intensity of the shock wave and define, as usual, the adiabatic sound speed by the relation $a^2 = (dp/d\rho)_S$.

PROPERTY 1. Along the Hugoniot adiabatic for media for which the inequality $d^2V/dp^2 > 0$ is satisfied the entropy increases monotonically together with the pressure p. In accordance with the second law of thermodynamics this implies that the shock wave is a compression jump, and for it $p_2 > p_1$ and $\rho_2 > \rho_1$. This conclusion is one of the fundamental facts in the theory of strong discontinuities.

PROPERTY 2. A jump in entropy in a shock wave is a quantity of third order of smallness with respect to the intensity of the discontinuity, i.e.

$$\lim_{\rho_2 \to \rho_1} \frac{S_2 - S_1}{(\rho_2 - \rho_1)^3} = B > 0.$$

PROPERTY 3. The absolute value of the normal component of the velocity of motion of the gas with respect to the shock wave is greater than the sound speed before the front and less than the sound speed behind the front, i.e. $|v_{n1} - D| > a_1$ and $|v_{n2} - D| < a_2$. This conclusion is called Zemplén's theorem, and it is closely related to the condition of entropy increase in a shock wave.

PROPERTY 4. The tangential component of the velocity v suffers no discontinuity in passing through a shock wave.

DEFINITION. We shall call a shock wave *strong* if $[p]/p_1 \gg 1$ and *weak* if $[p]/p_1 < 1$.

PROPERTY 5. For a weak shock wave there is the identity

$$\lim_{\rho_2 \to \rho_1} \frac{p_2 - p_1}{\rho_2 - \rho_1} = a_1^2 = a_2^2.$$

Property 5 shows that when the intensity of the discontinuity tends to zero the relative velocity of the gas in the direction normal to the surface of the shock wave tends to the speed of sound.

Properties of discontinuity surfaces when $Q \neq 0$ are not considered here. We shall consider this case in Chapter 6 during the study of specific problems.

2.3. *Shock waves for one-dimensional flows.* For one-dimensional gas flows only plane, cylindrical or spherical shock waves are possible in agreement with the types of one-dimensional flows that we considered above. The conditions on the shock waves can be obtained either from the equations in integral form (1.52) for one-dimensional motions or from the general conditions on a shock wave (1.55).

Suppose that there is no heat flow ($q = 0$). In the case of one-dimensional motions the velocity vector is always directed along the normal to the surface of discontinuity, i.e. $v = v_n$. The conditions on a compression jump for one-dimensional motions with no heat supply can be written as follows:

$$\begin{gathered}\rho_1 (D - v_1) = \rho_2 (D - v_2),\\ \rho_1 v_1 (D - v_1) - p_1 = \rho_2 v_2 (D - v_2) - p_2,\\ \rho_1 (D - v_1)\left(\frac{v_1^2}{2} + \varepsilon_1\right) - p_1 v_1 = \rho_2 (D - v_2)\left(\frac{v_2^2}{2} + \varepsilon_2\right) - p_2 v_2,\end{gathered} \tag{1.60}$$

where, as before, the index 1 denotes quantities on one side of the discontinuity surface and the index 2 quantities on the other side.

If it is assumed that the medium in front of the shock wave is at rest, i.e. $v_1 = 0$, then conditions (1.60) simplify:

$$\begin{gathered}\rho_1 D = \rho_2 (D - v_2), \quad p_2 = \rho_2 v_2 (D - v_2) + p_1,\\ \rho_1 D \varepsilon_1 = \rho_2 (D - v_2)\left(\frac{v_2^2}{2} + \varepsilon_2\right) - p_2 v_2.\end{gathered} \tag{1.61}$$

In the case of a perfect gas, on introducing the sound speed in the quiescent gas $a_1 = (\gamma p_1/\rho_1)^{1/2}$ and replacing ϵ in accordance with (1.10), the conditions (1.61) can be written in the form

$$\begin{gathered}v_2 = \frac{2}{\gamma + 1}(1 - q)\, D, \qquad \rho_2 = \frac{\gamma + 1}{\gamma - 1}\rho_1\left[1 + \frac{2q}{\gamma - 1}\right]^{-1},\\ p_2 = \frac{2}{\gamma + 1}\rho_1 D^2\left[1 - \frac{\gamma - 1}{2\gamma} q\right],\end{gathered} \tag{1.62}$$

where $q = a_1^2/D^2$, and v_2, ρ_2, and p_2 represent the velocity, density and pressure directly behind the shock wave propagating through the quiescent gas with speed D.

Replacing D by q in (1.62), we obtain

$$v_2 = \frac{2}{\gamma + 1}\frac{1 - q}{\sqrt{q}} a_1, \qquad \rho_2 = \frac{\gamma + 1}{\gamma - 1 + 2q}\rho_1, \qquad p_2 = \frac{2\gamma - (\gamma - 1)\, q}{(\gamma + 1)\, q} p_1. \tag{1.63}$$

From the last relation of (1.63) we obtain

$$q = \frac{2\gamma}{(\gamma + 1)\dfrac{p_2}{p_1} + \gamma - 1}.$$

This implies that for large magnitudes of p_2/p_1, i.e. very intense shock waves, the quantity q and hence also a_1/D are small.

It is evident from (1.62) that for small q the quantities in square brackets are close to one. Thus, for $\gamma = 1.4$ even for $q < 0.01$ these quantities differ from unity by less than 0.05. If in the relations (1.62) we set $q = 0$ (this is equivalent to $p_1 = 0$), then for $q < 0.01$ the values of v_2, ρ_2, and p_2 admit an error of less than 5%. For $q = 0$ the conditions on the shock wave (1.62) assume the simple form

$$v_2 = \frac{2}{\gamma+1} D, \qquad \rho_2 = \frac{\gamma+1}{\gamma-1} \rho_1, \qquad p_2 = \frac{2}{\gamma+1} \rho_1 D^2. \tag{1.64}$$

In the sequel we shall frequently make use of conditions on a shock wave of the form (1.63) and (1.64). The conditions (1.60) and (1.61) have been derived assuming the absence of heat transfer in the flow by means of heat conduction. If heat conduction is reckoned with, then to the last equation of the system (1.60), which expresses the law of energy conservation, we must add the term $\kappa_2(\partial T/\partial r)|_2$ on the right and the term $\kappa_1(\partial T/\partial r)|_1$ on the left; these terms give the flow of heat to the shock wave (κ is the coefficient of thermal conductivity).

We point out that during the propagation of a shock wave through a quiescent gas conditions of the form (1.64) hold even for motions which are not one-dimensional, only in this case we must take $v_2 = v_{n2}$. This follows from the fact that the tangential component of the velocity is continuous, and since it is equal to zero in front of the wave it is also equal to zero behind the wave, i.e. $v_2 = nv_{n2}$.

§3. The equations of magnetohydrodynamics

3.1. *The system of differential equations of magnetohydrodynamics.* During the motion of an electrically conducting fluid or gas in electromagnetic fields both the electromagnetic forces and energy exchange between the moving gas and the electromagnetic field have an effect on the picture of gas flow. On the other hand, the motion of the fluid can substantially affect the parameters of the electromagnetic field.

In describing the motion of an electrically conducting gas taking the interaction with an electromagnetic field into account Euler and Lagrange variables may be used just as in ordinary gas dynamics. Essentially the only change is in the number of unknown functions, and the system of Maxwell's equations for the variation of electric and magnetic fields in moving conductors is added to the equations of gas dynamics.

The equations of magnetohydrodynamics consist of the system of Maxwell's equations and the equations of fluid motion (the equations of continuity, momentum, and energy) with consideration of the interaction between the electromagnetic field and the moving medium.

In describing the motion of electrically conducting gases we shall use the Gaussian system of electromagnetic units. We take Maxwell's equations in the form

$$\begin{aligned} \operatorname{curl} \boldsymbol{E} &= -\frac{1}{c}\frac{\partial \boldsymbol{B}}{\partial t}, & \operatorname{div} \boldsymbol{B} &= 0, \\ \operatorname{curl} \boldsymbol{H} &= \frac{4\pi}{c}\boldsymbol{j} + \frac{1}{c}\frac{\partial \boldsymbol{D}_u}{\partial t}, & \operatorname{div} \boldsymbol{D}_u &= 4\pi\rho_e, \end{aligned} \tag{1.65}$$

where $\boldsymbol{E}$ is the stress vector of the electric field, $\boldsymbol{H}$ is the stress vector of the magnetic field, $\boldsymbol{j}$ is the current density, $\boldsymbol{B}$ is the vector of magnetic induction, $\boldsymbol{D}_u$ is the vector of electric induction, ρ_e is the macroscopic density of electric charge, and c is the speed of light in a vacuum.

We shall explain some of the concepts introduced. The electric current arising in conductors, in particular in an ionized gas (plasma), represents the motion of charged particles. We shall denote by v_k the microscopic velocities of electrons and ions, and we denote their charges by e_k.

The current density j is formed as the mean, over a small volume Ω of the medium, of the sum of products of the charges by the velocities of the electrons and ions in Ω: $j = \Sigma e_k v_k / \Omega$.

The vector j is often written as the sum of two terms:

$$j = j^* + \rho_e v,$$

where j^* is the usual current both in moving and stationary conductors arising under the influence of an electromagnetic field and called the conduction current, while the part $\rho_e v$ represents the current related to the transport of macroscopic charge. Since for the majority of conductors $\rho_e v \ll j^*$, we shall, as a rule, neglect the convective part of the current and assume that $j = j^*$.

DEFINITION. The vector relation between the current density j and the characteristics of the magnetic field and the moving medium we shall call *Ohm's law*.

In its simplest form Ohm's law for moving conductors is written as

$$j^* = \sigma\left(E + \frac{1}{c}(v \times H)\right). \tag{1.66}$$

Here the scalar coefficient σ is called the conductivity of the medium and will be assumed to be a function of the thermodynamic parameters of the medium.

The relation between the induction vectors D_u and B and the field vectors E and H is determined by the properties of polarization and magnetization of the medium. For motions of media of the type of dielectrics and ferromagnets these relations can be rather complicated and nonlinear. In many cases for isotropic media these relations are written simply as $B = \mu H$ and $D_u = \epsilon_e E$, where μ is the coefficient of magnetic permeability and ϵ_e is the dielectric permeability (the dielectric constant). In magnetohydrodynamics properties of polarization and magnetization are usually not considered and it is assumed that $\mu = \text{const}$ and $\epsilon_e = \text{const}$ [9], [10] (in our system of units $\mu = \epsilon_e = 1$). We note that for the propagation of electromagnetic waves in a vacuum we have $B = H$ and $D_u = E$, i.e. $\mu = 1$ and $\epsilon_e = 1$. The second group of equations of magnetohydrodynamics is essentially a generalization of the equations of gas dynamics introduced above and without considering viscosity and nonelectromagnetic inertial forces can be written as follows:

$$\frac{\partial \rho}{\partial t} + \operatorname{div} \rho v = 0 \tag{1.67}$$

—the equation of continuity,

$$\frac{\partial \rho \boldsymbol{v}}{\partial t} + \operatorname{div} \hat{P} = \boldsymbol{F}_L \tag{1.67a}$$

—the equation for momentum, and

$$\frac{\partial}{\partial t}\left(\rho\varepsilon + \rho\frac{v^2}{2}\right) + \operatorname{div}\left[\rho\boldsymbol{v}\left(\varepsilon + \frac{v^2}{2}\right) + p\boldsymbol{v} + \boldsymbol{q}\right] = \boldsymbol{jE} \tag{1.67b}$$

—the energy equation.

In the system of equations (1.67)–(1.67b) $\boldsymbol{F}_L$ is the Lorentz force or inertial pondermotive force of the action of the electromagnetic field on the medium:

$$\boldsymbol{F}_L = \rho_e \boldsymbol{E} + \frac{1}{c}(\boldsymbol{j} \times \boldsymbol{H}),$$

$\hat{P}$ is the tensor with components $P_{ik} = -\delta_{ik}p + \rho v_i v_k$, $\boldsymbol{q}$ is the vector of heat flow, and the product $\boldsymbol{jE}$ in (1.67b) characterizes the change in energy of the fluid caused by the presence of the electromagnetic field.

The system of equations (1.67) can be reduced to divergence form. Using Maxwell's equations, the system (1.67) can be written in the form [10]

$$\begin{gathered}
\frac{\partial \rho}{\partial t} + \operatorname{div}(\rho\boldsymbol{v}) = 0, \\
\frac{\partial}{\partial t}\left(\rho\boldsymbol{v} + \frac{1}{4\pi c}(\boldsymbol{E} \times \boldsymbol{H})\right) + \operatorname{div}(\rho v_i v_k + p\delta_{ik} - \tau_{ik}) = 0, \\
\frac{\partial}{\partial t}\rho\left[\varepsilon + \frac{v^2}{2} + \frac{H^2 + E^2}{8\pi}\right] + \operatorname{div}\left[\frac{c}{4\pi}(\boldsymbol{E} \times \boldsymbol{H}) + \boldsymbol{q}\right. \\
\left. + p\boldsymbol{v} + \rho\boldsymbol{v}\left(\varepsilon + \frac{v^2}{2}\right)\right] = 0, \\
\tau_{ik} = \frac{1}{4\pi}(H_i H_k + E_i E_k) - \frac{1}{8\pi}(H^2 + E^2)\delta_{ik}.
\end{gathered} \tag{1.68}$$

Equations (1.68) together with (1.65) and (1.66) constitute the full system of equations for the motion of a compressible, electrically conducting fluid.

We shall not discuss here the question of the limits of applicability of these equations in describing the motion of ionized gases. We mention only that these equations for the motion of an electrically conducting fluid are applied very often in considering non-relativistic astrophysical phenomena such as the motion of ionized gases in the sun's atmosphere.

We now turn to a simplified system of equations which are also often called the system of equations of magnetohydrodynamics or, more precisely, the equations of magneto-gas dynamics.

We neglect the displacement currents in Maxwell's equations, i.e. $\partial E/\partial t$, and the convective current $\rho_e v$; this is justified for cases of observable conduction currents j^* and if there are no very high frequency oscillations [10]. Estimates given in [9]–[11] indicate that for conducting media the ratio of the electric energy per unit volume to the corresponding magnetic energy is of the order v^2/c^2, which is a very small quantity, since for the majority of phenomena $(v/c) \ll 1$.

Under the conditions indicated the equations of magnetohydrodynamics take the form

$$\frac{\partial \rho}{\partial t} + \operatorname{div} \rho \boldsymbol{v} = 0, \tag{1.69}$$

$$\frac{\partial \rho v_i}{\partial t} = -\frac{\partial}{\partial x_k}\left[\rho v_i v_k + p\delta_{ik} - \frac{1}{4\pi}\left(H_i H_k - \frac{1}{2} H^2 \delta_{ik}\right)\right], \tag{1.70}$$

$$\frac{\partial}{\partial t}\left(\frac{\rho v^2}{2} + \rho\varepsilon + \frac{H^2}{8\pi}\right) = -\operatorname{div}\left\{\rho \boldsymbol{v}\left(\frac{v^2}{2} + \varepsilon + \frac{p}{\rho}\right) + \frac{1}{4\pi}[\boldsymbol{H} \times (\boldsymbol{v} \times \boldsymbol{H})]\right\}, \tag{1.71}$$

$$\frac{\partial \boldsymbol{H}}{\partial t} = \operatorname{curl}\,[\boldsymbol{v} \times \boldsymbol{H} - \nu_m \operatorname{curl} \boldsymbol{H}], \tag{1.72}$$

$$\operatorname{div} \boldsymbol{H} = 0, \tag{1.73}$$

where $\nu_m = c^2/4\pi\sigma$ is the magnetic viscosity.

In the system (1.69)–(1.72) equation (1.72) is a consequence of Maxwell's equations and Ohm's law. Equation (1.72) is called the induction equation (as in the system of Maxwell's equations). Equation (1.73) occupies a special place in this system. Indeed, if we apply the divergence operation to the induction equation, we find that in magnetohydrodynamics

$$\frac{\partial}{\partial t} \operatorname{div} \boldsymbol{H} = 0.$$

Therefore, in nonstationary problems, in order that a solution H satisfy (1.73) it is sufficient that the initial conditions satisfy this equation. Thus the equation div $H = 0$ is satisfied by virtue of the induction equations and the initial conditions. However, in solving problems it is convenient to use this equation in place of one of the projections of the induction equation.

DEFINITION. A gas is called *ideally* (or *infinitely*) *conducting* if the condition $\nu_m = 0$ is satisfied. From the definition of the magnetic viscosity it follows that this is satisfied when $\sigma \to \infty$. The equations of magentohydrodynamics for an ideally conducting gas are easily obtained from the system (1.69)–(1.73) when it is assumed that $\nu_m = 0$.

We mention that for an ideally conducting gas Ohm's law takes the simple form

$$\boldsymbol{E} = -\frac{1}{c}(\boldsymbol{v} \times \boldsymbol{H}). \tag{1.74}$$

Let L be a characteristic linear dimension of the region in which the flow of an electrically conducting gas takes place, and let v_* be a characteristic velocity of the motion of the gas. In analogy with the usual Reynolds number we introduce the dimensionless number

$$R_m = \frac{v_* L}{\nu_m}. \tag{1.75}$$

The number R_m is called the magnetic Reynolds number. If $R_m \ll 1$ we speak of motions with small Reynolds numbers; if $R_m \gg 1$ we speak of motions with large Reynolds numbers. If $R_m \ll 1$, the effect of the motion of the gas on the electromagnetic field may be neglected [12]. For $R_m \gg 1$ the equations of motion may be taken to be the equations of magnetohydrodynamics in the approximation of an ideally conducting gas. Indeed, if in the induction equation and Ohm's law we go over to dimensionless quantities and then let R_m tend to infinity, then, assuming that the other quantities remain finite, we obtain the corresponding equations for an ideally conducting gas. If the numbers R_m are finite but sufficiently large, the model of an infinitely conducting gas can be used as an approximate model of the phenomenon.

If the magnetic Reynolds numbers are not large and are of order one (which is very often the case in real processes), then the complete system of equations (1.69)–(1.73) must be used in the theoretical solution of problems.

The quantity $H^2/8\pi$ we call the magnetic pressure. If for the motion of the gas the relation $(H^2/8\pi) \ll p$ $((H^2/8\pi) \ll \rho v^2)$ holds, we consider the magnetic field to be weak. If the numbers R_m are not small, then for weak fields it is possible to neglect the Lorentz force and the electromagnetic energy but take into account the effect of the motion of the gas on the variation of the parameters of the electromagnetic field. The approximate description of the motions indicated above we shall call the *weak field approximation.*

We now consider briefly the symmetry properties of gas flows in a magnetic field. For $H \neq 0$ the equations of magnetohydrodynamics do not admit spherically symmetric one-dimensional motions for which there is interaction of the medium and the field. However, one-dimensional motions with plane and cylindrical symmetry are altogether possible, and we shall consider these in the sequel. It is natural that plane motions and motions with axial symmetry are encountered very frequently in problems of magnetohydrodynamics.

To conclude this subsection we mention some important properties of the equations of motion of an ideally conducting liquid. As already noted, the induction equation in this case has the form

$$\frac{\partial \boldsymbol{H}}{\partial t} = \operatorname{curl} (\boldsymbol{v} \times \boldsymbol{H}). \tag{1.76}$$

This equation is identical to the equation for the vortex velocity in the hydrodynamics of a nonviscous fluid; as is known, it implies that the vorticity lines move together with the fluid.

Thus it follows from (1.76) that the magnetic field varies as if the magnetic force lines were rigidly coupled with matter. It can be shown (cf., for example, [9]) that (1.76) implies the constancy of the magnetic current penetrating a contour each point of which moves with local velocity $\boldsymbol{v}$. Motion along the force lines has no effect on the field; for motion in a transverse direction the force lines are displaced together with the matter. This property of the motion of an ideally conducting fluid is called the property of frozen force lines of the magnetic field. Following [9], [13], we now present certain general results related to the property of the force lines being frozen. Using (1.69) and (1.73), equation (1.76) determining the variation of the field can be brought to the form

$$\frac{d}{dt}\left(\frac{\boldsymbol{H}}{\rho}\right)=\left(\frac{\boldsymbol{H}}{\rho}\operatorname{grad}\right)\boldsymbol{r}, \tag{1.77}$$

where d/dt is the total derivative.

Suppose that the radius vector $\boldsymbol{r}_0$ of some particle over a certain interval of time becomes $\boldsymbol{r}$ and that $\boldsymbol{H}_0$ and ρ_0 assume correspondingly the values $\boldsymbol{H}$ and ρ.

Equation (1.77) implies a relation between $\boldsymbol{r}_0$, $\boldsymbol{H}_0$, ρ_0 and $\boldsymbol{r}$, $\boldsymbol{H}$, ρ, called the integral of the frozen characteristic [9], [13],

$$\frac{\boldsymbol{H}}{\rho}=\left(\frac{\boldsymbol{H}_0}{\rho_0}\operatorname{grad}\right)\boldsymbol{r} \tag{1.78}$$

or in coordinate form

$$\frac{H_i}{\rho}=\frac{H_{0j}}{\rho_0}\frac{\partial x^i}{\partial x_0^j}=\frac{\partial x^i}{\partial x_0^j}\frac{H_{0j}}{\rho\left|\dfrac{\partial x^i}{\partial x_0^j}\right|}, \tag{1.79}$$

where $|\partial x^i/\partial x_0^j|$ denotes the determinant of the matrix $\|\partial x^i/\partial x_0^j\|$.

We consider one very important particular form of relation (1.79). Suppose that the motion is planar and the vector $\boldsymbol{H}$ is perpendicular to the plane of the motion. Then from (1.77) or (1.78) it follows that

$$\frac{H}{\rho}=\frac{H_0}{\rho_0}, \tag{1.80}$$

since in this case the length of an element of a magnetic force line does not change.

Formally this is easy to see if (1.77) is transformed using the condition $\operatorname{div}\boldsymbol{H}=0$ into

$$\frac{d}{dt}\frac{\boldsymbol{H}}{\rho}=\frac{\nabla\boldsymbol{H}\boldsymbol{v}}{\rho}.$$

This implies the integral mentioned above, since $\boldsymbol{H}\cdot\boldsymbol{v}=0$. When studying plane motions of an ideally conducting gas in a magnetic field perpendicular to the plane of motion the relation (1.80) makes it possible to eliminate H from the equations of magnetohydrodynamics. Moreover, if in place of the pressure and internal energy we introduce the functions $p^* = p + (H^2/8\pi)$ and $\epsilon^* = \epsilon + (H^2/8\pi)$, then the equations of magnetohydrodynamics have the same form as the equations of ordinary gas dynamics. This implies that solutions of problems of magnetohydrodynamics in this case can be obtained by conversion of the corresponding problems of ordinary gas dynamics. It is only necessary to alter the boundary and initial conditions of the problems.

We also mention here the following important property for a perfect gas with $\gamma = 2$: all thermodynamic formulas remain unchanged if p is replaced by p^* therein.

3.2. *Equations of the magnetohydrodynamic approximation for the motion of a rarefied plasma.* Equations analogous to the hydrodynamical equations can be written down also for the motion of a strongly rarefied plasma in a magnetic field. From the kinetic Boltzmann-Vlasov equation and Maxwell's equations [14], [15] it is possible to obtain the following equations of the hydrodynamical approximation:

$$\frac{\partial \rho}{\partial t} + \operatorname{div} \rho \boldsymbol{v} = 0, \tag{1.81}$$

$$\frac{\partial \rho v_i}{\partial t} + \operatorname{div} \hat{T} = 0, \tag{1.82}$$

$$\frac{\partial}{\partial t}\left(\varepsilon + \frac{\rho v^2}{2}\right) + \operatorname{div} \boldsymbol{Q} = 0, \tag{1.83}$$

$$\frac{\partial}{\partial t}(p_\perp/B) + \operatorname{div}(p_\perp \boldsymbol{v}/B) = 0, \tag{1.84}$$

$$\frac{\partial \boldsymbol{B}}{\partial t} + \operatorname{curl}(\boldsymbol{B} \times \boldsymbol{v}) = 0, \quad \operatorname{div} \boldsymbol{B} = 0. \tag{1.85}$$

Here $\hat{T}$ is the tensor of momentum current with components

$$T_{ik} = \rho v_i v_k + p_\perp \delta_{ik} + \frac{p_\parallel - p_\perp}{B^2} B_i B_k - \frac{1}{4\pi}\left(B_i B_k - \frac{1}{2} B^2 \delta_{ik}\right),$$

Q is the vector of energy current

$$\boldsymbol{Q} = \rho \boldsymbol{v}\left(\frac{v^2}{2} + \varepsilon + \frac{p_\perp}{\rho}\right) + \frac{p_\parallel - p_\perp}{B^2} \boldsymbol{B}(\boldsymbol{v}\boldsymbol{B}) + \frac{1}{4\pi}(\boldsymbol{B} \times (\boldsymbol{v} \times \boldsymbol{B})),$$

$\epsilon = (1/\rho)(p_\perp + p_\parallel)$ has the interpretation of the internal energy of the plasma, and the quantities $p_\perp$ and $p_\parallel$ which enter in the stress tensor, the vector of energy current, and the expression for ϵ are called the transverse and longitudinal pressures. In considering questions of the motion of a rarefied plasma, for the characteristic of the magnetic field we shall use the magnetic induction vector $\boldsymbol{B}$, this being the quantity most frequently used in the corresponding literature.

The system (1.81)–(1.85) provides a rough description of the behavior of a rarefied plasma. However, it is frequently used for qualitative investigation of the motion of a plasma of very low density in the presence of a magnetic field. There is a discussion of the applicability of these equations to physical problems in [16].

Without going into this, we mention that the system (1.81)–(1.85) can be used only for the description of motions of a rarefied plasma in the presence of a magnetic field and cannot be applied to describe the motion of a rarefied gas which is not electrically conducting. We note further that the energy equation (1.84) can be transformed by means of the other equations of the system (1.81)–(1.85) to the form

$$\frac{d}{dt}\left(\frac{p_{\parallel} B^2}{\rho^3}\right) = 0. \tag{1.86}$$

Equation (1.86) may be used in the system (1.81)–(1.85) in place of the energy equation.

The system of equations (1.81)–(1.85), or one equivalent to it, we shall henceforth call the *system of magnetohydrodynamic equations* for a rarefied plasma.

The systems introduced are basic for the investigation of many problems of magnetohydrodynamics and the dynamics of plasmas. In these equations an ionized gas is considered as a unified electrically conducting fluid; therefore, equations of the type considered are called the equations in the single-liquid approximation.

It is possible to introduce two-liquid or two-component models of the motion of completely ionized gases, i.e. to introduce an electron liquid for the flow of electrons, an ion liquid for the motion of ions, and forces which take into account their interaction and also the interaction with electromagnetic fields.

If the gas is not completely ionized, a three-liquid or three-component model is used to describe the motion of a gas consisting of electrons, ions (of a single sort), and neutral particles.

We shall not consider here two- or multiple-component MHD equations. For sufficiently dense gases these equations are derived in [17], [18]. For a strongly rarefied plasma they are presented, for example, in [16], [19].

In addition to the equations in the form of conservation laws, in plasma physics one utilizes the consideration of the behavior of the trajectories of individual particles in electromagnetic fields, and on the basis of the composite data on this motion the currents and charges are found which are used in Maxwell's equations describing the variation of the parameters of the electromagnetic field. Here there has been developed a special technique of averaging the motions and estimating the variation of the electromagnetic fields caused by the motion of the particles [20]–[25]. We shall present certain facts from the theory of motion of a charged particle in electromagnetic fields.

The motion of a particle in an external electromagnetic field is described by the equations

$$\frac{d\boldsymbol{r}}{dt} = \boldsymbol{v}, \qquad \frac{d\,(m\boldsymbol{v})}{dt} = e\left(\boldsymbol{E} + \frac{1}{c}\,\boldsymbol{v}\times\boldsymbol{B}\right), \tag{1.87}$$

where $\boldsymbol{r}$ is the radius vector of the particle, e is its charge, and m its mass. For relativistic velocities the mass m must be considered a variable quantity:

$$m = \frac{m_0}{\sqrt{1 + v^2/c^2}}, \tag{1.88}$$

where m_0 is the rest mass.

If $E = 0$, then the force acting on the particle $F = (e/c)(v \times B)$ has no component parallel to B, and therefore the momentum $p_e = mv$ has a constant component parallel to B. Further, scalar multiplication of (1.87) by p_e gives $p_e(dp_e/dt) = 0$ and $p_e^2 = \text{const}$. From this and (1.88) it follows that $m = \text{const}$ and the length of the vector v is constant.

Suppose that the magnetic field is constant and homogeneous. We decompose the total velocity v into the velocity $v_\parallel$ parallel to the vector B and the velocity $v_\perp$ perpendicular to B. In a coordinate system moving with velocity v the trajectory of the particle is a circle. The motion of the particle perpendicular to the magnetic field is a uniform rotation about a circle with angular velocity

$$\omega = -\frac{eB}{mc}, \tag{1.89}$$

which is called the Larmor or cyclotron frequency. The radius of the circle and the circle itself are called the Larmor radius and circle. For the radius of the Larmor circle r_L we have

$$r_L = \left|\frac{v_\perp}{\omega}\right| = \left|\frac{mcv_\perp}{eB}\right|. \tag{1.90}$$

For a positively charged particle the angular velocity vector ω is directed opposite to B, and for a negatively charged particle it is directed along B.

We now make the following

DEFINITION. The *magnetic moment* of the particle is the vector quantity

$$\mu = -\frac{mv_\perp^2}{2B^2} B. \tag{1.91}$$

The magnetic moment μ has the interpretation of the product of the mean current created by the rotating particle and the vector area of the Larmor circle: $\mu = Is/c = (1/c)Is\omega/|\omega|$.

Equation (1.91) can be obtained from this by taking

$$I = \frac{e}{2\pi}\omega, \qquad s = \pi r_L^2.$$

In a constant and homogeneous magnetic field the particle moves along a cylindrical screw line (a spiral). The axis of the screw line is parallel to B. The magnetic moment of the particle μ remains constant during such motion. The picture of the motion of the particle becomes more complicated if $E \neq 0$ and $B \neq 0$, and E and B are variable.

We shall not discuss in detail questions of the motion of a particle in variable electromagnetic fields. We mention, however, that the quantity $m\mu$ for relativistic velocities and the quantity μ for nonrelativistic velocities change very little during the process of motion for weak electric fields and for slow spatial and time variations of $\boldsymbol{B}$. In other words, these quantities are approximate integrals of the motion of the particle or adiabatic invariants.

Returning to the equations of MHD for a rarefied plasma, we shall clarify the physical interpretation of equations (1.81)–(1.86). We point out, first of all, that the derivation of these equations proceeds from the Boltzmann-Vlasov equation under the assumption that the characteristic Larmor radius of motion of the particles is considerably less than the characteristic length of inhomogeneities of the plasma and the corresponding Larmor frequencies are greater than the characteristic time frequencies of the processes considered.

Further, the quantity $p_\perp/\rho B$ has the interpretation of the mean magnetic moment of particles calculated per unit mass. The velocity $\boldsymbol{v}$ entering in the system of equations has the meaning of the mean velocity per unit volume of ncninteracting (in the sense of short-range interactions or collisions) particles. Therefore, if for the moment we call an element of volume moving with velocity $\boldsymbol{v}$ an individual (or fluid) volume, equation (1.84) means that the average magnetic moment $p_\perp/\rho B$ is conserved in an individual volume. During compression of the plasma in the direction of the magnetic field the quantities $\boldsymbol{B}$ and $p_\perp$ do not change, while the quantities $p_\parallel$ and ρ are related by an adiabatic dependence with exponent $\gamma = 3$. Thus (1.86) plays the role of an adiabatic condition in an individual volume during the motion of the medium along the magnetic force lines. We mention also that if we formally set $p_\parallel = 0$, then the system (1.81)–(1.85) goes over into the system of equations of ideal MHD for a perfect gas with $\gamma = 2$, where $p_\perp$ plays the role of the usual pressure.

3.3. *The linearized system of equations of magnetohydrodynamics.* By using the procedure for linearization of the equations of hydrodynamics considered above, the system of MHD equations can be linearized about the states where $\boldsymbol{v}_0 = 0$, $\rho =$ const and $\boldsymbol{H}_0 =$ const (a homogeneous field). Suppose that the vector $\boldsymbol{H}_0$ is directed along the z axis and that the conductivity is infinite. Then by transformation of the system of equations (1.69) we find that

$$\frac{\partial^4\rho}{\partial t^4} = L\Delta\rho, \tag{1.92}$$

$$\frac{\partial^2}{\partial t^2} L_H \boldsymbol{H} = \frac{\boldsymbol{H}_0}{\rho_0}\frac{\partial^4\rho}{\partial t^4} - \frac{\boldsymbol{H}_0}{\rho_0}\nabla\left(L\frac{\partial\rho}{\partial z}\right), \tag{1.93}$$

$$\frac{\partial \boldsymbol{v}}{\partial t} = -\frac{a_0^2}{\rho_0}\nabla\rho - \frac{H_0}{4\pi\rho_0}\left(\nabla H_z - \frac{\partial}{\partial z}\boldsymbol{H}\right), \tag{1.94}$$

where the differential operators L and L_H are given by

$$L = \left(a_0^2 + a_A^2\right)\frac{\partial^2}{\partial t^2} - a_0^2 a_A^2 \frac{\partial^2}{\partial z^2}, \tag{1.95}$$

$$L_H = \frac{\partial}{\partial t^2} - a_A^2 \frac{\partial^2}{\partial z^2}, \tag{1.96}$$

a_0^2 is the usual speed of sound, and $a_A = H_0/\sqrt{4\pi\rho_0}$ is the Alvén velocity.

We note that if $a_A = 0$, in place of (1.92) we have $(\partial^2/\partial t^2 - a_0^2\Delta)\rho = 0$, i.e. the usual wave equation.

If $a_0 = 0$, from (1.92) we also obtain the wave equation, where in place of a_0^2 we have a_A^2.

In the particular case of an incompressible fluid ($\rho = \text{const}$), from (1.93) we obtain

$$L_H \boldsymbol{H} = 0. \tag{1.97}$$

Equation (1.97) gives us the equation of propagation of MHD or Alvén waves. As is evident from (1.97), these waves are connected with the propagation of disturbances along the force lines of the field frozen into the incompressible medium.

Analysis of the equations for the propagation of small disturbances along the z axis shows [9], [10] that there are three characteristic propagation speeds a_A, a_+, a_-, where for the latter we have

$$a_{+,-}^2 = \frac{1}{2}\left(a_0^2 + \frac{H_0^2}{4\pi\rho_0} \pm \sqrt{\left(a_0^2 + \frac{H_0^2}{4\pi\rho_0}\right) - \frac{a_0^2 H_{z0}^2}{\pi\rho_0}}\right), \tag{1.98}$$

where a_+ is the speed of the fast magnetosonic wave and a_- the speed of the slow magnetosonic wave.

§4. Conditions on magnetohydrodynamic shock waves

4.1. *Magnetohydrodynamic discontinuities.* The conditions on MHD shock waves can be obtained by standard methods from the integral conservation laws written both for the moving fluid and for the electromagentic field. We shall not discuss in detail the derivation of these conditions for ordinary MHD, since the derivation can be carried out in analogy with that of §2 for the case of shock waves of gas dynamics. A derivation of these conditions is given, for example, in [6], [9], [10], [21].

For the parameters of the electromagnetic field there are the conditions [6], [9], [26]

$$\boldsymbol{n} \times [\boldsymbol{H}] - \frac{1}{c} D[\boldsymbol{D}_u] = \boldsymbol{i}, \tag{1.99}$$

$$\boldsymbol{n} \times [\boldsymbol{E}] - D[\boldsymbol{B}] = 0, \tag{1.100}$$

$$\boldsymbol{n} \times [\boldsymbol{B}] = 0, \tag{1.101}$$

where D denotes the speed of the shock wave, $\boldsymbol{n}$ is the unit normal, and $\boldsymbol{i}$ is the surface current. If displacement currents are neglected and it is assumed that $B = H$ and $D_u = E$, then in correspondence with Maxwell's equations and Ohm's law (1.66) we have

$$\boldsymbol{E} = \frac{1}{c} (\nu_m \operatorname{curl} \boldsymbol{H} - \boldsymbol{v} \times \boldsymbol{H}). \tag{1.102}$$

If it is assumed that over the surface of the shock wave no concentrated current flows, then (1.99) together with (1.101) leads to the continuity of the magnetic field, since in addition to (1.101) the difference of the tangential components of the field H is also equal to zero [9]. Thus, in the case of finite conductivity (1.100), (1.102) give

$$\boldsymbol{n} \times (\nu_m \operatorname{curl} \boldsymbol{H} - \boldsymbol{v} \times \boldsymbol{H}) = 0. \tag{1.103}$$

Since the magnetic field is continuous in the case of finite conductivity, the hydrodynamical conditions on the surface of discontinuity reduce to the conditions (1.55) already considered. However, the additional condition (1.103) must be considered when solving problems on the motion of an electrically conducting gas of finite conductivity. For finite conductivity the conditions on the discontinuities are simply obtained from the equations of MHD in integral form for an arbitrary moving volume $\Omega(t)$. These equations have the form

$$\frac{d}{dt} \int_\Omega \rho \, d\Omega + \oint_\Sigma \rho (v_n - D) \, d\Sigma = 0, \tag{1.104}$$

$$\frac{d}{dt} \int_\Omega G \, d\Omega + \oint_\Sigma (\hat{T} \boldsymbol{n}) (v_n - D) \, d\Sigma = 0, \tag{1.105}$$

$$\frac{d}{dt} \int_\Omega W \, d\Omega + \int_\Sigma S_n (v_n - D) \, d\Sigma = 0. \tag{1.106}$$

Equations (1.104)–(1.106) follow from the analogous relations for a constant volume obtained in [9] and from the formula (1.46) for the differentiation of integrals. In these equations $\hat{T}$ is the tensor of the density of the total momentum current with components

$$T_{ik} = \rho v_i v_k + p \delta_{ik} - \frac{1}{4\pi} (H_i H_k + E_i E_k) - \frac{1}{8\pi} (H^2 + E^2) \delta_{ik},$$

W is the density of total energy,

$$W = \rho \left(\varepsilon + \frac{v^2}{2} + \frac{1}{8\pi} (H^2 + E^2) \right),$$

S is the density of total energy current,

$$S_n = \rho \left(\varepsilon + \frac{v^2}{2} \right) v_n + q_n - (\hat{T} \boldsymbol{v}) \boldsymbol{n} + \frac{c}{4\pi} (\boldsymbol{E} \times \boldsymbol{H}) \boldsymbol{n},$$

D is the velocity of the boundary Σ, and $S_n = \boldsymbol{S} \cdot \boldsymbol{n}$.

Conditions on the discontinuities for ideally conducting media are easily obtained from (1.104)–(1.106).

If it is recalled that $E = (1/c)(v \times H)$ and use is made of the electrodynamic relations (1.99)–(1.102), then for $\mathbf{q} = 0$ the complete system of conditions on a surface of discontinuity takes the form [9], [11]

$$[H_n] = 0, \tag{1.107}$$

$$[(v_n - D)\,\boldsymbol{H} - H_n \boldsymbol{v}] = 0, \tag{1.108}$$

$$[(v_n - D)\,\rho] = 0, \tag{1.109}$$

$$\left[(v_n - D)\,\rho \boldsymbol{v} + \left(p + \frac{H^2}{8\pi}\right)\boldsymbol{n} - \frac{H_n}{4\pi}\boldsymbol{H}\right] = 0, \tag{1.110}$$

$$\left[(v_n - D)\left(\frac{\rho v^2}{2} + \rho\varepsilon + \frac{H^2}{8\pi}\right) + v_n\left(p_n + \frac{H^2}{8\pi}\right) - \frac{H_n}{4\pi}(\boldsymbol{H} \cdot \boldsymbol{v})\right] = 0. \tag{1.111}$$

Here the first relation is a consequence of the absence of magnetic charges, and the second follows from the field equations (1.100) and the formula for E. The relation (1.108) is essentially an expression of the frozen-in law. The remaining relations are respectively consequences of the laws of conservation of mass, momentum, and energy. The relations (1.107)–(1.111) are substantially more complicated than the analogous relations for ordinary hydrodynamics, and they provide a greater variety of possible discontinuities. A detailed classification of MHD discontinuities is given in [9], [10], [26], [28].

If there is no flow of matter through the discontinuity surface, i.e. $m = \rho_1(v_{n1} - D) = \rho_2(v_{n2} - D) = 0$, then there is either a tangential discontinuity ($H_n = 0$) or a contact discontinuity ($H_n \neq 0$), $[p] = 0$, $[v_n] = 0$, $[H] = 0$. For $m \neq 0$ there are rotational discontinuities and shock waves. A discontinuity is called rotational if the condition $[v_n] = [\rho] = 0$ is satisfied. For this discontinuity from the relations (1.107)–(1.111) we obtain

$$m = \frac{H_n \sqrt{\rho}}{\sqrt{4\pi}}, \qquad v_n - D = \frac{H_n}{\sqrt{4\pi\rho}}. \tag{1.112}$$

All thermodynamic parameters are continuous at a rotational discontinuity. Moreover, we have $[H^2] = 0$, i.e. the modulus of the vector H is continuous. The vector H changes its direction on passing through the discontinuity, and $[H] = km\frac{1}{2}(H_1 + H_2) \times \boldsymbol{n}$, where k is a constant.

Surfaces of the discontinuity for which $m \neq 0$ and $[\rho] \neq 0$ are called shock waves. The following types of shock waves occur: a fast shock wave, a slow shock wave, a switch-on shock wave, and a switch-off shock wave. The last two are limiting cases of the first two types.

The fast shock wave is the analogue of the ordinary gas-dynamic wave, and the strength of the magnetic field increases on passing through the discontinuity. In the slow wave the magnetic field decreases. We mention the basic properties of fast and slow shock waves.

PROPERTY 1. Shock waves are compression jumps. Rarefaction jumps are impossible, since these correspond to transitions with entropy decrease [29], [30].

PROPERTY 2. The change in entropy is a quantity of third order relative to the intensity of the shock wave, and the following formula holds:

$$T(S_2 - S_1) = \frac{1}{12}\left(\frac{\partial^2 V}{\partial p^2}\right)_S (p_2 - p_1)^3 - \frac{1}{16\pi}\left(\frac{\partial V}{\partial p}\right)_S (p_2 - p_1)(\boldsymbol{H}_{\tau 2} - \boldsymbol{H}_{\tau 1})^2. \quad (1.113)$$

Here T is the temperature, S is the entropy, $V = 1/\rho$ is the specific volume, and H_τ is the magnitude of the tangential component of the magnetic field H.

PROPERTY 3. For slow and fast waves respectively there are the inequalities

$$a_{-,1} \leqslant D - v_{n1} < a_{A1}, \qquad D - v_{n2} \leqslant a_{-,2}, \qquad a_{+,1} \leqslant D - v_{n1},$$
$$a_{A2} < D - v_{n2} \leqslant a_{+,2}. \quad (1.114)$$

Conditions (1.114) follow from the evolutionary conditions of a shock wave. In (1.114) the equality signs correspond to shock waves of infinitely small intensity. We now make the following

DEFINITION. The *equation of a shock adiabatic* is

$$\varepsilon_2 - \varepsilon_1 + \frac{p_2 + p_1}{2}(V_2 - V_1) + \frac{1}{16\pi}(V_2 - V_1)(\boldsymbol{H}_{\tau 2} - \boldsymbol{H}_{\tau 1})^2 = 0. \quad (1.115)$$

We mention one more important property of shock waves for which $H_n = 0$, i.e. the magnetic field is parallel to the wave front. For such shock waves we find from (1.108) and (1.109) that $[H/\rho] = 0$.

It is customary to call *switch-on shock waves* those waves for which $H_{\tau 1} = 0$ but $H_{\tau 2} \neq 0$. After the passage of such a wave a tangential component of the magnetic field appears (is switched on) which was absent before the wave. A switch-on shock wave is essentially a fast shock wave whose intensity must be lower than some critical value and in front of which the velocity of the Alvén wave is supersonic.

We point out that in a switch-on wave together with the tangential field component there also appears a tangential component of the gas velocity. A switch-on shock wave can occur if the following inequality is satisfied [21]:

$$p_1 < \frac{H_1^2}{4\pi\gamma}. \quad (1.116)$$

Switch-off shock waves are characterized by the fact that for them $H_{\tau 2} = 0$ and $H_{\tau 1} \neq 0$, i.e. during the propagation of this wave after its passage the tangential component of the magnetic field which existed before its arrival vanishes (is switched off). The switch-off wave is a slow wave whose intensity reaches some characteristic value, while the velocity of the Alvén wave behind the shock wave is supersonic.

Questions of the effect of finite conductivity on the propagation of shock waves have been considered in a number of works (cf., for example, [31]).

In MHD so-called shock waves with a conductivity jump are also possible. These shock waves arise in media for which the conductivity in front of the shock wave can be assumed to be zero while behind the front of the shock wave it is nonzero and may even be infinite [32], [33]. Shock waves with a conductivity jump can occur in gases under the condition of propagation of a strong shock wave which ionized the gas. Conditions on shock waves of this type can be rather diverse depending on the relations between the dissipative coefficients (i.e. the coefficients of viscosity, thermal conductivity, and electrical conductivity) and their behavior on passage through the discontinuity. We mention that if the dominant dissipative coefficient is the coefficient of magnetic viscosity, then the supplementary condition on fast ionizing shock waves should be [33], [34]

$$[\boldsymbol{H}]=0.$$

We shall consider this type of wave also in the solution of problems in the theory of explosions.

4.2. *Conditions on shock waves for hydrodynamical models of a rarefied plasma.* The study of the structure of shock waves in ordinary hydrodynamics from the point of view of the kinetic theory of gases or by application of the complete equations of motion for a viscous, heat-conducting fluid indicates that the width of shock waves (i.e. the zone of abrupt gradients of velocities and thermodynamic parameters) is of the order of several lengths of the molecular mean free path. An analogous situation occurs in a dense plasma with a relatively small molecular mean free path. In a plasma of very low density the length of the free path of molecules is extremely large (greater than the characteristic dimensions over which the average parameters of the plasma vary), and from the molecular-kinetic point of view in the formation of zones with rather large gradients of the average plasma parameters an important role is played by the collective properties of the plasma by means of which interactions of the particles through electromagnetic fields and instability mechanisms of the moving ensemble of particles are realized.

Investigations carried out here (cf. [35], [36]) indicate that shock waves can exist with a width much less than the length of the mean free path of the particles. There are now no standard relations on shock waves analogous to the classical Rankine-Hugoniot conditions.

In considering problems concerning a rarefied plasma we take as basis the macroscopic equations of the MHD of a rarefied plasma (1.81)–(1.85). Because of their nonlinearity the possibilities of occurrence of discontinuities here follow from results in the general theory of partial differential equations [37]–[39] and were predicted for the hydrodynamical equations already by Riemann. Since the derivation of conditions on shock waves of this type are but weakly illuminated in the literature, we shall give a brief derivation here. We proceed from the equations of MHD for a rarefied plasma in integral form.

We point out that the electrodynamical part of the conditions will have the form (1.107), (1.108), since we begin here also with Maxwell's equations and the condition of infinite conductivity. The hydrodynamical part of the equations in integral form consists of the equation of mass conservation (1.104), the momentum equation (1.105) (where in place of the tensor $\hat{T}$ we must take the tensor $\hat{T}$ entering in (1.82)), and an energy equation of the form (1.106) (where we must take into account the dependence of the energy ϵ on $p_\perp$ and $p_\parallel$, and in place of the vector $\boldsymbol{S}$ we must take the vector $\boldsymbol{Q}$ of energy flow entering in (1.83)). Moreover, to the equations of the form (1.104)–(1.106) there is added a new equation

$$\frac{d}{dt}\int\limits_{\Omega}(p_\perp/B)\,d\Omega+\oint\limits_{\Sigma}(p_\perp/B)(v_n-D)\,d\Sigma=0, \tag{1.117}$$

which corresponds to the law of conservation of the mean adiabatic invariant $p_\perp/B$. Applying now to the integral conservation laws indicated above the standard procedure described in §2 and using conditions (1.107) and (1.108) for the electromagnetic field, we find

$$[B_n]=0, \tag{1.118}$$

$$[(v_n-D)\,\boldsymbol{B}-B_n\boldsymbol{v}]=0, \tag{1.119}$$

$$[(v_n-D)\rho]=0, \tag{1.120}$$

$$\left[\rho\boldsymbol{v}(v_n-D)+\left(p_\perp+\frac{B^2}{8\pi}\right)\boldsymbol{n}+\left(\frac{p_\parallel-p_\perp}{B^2}-\frac{1}{4\pi}\right)B_n\boldsymbol{B}\right]=0, \tag{1.121}$$

$$\left[(v_n-D)\left(\frac{\rho v^2}{2}+\rho\varepsilon+\frac{B^2}{8\pi}\right)+v_n\left(p_\perp+\frac{B^2}{8\pi}\right)-\right.$$
$$\left.-\left(\frac{p_\parallel-p_\perp}{B^2}-\frac{1}{4\pi}\right)B_n(\boldsymbol{B}\boldsymbol{v})\right]=0, \tag{1.122}$$

$$\left[\frac{p_\perp}{B}(v_n-D)\right]=0. \tag{1.123}$$

The conditions (1.118)–(1.123) make it possible to analyze the change in the basic quantities on passage through a discontinuity. In deriving these conditions a basic assumption was made to the effect that in a shock wave the mean magnetic moment $p_\perp/B\rho$ is conserved. This condition was proposed by R. V. Polovin [40]. The utility of this assumption is pointed up by the fact that in a compressive wave the evolutionary conditions are satisfied and also the fact that the adiabatic invariant of a particle is conserved with great accuracy during its motion in electromagnetic fields across a surface where the field parameters have a discontinuity of first kind [41]–[43].

We shall call (1.118)–(1.123) the conditions on a MHD shock wave in a rarefied plasma. We shall mention certain properties of shock waves in a rarefied plasma which follow from these conditions. We note that these conditions also admit discontinuities of the tangential, contact, and rotational types similar to those for the case of ideal MHD.

Here there may be fast as well as slow shock waves and also switch-on waves. All waves for which $B_{n1} \neq 0$ and $B_{\tau 1} = 0$ are switch-on waves (behind them $B_\tau \neq 0$) [43]. If for the entropy we take the formula [40] $S = (k/2)\ln(p_\parallel p_\perp^2/\rho^5)$, where k is the Boltzmann constant, then there is an increase in entropy in the shock wave.

The motion of the gas in the presence of shock transitions is most stable under the condition $p_\parallel - p_\perp > 0$ [14], [43].

The properties of MHD shock waves in a rarefied plasma have been considerably less studied than the properties of analogous discontinuities in ordinary MHD.

§5. Elements of dimension theory and the concept of self-similar motions of the medium

In the study of problems in the theory of point explosions we shall make extensive use of the results and methods of dimension and similarity theory. Below we present the background from dimension theory which is needed below and introduce the concept of self-similar motions [46].

The units for measuring physical quantities are divided into basic and derived units. Various physical quantities may be connected by definite relations which express physical laws. If some of these quantities are taken as basic and units of measurement are set up for them, the units of measurement for the remaining quantities will be expressed in terms of the units of measurement of these basic quantities, i.e. will be derived units of measurement. The expression for the derived unit of measurement in terms of the basic units is call the dimension.

In studying mechanical, heat, and certain other physical phenomena it suffices to introduce only three basic independent units of measurement: the units for length, mass, and time. We shall mainly use this system of basic units of measurement in studying explosion phenomena.

The dimension of quantities is written symbolically in the form of a formula in which the symbol for the unit of length is denoted by the letter L, the unit of mass by the letter M, and the unit of time by the letter T. The symbol $[a]$ denotes the dimension of any quantity a.

The dimension of all physical quantities which we shall use in the sequel can be expressed by a formula of the form $[a] = M^{n_1} L^{n_2} T^{n_3}$.

A dimensional quantity is said to have dimension independent of the remaining quantities entering in the problem if the formula expressing its dimension cannot be represented as a combination in the form of a monomial of the dimensional formulas for the other quantities. Thus, for example, the dimensions of density $[\rho] = ML^{-3}$ and energy $[E] = ML^2T^{-2}$ are independent; the dimensions of length L, velocity LT^{-1}, and acceleration LT^{-2} are dependent.

In studying problems of physics and mechanics we often encounter the case in which some dimensional quantity is a function of other dimensional quantities. Suppose that we have a dimensional quantity a which is a function of the dimensional quantities $a_1, \cdots, a_n$,

$$a = \varphi(a_1, a_2, \ldots, a_n). \tag{1.124}$$

We shall assume that the dependence (1.124) expresses a physical law; that is, a fixed physical relation which does not depend on the choice of the system of units.

Suppose that among the dimensional quantities only the first k quantities, $k \leqslant n$, have independent dimensions. Since in problems of interest to us the number of basic units of measurement is equal to three, in our case $k \leqslant 3$. Any physical relation of the type (1.124) can be considered as a relation between dimensionless quantities. If it is known that this dimensionless quantity is a function of a number of dimensional quantities, then this function can depend only on dimensionless combinations composed of the defining dimensional quantities.

According to the π theorem of dimension theory a functional relation between $n + 1$ dimensional quantities which does not depend on the system of units takes the form of a relation between $n + 1 - k$ dimensionless quantities representing the dimensionless combinations of the $n + 1$ dimensional quantities.

Naturally, the less the number of parameters determining the quantity under study, the more restricted the functional relation of the type (1.124) and the easier the quantity is to study. Thus, if the number of basic units of measurement is equal to the number of defining parameters with independent dimensions, this dependence is determined up to a constant factor. Indeed, if $n = k$, it is not possible to form a dimensionless combination from the quantities $a_1, \cdots, a_n$. For the dimension of the quantity a we have the relation $[a] = [a_1]^{k_1}[a_2]^{k_2} \cdots [a_n]^{k_n}$ which is satisfied when $a = c_k a_1^{k_1} \cdots a_n^{k_n}$, where c_k is a dimensionless constant and the exponents $k_1, \cdots, k_n$ are easily found by comparing the dimensions of the right and left sides of the equation.

When starting to solve some physical or mechanical problem, we begin by distinguishing the basic factors, essential variables, and constant parameters which determine the quantities of interest to us. One of the first steps in investigating the functional dependencies of dimensional quantities consists in forming a table of the basic parameters determining the phenomenon in question.

Dimension theory imposes on the system of defining parameters the requirement that there be among the defining parameters dimensional quantities in terms of which the dimensions of all dependent quantities can be expressed. Thus, for example, it cannot be asserted that the static state of a perfect gas is determined by only the two dimensional quantities temperature T $([T] = C^\circ)$ and density ρ, since the dimension of the pressure cannot be expressed in terms of $[T]$ and $[\rho]$. To these quantities it is necessary to add still another dimensional constant R (R is the gas constant).

We proceed to apply the conclusions of dimension theory mentioned above to problems concerning the motion of continuous media. In order to solve problems on the spatial motions of a compressible fluid it is necessary to determine the pressure p, the density ρ, and the velocity components v^i along the axes x^i as functions of time, coordinates, and constant parameters entering in the conditions of the problem. In the general case of spatial motions the system of defining parameters can always be brought to the form

$$x^i, \quad t, \quad a, \quad b, \quad c, \quad \alpha_1, \quad \alpha_2, \ldots, \alpha_n, \tag{1.125}$$

where a, b and c are constants with independent dimensions and $\alpha_1, \cdots, \alpha_n$ are dimensionless constants.

A quantity χ defining some characteristic of motion of the medium will be a function of the parameters (1.125) not depending on the system of units of measurement. We apply the π theorem to this function. We have $n = 7$ and $k = 3$. Forming a dimensionless combination from (1.125) and χ, we can introduce a dimensionless function $\tilde{\chi}$ which according to the π theorem can depend in this case on four dimensionless variables. This also follows directly from the fact that from the parameters (1.125) it is possible to form four independent dimensionless variable combinations. Thus in the general case we have

$$\tilde{\chi} = \psi(\tilde{x}^i, \tilde{t}, \alpha_1, \alpha_2, \ldots, \alpha_n),$$

where $\tilde{x}^i$ and $\tilde{t}$ are dimensionless coordinates and dimensionless time. The number of independent variables does not decrease and is equal to four.

Suppose now that $c = 0$, and we have in the system of defining parameters (1.125) two constants a and b with independent dimensions

$$[a] = ML^k T^s, \qquad [b] = LT^{-\delta} \qquad (\delta \neq 0). \tag{1.126}$$

Now $n = 6$, $k = 3$ and $n - k = 3$, i.e. the dimensionless function $\tilde{\chi}$ will depend only on three dimensionless variable combinations. These dimensionless combinations have the form $x^1/bt^\delta, x^2/bt^\delta, x^3/bt^\delta$; it is not possible to form any other independent dimensionless combinations. Thus

$$\tilde{\chi} = \psi\left(\frac{x^i}{bt^\delta}, \alpha_1, \alpha_2, \ldots, \alpha_n\right).$$

In the case $\delta = 0$ we have $[b] = L$; then either a has dimension $L^k T^s$ or there is still another dimensional constant with dimension not depending on $[a]$ and $[b]$. The second possibility leads to the general case. This follows from the fact that we must form a dimensionless combination with the time t on which $\widetilde{\chi}$ depends. If $[a] = L^k T^s$, then in the system of parameters $x^i, t, a, b, \alpha_1, \cdots, \alpha_n$ only the two quantities a and b have independent dimension, and we have four dimensionless combinations on which $\widetilde{\chi}$ will depend.

We make the

DEFINITION. Transient motions of a continuous medium in which all dimensionless characteristics depend on variable combinations of x^1/bt^δ, x^2/bt^δ and x^3/bt^δ, $[b] = LT^{-\delta}$, are called *self-similar.* In this particular case of one-dimensional self-similar motions the dimensionless characteristics depend on one dimensionless variable combination r/bt^δ.

From the preceding discussion it follows that for self-similarity of a problem it is sufficient that the system of dimensional defining parameters given by the equations and auxiliary conditions (in particular, boundary and initial conditions) should contain no more than two constants with independent dimensions distinct from the length (and time), i.e. the system of defining parameters should have the form $x^1, x^2, x^3, t, a, b, \alpha_1, \cdots, \alpha_n$, where $\alpha_1, \cdots, \alpha_n$ are dimensionless constants and formulas (1.126) hold for $[a]$ and $[b]$, wherein k and s are arbitrary constants. In place of a it is possible to take the constant $\widetilde{a}$ with $[\widetilde{a}] = ML^{\omega-3}$, where ω is an arbitrary number.

For one-dimensional self-similar motions the system of defining parameters must have the form $r, t, a, b, \alpha_1, \cdots, \alpha_n$. We remark that in the initial formulation of the problem there may be many dimensional, constant defining parameters, but only two of them, which we take to be a and b, must have independent dimensions. We can bring all the remaining dimensional constants to dimensionless form using a and b and include them among the parameters $\alpha_1, \cdots, \alpha_n$. Since one-dimensional self-similar motions depend only on one dimensionless variable $\lambda = r/bt^\delta$, for one-dimensional self-similar motions the system of partial differential equations after transformation to dimensionless form is replaced by an equivalent system of ordinary differential equations which considerably simplifies solution of the problems posed.

Let us consider some examples of self-similar one-dimensional motions of a perfect gas.

1) *The propagation of the front of a flame or detonation with zero thickness of the combustion zone.* A homogeneous combustible mixture filling space with constant pressure p_1 and constant density ρ_1 is ignited at the moment $t = 0$ at a point (spherical symmetry), along a line (cylindrical symmetry) or on a plane (plane symmetry). A flame or detonation front will propagate through the mixture. Suppose that the conditions for propagation of the flame or detonation are such that it is possible to neglect the energy of initiation of the combustion process and to assume that

matter burns instantaneously on the front of the flame or detonation with the release of energy.

The defining parameters will be $r, t, \rho_1, p_1, \gamma$, the amount of heat Q released during combustion of a unit mass of the gas, and in the case of burning also the known velocity of the front of the flame U_1. The dimension of Q in mechanical units can be expressed in terms of the dimensions of ρ_1 and p_1. Thus the number of constants with independent dimensions is equal to two, i.e. the problem is self-similar. The solution of this problem is known [46], [47]. Other examples of self-similar problems related to the propagation of waves in a combustible mixture of gases are considered in Chapter 6.

2) *The problem of the inverse pinch effect.* Suppose that in an infinite conducting medium along a line there passes a current I which varies with time: $I = \sigma_1 t$ (σ_1 = const). The initial density of the gas ρ_1 and initial pressure p_1 are constant. The initial velocity $v_1 = 0$. In the medium there is a frozen initial field H_1 = const which is directed parallel to the line along which the current flows.

Because of the frozen state the magnetic pressure of the field of the linear current

$$\frac{H_\varphi^2}{8\pi} = \frac{I^2}{2cr_0^2}$$

(r_0 is the distance from the axis of symmetry to the point considered) will "eject" gas from the region next to the axis and will play the role of a piston of radius $r_0(t)$ which expands with constant velocity [44]. The process of compression of the gas along the line which takes place here has been called the inverse pinch effect. Since the dimension of the magnetic field H_1 is expressed in terms of the dimension of the pressure p_1 ($[H_1^2] = [p_1]$), the problem is self-similar. We have considered the solution to this problem in [48].

Another important example of self-similar problems of gas dynamics and MHD is the problem of a strong point explosion, which we shall consider in detail below.

§6. Invariant and self-similar solutions of the equations for one-dimensional motions

6.1. *On the relation between dimensional analysis and the theory of transformation groups.* From a mathematical point of view dimensional analysis is closely connected with results from the theory of continuous transformation groups. We shall present a number of known facts from the theory of Lie groups [49]–[51].

DEFINITION. A group G is called a *Lie group* if it contains a subset G_r satisfying the following conditions:

1. G_r contains the identity element e of G.
2. Products of elements of G_r generate G.
3. The elements of G_r can be put in one-to-one correspondence with the points

α of an open r-dimensional sphere Q in Euclidean space, whereby the center O of the sphere corresponds to e. This condition introduces coordinates in G_r. We shall denote the element $g \in G_r$ corresponding to $\alpha \in Q$ by g_α.

4. The product $g_\alpha \cdot g_\beta \in G_r$ is some element g_γ corresponding to the point $\gamma \equiv \varphi(\alpha, \beta)$, where the notation $\gamma = \varphi(\alpha, \beta)$ means $\gamma^\lambda = \varphi^\lambda(\alpha^1, \cdots, \alpha^r; \beta^1, \cdots, \beta^r)$, $\lambda = 1, \cdots, r$.

5. In Q^r the function $\varphi(\alpha, \beta)$ defined in some neighborhood of the points α and β is thrice continuously differentiable.

The subset G_r is called a group kernel or a local Lie group. The dimension r of the sphere Q is called the order of the local Lie group G_r. The associativity property for the Lie group has the form $\varphi[\varphi(\alpha, \beta), \gamma] = \varphi[\alpha, \varphi(\beta, \gamma)]$. The study of the properties of a local Lie group reduces to the study of the properties of the functions $\varphi^\lambda(\alpha, \beta)$ which are independent of the coordinate system and give the multiplication laws.

We introduce the concept of isomorphism: two groups G and H are called isomorphic if to each element $A, B, C, \cdots$ of the first there can be assigned in a one-to-one manner an element $a, b, c, \cdots$ of the second such that $ab = c$ whenever $AB = C$.

DEFINITION. A one-to-one mapping of a set E into itself is called a *transformation* of the set E.

If T_1 and T_2 are two transformations of a set E, their product $T_1 T_2 = T_3$ is defined by the relation

$$T_3(M) = T_1(T_2(M)) \tag{1.127}$$

for arbitrary $M \in E$. The role of the unit element is played by the identity transformation $e(M) = M$, $M \in E$, so that $eT = Te = T$. It can be shown [49], [50] that the product of transformations defined above has the associativity property. The transformation T^{-1} inverse to T is defined by the property that it takes any element $T(M) \in E$ into the element M.

Thus the set G of transformations of a set E which contains with any two transformations also their product and together with any transformation its inverse forms a group under the multiplication law (1.127). Any such group is called a transformation group.

There is the principle of Weyl: any group can be represented by means of some transformation group, i.e. there is a transformation group isomorphic to the given group.

We shall consider a representation of a Lie group by means of transformation groups. For example, the set G of all nondegenerate square matrices of order n is a group under ordinary matrix multiplication. The identity is the matrix $e = \|\delta^i_j\|$, where δ^j_i is the Kronecker delta. To the group of matrices $\|a^i_j\|$ there correspond transformations $x' = \|a^i_j\|x$, wherein to the product of matrices there corresponds a sequence of transformations. Thus, the group of matrices is represented by linear transformations. Suppose now that in the general case we have a local Lie group G_r and a set $E(x^1, \cdots, x^n)$. In this set we shall single out a subset–the nucleus. Suppose that we have constructed transformations by means of the elements of G_r which "shift" the nucleus of E. This means that in the transformation formulas of the given set there enter the coordinates of the elements of G_r, i.e. to each point α there corresponds a transformation

$$x'^i = f^i(x^1, \ldots, x^n, \alpha^1, \ldots, \alpha^r). \tag{1.128}$$

In what follows we shall take the set E to be the n-dimensional Euclidean space E_n of points x with coordinates $x^1, \cdots, x^n$. If in (1.128) f^i is thrice continuously differentiable and the transformation is invertible while $f(x, 0) = x$ and the property $f(f(x, \alpha), \beta) = f(x, \varphi(\alpha, \beta))$ holds, then the set of transformations (1.128) forms a local Lie group of point transformations of E_n. This group provides a representation of G_r. We shall denote the transformation group by G^n_r.

Then for $T_\alpha, T_\beta \in G^n_r$ to the formula $T_\beta T_\alpha = T_{\varphi(\alpha,\beta)}$ there corresponds the formula

$$f^i(f(x, \alpha), \beta) = f^i(x, \varphi(\alpha, \beta)).$$

Under the action of the transformation T_α a function $F(x)$ defined on E_n is transformed into a new function: $F'(x) = T_\alpha F(x) = F(T_\alpha x)$.

DEFINITION. A function $F(x) \not\equiv \text{const}$ is called an *invariant* of the group G_r if it is not changed by the action of any transformation $T_\alpha \in G^n_r$, i.e. $T_\alpha F(x) \equiv F(x)$ or $F(T_\alpha x) \equiv F(x)$.

We introduce the auxiliary functions

$$\xi^i_s(x) = \left.\frac{\partial f^i(x, \alpha)}{\partial \alpha^s}\right|_{\alpha=\alpha_0}, \quad \begin{matrix} i = 1, 2, \ldots, n, \\ s = 1, 2, \ldots, r, \end{matrix} \tag{1.129}$$

and the matrix $A = \|\xi^i_j\|$ ($\alpha = \alpha_0$ corresponds to the identity of the group).

Suppose that the common rank of the matrices A is R.

THEOREM [51]. *The group G^n_r has invariants if and only if $R < n$. If this inequality is satisfied, then there exist $t = n - R$ functionally independent invariants $I^k(x)$ $(k = 1, \cdots, t)$ of G^n_r such that any invariant of the group is a function of these.*

This means that if $F(x)$ is an invariant, then

$$F(x) = F(I^1(x), I^2(x), \ldots, I^t(x)).$$

From this one can deduce the generalized π theorem [52].

Suppose that the equation $F(x^i) = 0$ is invariant under a group G_r^n which takes an l-dimensional manifold into itself, the latter being an l-dimensional subspace of E_n. Then the equation $F = 0$ is equivalent to an equation

$$\Phi(I^1, I^2, \ldots, I^{n-l}) = 0 \tag{1.130}$$

involving $n - l$ functions $(1 \leqslant l \leqslant r)$ $I^1, \cdots, I^{n-l}$, each of which is invariant under G_r^n. We point out that the dimension l coincides with the rank of the matrix A.

We mention that a generalization of these results to tensor functions and applications of Lie groups to the theory of crystals are given in [53].

In dimensional analysis it is postulated that certain basic units $q_1, \cdots, q_r$ may be varied in arbitrary positive ratios by transformations of the form

$$T_\alpha(q_i) = q_i' = \alpha_i q_i \qquad (\alpha_i > 0, \quad i = 1, \ldots, r), \tag{1.131}$$

which are called scale changes. Homogeneous quantities $Q_1, \cdots, Q_n$ are also considered which are transformed by (1.131) in the following manner:

$$T_\alpha(Q_j) = Q_j' = \alpha_1^{b_{j1}} \ldots \alpha_r^{b_{jr}} Q_j. \tag{1.132}$$

Thus each of the Q_j has the dimension $q_1^{b_{j1}} \cdots q_r^{b_{jr}}$. We note that (1.131) is the special case of (1.132) in which $b_{ik} = \delta_k^i$ and $r = n$.

To the transformations T_α there correspond in a one-to-one manner vectors $\alpha = (\alpha_1, \cdots, \alpha_r)$ with positive components. Moreover, if we introduce the following actions:

$$\alpha\beta = (\alpha_1 \beta_1, \ldots, \alpha_r \beta_r), \qquad \alpha^{-1} = (\alpha_1^{-1}, \ldots, \alpha_r^{-1}), \tag{1.133}$$

then clearly the following equations hold:

$$T_\alpha(T_\beta(Q_j)) = T_\beta(T_\alpha(Q_j)) = T_{\alpha\beta}(Q_j), \qquad T_{\alpha^{-1}}(T_\alpha(Q_j)) = Q_j.$$

From this it follows that equations (1.132) provide a representation of the group of positive r-vectors defined by (1.133), while equations (1.131) define a group of linear transformations of the vector space Q.

If we form the matrix A according to (1.129), then its rank will be equal to the rank of the matrix $\|b_{ik}\|$, i.e. to the number of quantities with independent dimensions. Using the generalized π theorem of group theory mentioned above, we arrive at the π theorem of dimension and similarity theory formulated in the preceding section. We shall call the group (1.132) the similarity group.

Applications of dimension theory are of considerable use in considering concrete problems of gas dynamics, and extensive use will be made of this fact below. The great generality of group-theoretic methods makes it possible to extend results and methods of dimension theory to broader classes of problems and to obtain new particular solutions of the equations of gas dynamics.

6.2. *Invariant solutions and their relation to self-similar solutions.* Suppose that the system of equations of hydrodynamics is invariant under a group G_r^9 of transformation in the space of v^i, ρ, p, x^i, t, i.e. the system of equations of hydrodynamics is a differential invariant of G_r^9. In this case the system of equations of hydrodynamics is said to admit the group G_r^9.

In the theory of invariant solutions it is possible to distinguish two directions:

1) the use of known groups (for example, the similarity group) to construct invariant solutions [45], [46];

2) the search for the largest Lie group of transformations admitted by the system of equations of hydrodynamics and construction of a complete set of essentially distinct invariant solutions [51].

We shall not consider these questions in detail here, but we will prove that the invariant solutions of the equations of one-dimensional hydrodynamics can be obtained from self-similar solutions by elementary transformations and limiting processes.

It is shown in [51] that for arbitrary values of the ratio of the specific heat capacities γ the system of equations (1.118)–(1.120) for one-dimensional motions with cylindrical and spherical symmetry has only the following invariant solutions:

$$v = u(r), \qquad \rho = e^{\beta t} R(r), \qquad p = e^{\beta t} P(r), \tag{1.134}$$

$$v = r + e^t u(\lambda), \qquad \rho = e^{(\beta-2)t} R(\lambda), \qquad p = e^{\beta t} P(\lambda), \qquad \lambda = re^{-t}. \tag{1.135}$$

$$v = \frac{1}{t} u(r), \qquad \rho = t^{2\beta+2} R(r), \qquad p = t^{2\beta} P(r). \tag{1.136}$$

$$v = \frac{r}{t} u(\lambda), \qquad \rho = r^{\beta+2\alpha-2} R(\lambda), \qquad p = r^{\beta} P(\lambda), \qquad \lambda = tr^{-\alpha}. \tag{1.137}$$

Invariant solutions of the form (1.135), (1.137) have been studied by K. P. Stanjukovič [45].

For the case of motions with plane waves, to these solutions there are added the following:

$$v = u(t), \qquad \rho = e^{\beta r} R(t), \qquad p = e^{\beta r} P(t), \tag{1.138}$$

$$v = r + u(\lambda), \qquad \rho = e^{\beta t} R(\lambda), \qquad p = e^{\beta t} P(\lambda), \qquad \lambda = r - \frac{t^2}{2}, \tag{1.139}$$

$$v = \frac{1}{t} u(\lambda), \qquad \rho = t^{2\beta+2} R(\lambda), \qquad p = e^{2\beta} P(\lambda), \qquad \lambda = te^{-r}, \tag{1.140}$$

$$v = \frac{r}{t} + u(\lambda), \qquad \rho = t^{\beta} R(\lambda), \qquad p = t^{\beta} P(\lambda), \qquad \lambda = te^{-r/t}, \tag{1.141}$$

$$v = \frac{r}{t} + u(t), \qquad \rho = e^{\beta r/t} R(t), \qquad p = e^{\beta r/t} P(t). \tag{1.142}$$

In the solutions (1.134)–(1.142), α and β are constants.

Using the methods of dimensional analysis, L. I. Sedov indicated a class of self-similar solutions of the equations of one-dimensional gas dynamics of the form

$$v=\frac{r}{t}V(\lambda),\qquad \rho=\frac{a}{r^{k+3}t^{s}}R(\lambda),\qquad p=\frac{a}{r^{k+1}t^{s+2}}P(\lambda),\qquad \lambda=\frac{r}{bt^{\delta}}. \tag{1.143}$$

Here a and b are dimensional constants, and k, s and δ are numbers.

We shall establish a relation between the solutions (1.143) and the invariant solutions. We shall call *limits of self-similar solutions* [46], [51] those solutions which are obtained from self-similar solutions by applying the transformation of translation in time (or space) and passing to the limit with respect to parameters entering in the self-similar solution (1.143). We have the following theorem [55].

THEOREM 1. *All the invariant solutions of the equations of gas dynamics for spherical and cylindrical symmetry are either self-similar solutions or limits of self-similar solutions.*

PROOF. The solutions (1.137) are essentially self-similar, since if in (1.143) we introduce $\widetilde{\lambda}=(\lambda/b)^{-1/\delta}$ and transform and redesignate the functions $R(\lambda)$, $P(\lambda)$, $V(\lambda)$ and the factors in front of them, we arrive at a solution of the form (1.137).

Further, if we make the change $t\longrightarrow t+t_0$, $t_0=\delta\tau$, $b=r_0(b\tau)^{-\delta}$, $V=\delta\widetilde{V}$, $R=\delta^s\widetilde{R}$, $P=\delta^{s+2}\widetilde{P}$ and pass to the limit as $\delta\longrightarrow\infty$, we obtain [46], [54] solutions of the form

$$v=\frac{r}{t}\widetilde{V}(\lambda),\qquad \rho=\frac{a}{r^{k+3}\tau^{3}}\widetilde{R}(\lambda),\qquad p=\frac{a}{r^{k+1}\tau^{s+2}}\widetilde{P}(\lambda),\qquad \lambda=\frac{r}{r_0}e^{-t/\tau}.$$

These solutions are easily reduced to the form (1.135).

If in (1.143) we set $\delta=0$, $s=-2\beta-2$, $a=b=1$ and redesignate the functions V, P and R, we arrive at solutions of the form (1.136). It remains to obtain solutions of the form (1.134). For this again in the solution (1.143) we replace t by $t+t_0$ and set $t_0=s\tau$, $a=a_0(\tau s)^s$, $P=(\tau s)^2\hat{P}$, $V=s\hat{V}$ and $R=\hat{R}$. Then we find that

$$v=\frac{r\hat{V}}{\frac{t}{s}+\tau},\qquad \rho=\frac{a_0r^{-(k+3)}}{\left(\frac{t}{\tau s}+1\right)^{s}}\hat{R}(\lambda),\qquad p=\frac{a_0r^{-(k+1)}}{\left(\frac{t}{s}+\tau\right)^{2}\left(\frac{t}{\tau s}+1\right)^{s}}\hat{P}(\lambda)$$

and, passing to the limit as $s\longrightarrow\infty$ and $\delta\longrightarrow 0$, we obtain solutions of the form (1.134). This completes the proof of the theorem.

This theorem is also true for solutions of the equations of one-dimensional cylindrically symmetric motions of ideal MHD. The corresponding invariant solutions have been considered in work of the author [55].

For the case of plane symmetry solutions of the form (1.138), (1.140) can be obtained as limits of self-similar solutions by using transformations with respect to the coordinate r [46], [54]. In order to obtain the other invariant solutions, it would be necessary to make use of the Galilean group. We shall not consider these limits in detail or analyze obtaining all the invariant solutions for the case of plane symmetry.

6.3. *Integrals of adiabatic motions and the energy integral.* We shall consider one-dimensional motions of a gas without taking into account dissipative processes. We shall assume that the motion is self-similar.

Suppose that there is an integral conservation law of the form (1.48)

$$\frac{d}{dt}\int_{\Omega}\Phi\rho\, d\Omega=\oint_{\Sigma}\rho\Phi\,(D-v_n)\, d\Sigma,$$

so that

$$\frac{d}{dt}\int_{\Omega^*}\Phi\rho\, d\Omega=0.$$

For one-dimensional motions we have

$$\frac{d}{dt}\sigma_\nu\int_{r'}^{r''}\Phi\rho r^{\nu-1}dr=\left[\Phi\rho r^{\nu-1}\sigma_\nu\left(\frac{dr}{dt}-v\right)\right]\Big|_{r'}^{r''}. \tag{1.144}$$

Let a and b be the essential dimensional constants entering in the problem, $[a]=ML^kT^s$ and $[b]=LT^{-\delta}$.

For the function Φ we can write

$$\Phi=a^{\eta}b^{\xi}G(\lambda),\quad \lambda=\frac{r}{bt^{\delta}}.$$

For the integral

$$\mathcal{J}=\int_{r'}^{r''}\Phi\rho r^{\nu-1}dr,\quad r'=bt^{\delta}\lambda',\quad r''=bt^{\delta}\lambda'' \tag{1.145}$$

suppose that we have

$$\mathcal{J}=a^x b^y t^{\mu}K(\lambda',\lambda'').$$

For the functions v, p and ρ we introduce dimensionless variables V, P and R according to (1.143).

If Φ is such that $\mu=0$, then (1.144) implies the adiabatic integral [56]

$$G(\lambda)\,R(\lambda)\,[V-\delta]\,\lambda^{\nu-(k+1)}=\text{const.} \tag{1.146}$$

For the adiabatic motion of a perfect gas we take for Φ the quantity $(p/\rho^{\gamma})^z=\exp(sz/c_V)$, where z is some number. From (1.145) by comparison of dimensions we obtain a system of equations for the choice of z:

$$(1-\gamma)z+1=x,\quad (3\gamma-1)z=kx+y+3-\nu,\quad -2z=sx-\delta y.$$

The integral (1.146) assumes the form

$$\left(\frac{P}{R^{\gamma}}\right)^z R(-\delta+V)\lambda^{\omega_1}=\kappa_1,\quad \omega_1=\nu-(k+1)[1+z(1-\gamma)]$$
$$+1-(k+1)[1+(1-\gamma)z]. \tag{1.147}$$

This integral was obtained by M. L. Lidov [57]. We proceed to the equations of MHD for $\nu = 1, 2$. Let

$$h_z = \frac{H_z^2}{8\pi} = ar^{-(k+1)}t^{-(s+2)}G_z,$$
$$h_\varphi = \frac{H_\varphi^2}{8\pi} = ar^{-(k+1)}t^{-(s+2)}G_\varphi, \quad \boldsymbol{H} = (H_\varphi, H_z), \tag{1.148}$$
$$h_\varphi = 0 \quad (\nu = 1).$$

Then in the case of infinite conductivity from (1.146) we obtain integrals corresponding to the frozen-in characteristic of the field lines; these were first obtained in our work [58]:

$$G_\varphi^m (\lambda R)^{-2m} = \kappa_2 [R(V-\delta)]^{2-s-\delta(k+3)} \lambda^{-(k+5)m+(\nu-k+3)[2-s(k+3)]}, \tag{1.149}$$
$$G_z R^{-2m} = \kappa_3 [R(V-\delta)]^{2-s-\delta(k+5)} \lambda^{-(2+\nu)s-2(k+3-\nu)},$$
$$m = s + \delta(k+3-\nu) \qquad (\nu = 1, 2). \tag{1.150}$$

We now consider the equations of the MHD of a rarefied plasma for the case of one-dimensional cylindrical motions. Naturally the integrals (1.149) and (1.150) hold in this case. There are still two more integrals of adiabatic invariants.

Suppose that the system of dimensionless variables has the form (1.143), (1.148), where in place of the formula for p there are now analogous relations for $p_\perp$ and $p_\|$. The integrals of adiabatic invariance have the form

$$P_\perp^m R^{-m} G^{-m/2} = \kappa_4 [R(V-\delta)]^{-1/2[s+2+\delta(k+s)]} \lambda^{-(s+k+1)}, \tag{1.151}$$
$$(P_\| R^{-3} G)^m = \kappa_5 [R(V-\delta)]^{4-s-\delta(k+7)} \lambda^{-(6s+4k+4)}, \qquad G = G_\varphi + G_z. \tag{1.152}$$

If $s + 2 - \delta(\nu - 1 - k) = 0$, systems of self-similar equations admit an energy integral.

For the equations of MHD this integral has the form [58]

$$\lambda^{\nu+2}\left[(P+G)V + (V-\delta)\left(\frac{1}{2}RV^2 + P\frac{1}{\gamma-1} + G\right)\right] = \kappa_6. \tag{1.153}$$

For a rarefied plasma the integral (1.153) $(\nu = 2)$ takes the form

$$\lambda^4\left[(P_\perp + G)V + (V-\delta)\left(\frac{1}{2}RV^2 + P_\perp + \frac{1}{2}P_\| + G\right)\right] = \kappa_7. \tag{1.154}$$

Here the κ_i $(i = 1, \cdots, 7)$ denote arbitrary constants.

For the equations of gas dynamics the energy integral was first found by Sedov (cf. [46]).

Let $\delta = 2/(2 + \nu - \omega)$ and $s = 0$, $k = \omega - 3$. We shall write down the adiabatic and energy integrals for this case. For the adiabatic integral we have

$$\left[\frac{2}{\gamma-1}\left(\frac{\gamma+1}{2}-\frac{1}{\lambda}\frac{v}{v_2}\right)\right]^{1-\frac{\omega\gamma}{\nu}}\left(\frac{p}{p_2}\right)^{1-\frac{\omega}{\gamma}}\left(\frac{r}{r_2}\right)^{\nu-\omega\gamma}=\left(\frac{\rho}{\rho_2}\right)^{\gamma-1}, \qquad p_2=\frac{2\rho_1}{\gamma+1}D^2,$$
$$v_2=\frac{2D}{\gamma+1}, \qquad r_2=bt^\delta, \qquad D=\frac{dr_2}{dt}, \qquad \rho_2=\frac{\gamma+1}{\gamma-1}\rho_1. \tag{1.155}$$

Here we have chosen the arbitrary constant κ_1 such that the conditions (1.64) on a strong shock wave are satisfied.

The energy integral (1.153) can be brought to the form

$$\lambda^{\nu-1}\left[(\tilde{P}+\tilde{R}\tilde{V}^2)\left(\lambda-\frac{2}{\gamma+1}\tilde{V}\right)-\frac{2(\gamma-1)}{\gamma+1}\tilde{V}\tilde{P}\right]=\tilde{\kappa}_6,$$
$$\tilde{p}=\frac{p}{p_2}, \quad \tilde{V}=\frac{v}{v_2}, \quad \tilde{R}=\frac{\rho}{\rho_1}\frac{\gamma-1}{\gamma+1}, \quad \lambda=\frac{r}{r_2},$$
$$r_2=bt^\delta, \qquad D=\frac{dr_2}{dt}; \tag{1.156}$$

ρ_1 is a quantity with the dimension of density. We note that p_2, v_2 and ρ_2 are functions corresponding to the conditions on a strong shock wave if r_2 is the coordinate of the shock wave.

In the case $\tilde{\kappa}_6 = 0$ the integral (1.156) satisfies the conditions on a strong shock wave and takes the form

$$\gamma\tilde{P}\left(\frac{\tilde{V}}{\lambda}-\frac{\gamma+1}{2\gamma}\right)=\tilde{R}\tilde{V}^2\left(\frac{\gamma+1}{2}-\frac{\tilde{V}}{\lambda}\right). \tag{1.157}$$

We note also here that there will be adiabatic integrals and an energy integral [46] for motions which are limits of self-similar motions as well, i.e. for other invariant solutions of the equations of gas dynamics and MHD; these can be obtained, in particular, from (1.146) and (1.153) by limiting processes analogous to those used above in the study of the relation between self-similar and invariant solutions.

§7. Formulation of problems concerning point explosions

7.1. *Certain problems for linear equations of mathematical physics.*

A. The problem for the wave equation on the propagation of an instantaneous point disturbance. The change in pressure (or density) in a gas having weak compressibility will according to (1.32) and (1.33) be described by the wave equation, which for the case of one-dimensional motions can be written in the form (1.36):

$$\frac{1}{r^{\nu-1}}\frac{\partial}{\partial r}\left(r^{\nu-1}\frac{\partial p'}{\partial r}\right)=\frac{1}{a_0^2}\frac{\partial^2 p'}{\partial t^2}.$$

Suppose for simplicity that the internal energy of the medium is proportional to the pressure, i.e. $\epsilon = pf(\rho) + \text{const}$. Then, making $\epsilon' = \epsilon - p_0 f(\rho_0)$, we obtain

$$\varepsilon'=\left(f(\rho_0)+\left.\frac{df}{d\rho}\right|_{\rho=\rho_0}\frac{p_0}{a_0^2}\right)p',$$

i.e. for the perturbed internal energy ϵ' there is the equation

$$\frac{1}{r^{\nu-1}}\frac{\partial}{\partial r}\left(r^{\nu-1}\frac{\partial \varepsilon'}{\partial r}\right)=\frac{1}{a_0^2}\frac{\partial^2\varepsilon'}{\partial t^2}. \tag{1.158}$$

Suppose that in a medium with initial state $p_0, \rho_0, \epsilon_0(p_0, \rho_0)$ a finite amount of energy E_0 is released at the moment $t = +0$ (instantaneously) at a point, along a line, or over a plane, where in the latter two cases E_0 is reckoned per unit length or area respectively. We assume that the point, line, or plane of energy release has coordinate $r = 0$. The resulting perturbed state will have spherical $(\nu = 3)$, cylindrical $(\nu = 2)$ or plane $(\nu = 1)$ symmetry, respectively, of the released energy. We shall assume that the released energy E_0 goes entirely to change the internal energy ϵ, and we shall not consider the part of the energy going to the macroscopic motion of the medium, i.e. going to increase its kinetic energy. Suppose that the perturbation arising in the medium is subject to (1.158). Assuming that the law of energy conservation is satisfied, we have the equality

$$E_0=\int_{-\infty}^{+\infty}\rho_0\varepsilon' d\Omega, \tag{1.159}$$

where $d\Omega$ is the volume element and the integration extends over all of space.

This equality holds for any $t \geqslant 0$. Thus we arrive at the problem of finding a function with initial data $\epsilon' = 0$ $(r > 0, t = +0)$ and the additional integral condition (1.159). We first consider the three-dimensional case $(\nu = 3, d\Omega = dx^1dx^2dx^3)$. In terms of generalized functions it can be said that for this problem the initial data are given in the form of a singular generalized function, namely, the Dirac δ function (for the definition and properties of the δ function, cf., for example, [59]–[61])

$$\varepsilon'\big|_{t=+0}=\frac{1}{2}\frac{E_0}{\rho_0}\delta(\boldsymbol{r}), \tag{1.160}$$

where $\delta(\boldsymbol{r})$ is the three-dimensional δ function.

A generalized solution of the problem is easily written down in terms of generalized functions of the fundamental solution for the wave equation in the three-dimensional case and has the form [60]

$$\varepsilon'=\frac{\partial}{\partial t}\frac{E_0}{4\pi\rho_0}e(t)\frac{\delta(a_0^2t^2-r^2)}{a_0}, \tag{1.161}$$

where $e(t)$ is the unit step function $(e = 0, t < 0; e = 1, t \geqslant 0)$.

The solution (1.161) shows that the perturbation ϵ' will propagate in the form of a spherical wave with radius $r = a_0t$ which moves with velocity a_0 through the state $\epsilon' = 0$, while after passage of this wave the unperturbed state again ensues.

In the case of cylindrical symmetry $(\nu = 2)$ the solution has the form

$$\varepsilon' = \frac{\partial}{\partial t} \frac{E_0}{2\rho_0} \frac{e\,(a_0 t - r)}{2\pi a_0 \sqrt{a_0^2 t^2 - r^2}}. \tag{1.162}$$

It follows from this solution that the perturbation ϵ' from the instantaneouly moving point source $\frac{1}{2}(E_0/\rho_0)\delta(x^1, x^2)\delta(t)$ occupies at time $t > 0$ a cylindrical region of radius $r = a_0 t$. In contrast to the spatial case, the perturbation does not vanish behind the wave front but is rather preserved at a point of space after the passage of the wave. For the plane one-dimensional case $(\nu = 1)$ the solution of the problem takes the form

$$\varepsilon' = \frac{E_0}{2\rho_0} \delta\,(a_0 t - |\,x^1\,|) \quad (r = x^1). \tag{1.163}$$

From this solution it follows that at the moment of time $t > 0$ the perturbation from the source $(E_0/\rho_0)\delta(x^1)\delta(t)$ propagates in the form of a plane wave $|x^1| \leqslant a_0 t$ whose leading front $|x^1| = a_0 t$ moves with velocity a_0 in a direction perpendicular to the plane $x^1 = 0$. Here the leading front consists of two planes $x^1 = a_0 t$ and $x^1 = -a_0 t$ moving with speed a_0 respectively to the right and left of the initial plane.

It follows from (1.163) that the perturbation ϵ' at time t will be found only at the two points $x = \pm a_0 t$, so that after passage of the wave front the unperturbed state again ensues. By considering the equation relating the pressure and velocity, namely, $\rho_0(\partial v/\partial t) + (\partial p/\partial r) = 0$, it is possible to obtain the solution of the problem on the instantaneous release of energy also for the complete system of acoustic equations, taking into account the conservation of the total acoustic energy defined in terms of the quadratic energy E (formula (1.34)).

B. The problem of an instantaneous heat source. As was shown in §1, the flow of heat in a stationary medium is described by the heat equation (1.44) which for the case $\chi = \text{const}$ takes the form

$$\frac{\partial \varepsilon}{\partial t} = \chi \frac{1}{r^{\nu-1}} \frac{\partial}{\partial r}\left(r^{\nu-1} \frac{\partial \varepsilon}{\partial r}\right). \tag{1.164}$$

Suppose now that at the moment $t = 0$ a finite amount of heat E_0 is released instantaneously at a point, along a line, or over a plane, where, as in the problem of the preceding section, in the latter two cases E_0 is taken to be the specific amount of heat. If there are no other heat sources and $\epsilon = 0$ for $t = +0$ and $r > 0$, then by the law of energy conservation we have

$$E_0 = \rho_0 \int_0^{+\infty} \varepsilon\, d\Omega. \tag{1.165}$$

Thus we have again arrived at the problem of an instantaneous point perturbation of the type of a δ function:

$$\varepsilon\Big|_{t=+0} = \frac{E_0}{\rho_0}\,\delta(r). \tag{1.166}$$

The solution to this problem is well known and (for $t > 0$) is given by

$$\varepsilon = \frac{E_0}{\rho_0}\,e(t)\,\frac{1}{(2\sqrt{\chi\pi t})^{\nu}}\exp\left(-\frac{r^2}{4\chi t}\right). \tag{1.167}$$

From the solution (1.167) and the properties of the δ function [60] it follows that $\epsilon \longrightarrow E_0\delta(r)/\rho_0$ for $t \longrightarrow +0$, i.e. the initial condition (1.166) is satisfied. The relation (1.165) is also satisfied. Further, since $\epsilon(r, t) > 0$ for all r and $t > 0$, it follows that in the model under consideration heat propagates with infinite velocity. We remark that the solution (1.167) is self-similar.

We now make the following definition of a point explosion.

DEFINITION. We call a process of the instantaneous release of a finite amount of energy at a point of of finite specific energy along a line or over a plane in three-dimensional space E_3 a *point explosion.*

The problem of finding the perturbations caused by a point explosion we shall call the *point explosion problem* (PEP). The examples presented bespeak the fact that the PEP can be formulated for various equations and is a special Cauchy problem. Below we shall present the formulation of the PEP for the equations of gas dynamics and magnetohydrodynamics.

The PEP at a point $(\nu = 3)$, along a line $(\nu = 2)$, or over a plane $(\nu = 1)$ will be called the basic problem. In the sequel we shall not, as a rule, resort to the formalism of δ functions and unit step functions to write down the initial and boundary conditions and the solutions of the problems in question.

7.2. *Point explosion in an incompressible fluid.* The simplest example of a medium for which the basic PEP is solved rather simply is an incompressible fluid. We shall consider the case of spherical symmetry $(\nu = 3)$ and shall assume that the initial pressure and initial density are constant, with $\rho = \rho_1$. The formulations of this problem and the path to its solution were first indicated by Sedov [46]. Below we present the method of solving this problem used in [4].

If it is assumed that $\rho = \rho_1 = \text{const}$ and $\nu = 3$, then the system of equations (1.18)–(1.20) can be written as follows:

$$\frac{\partial v}{\partial t} + v\frac{\partial v}{\partial r} + \frac{1}{\rho}\frac{\partial p}{\partial r} = 0, \qquad \frac{\partial v}{\partial r} + 2\frac{v}{r} = 0, \qquad \rho = \rho_1. \tag{1.168}$$

This system is completely integrable; its solution depends on two arbitrary functions of time which we denote by $\psi_1(t)$ and $\psi_2(t)$:

$$v = \frac{\psi_1(t)}{r^2}, \qquad p = \psi_2(t) - \frac{1}{2}\rho_1 v^2 + \frac{\rho_1}{r}\frac{d\psi_1}{dt}. \tag{1.169}$$

The solution (1.169) can be used in the PEP in an incompressible fluid. In an incompressible fluid perturbations propagate with infinitely large velocity; therefore in this case solutions of the problem of explosion without a shock wave are possible. If at the initial moment of time the pressure is p_1 and the fluid is at rest, then at subsequent moments of time a pressure equal to p_1 is preserved at $r = \infty$.

Since in the problem we have three defining dimensional parameters ρ_1, E_0 and p_1, it follows that for $p_1 = 0$ the problem is self-similar. We shall consider this particular case. For $p_1 = 0$ the explosion's energy E_0 is entirely expended in increasing the kinetic energy of the liquid. In view of this we have the equality

$$E_0 = 4\pi \int_{r_*}^{\infty} \frac{\rho_1 v^2}{2} r^2 dr,$$

where r_* is a radius defining the region of the perturbed motion (we note that v^2 at the initial moment of time is the three-dimensional δ function). Substituting v from (1.169) here, we obtain

$$E_0 = 2\pi\rho_1 \frac{\psi_1^2}{r_*}. \tag{1.170}$$

From dimensional considerations it follows that

$$r_* = k\left(\frac{E_0}{\rho_1}\right)^{1/5} t^{2/5}. \tag{1.171}$$

We shall assume that r_* defines a sphere moving with the speed of the fluid particles. Then

$$\frac{dr_*}{dt} = v_* = \frac{\psi_1(t)}{r_*^2}. \tag{1.172}$$

From (1.171) and (1.172) we find that

$$\psi_1(t) = r_*^2 \frac{dr_*}{dt} = \frac{2}{5}\frac{r_*^3}{t}.$$

The following dependence can be found by using (1.170):

$$E_0 = \frac{8\pi\rho_1}{25}\frac{r_*^5}{t^2}.$$

Substituting r_* from (1.171), we find the constant

$$k = \left(\frac{25}{8\pi}\right)^{1/5}.$$

Finally, for r_* we have

$$r_* = \left(\frac{25}{8\pi}\frac{E_0}{\rho_1}\right)^{1/5} t^{2/5} = \left(\frac{5}{2}\right)^{2/5}\left(\frac{E_0}{2\pi\rho_1}\right)^{1/5} t^{2/5}. \tag{1.173}$$

Taking account of the boundary condition at infinity $p(\infty, t) = 0$, from (1.169) we find that $\psi_2 = 0$.

The pressure distribution is given by

$$\frac{p}{\rho_1} = \frac{2}{25}\frac{r_*^3}{t^2 r}\left[1 - \left(\frac{r_*}{r}\right)^3\right]. \tag{1.174}$$

For $r = r_*$ we have $p = 0$. This means that at the center of symmetry an empty sphere is formed—a cavity of radius r_* which expands in time according to the law (1.174). The maximum pressure p_2 in space is reached for $r = r_2 = 4^{1/3} r_*$. The dependence of p_2 on time is determined by the formula

$$p_2 = \frac{3}{50}\left(\frac{5}{2}\right)^{4/5} 4^{-1/3}\rho_1\left(\frac{E_0}{2\pi\rho_1}\right)^{2/5} t^{-6/5} = 4^{-4/3}\frac{3E_0}{\pi} r_2^{-3}. \tag{1.175}$$

We have thus obtained an exact solution to the problem of a strong explosion in an incompressible fluid.

Having the general solution (1.169) of the system (1.168), it is possible to obtain a solution of the self-similar PEP for $p_1 \neq 0$. For the solution of this problem we have the initial and boundary conditions: $p = p_1 = \text{const}, v = v_1 = 0, r_* = 0$ at $t = 0$ and energy E_0 is released at the center of the explosion;

$$\begin{aligned} &\text{for } r = r_* \quad p = p_* = 0, \quad v = v_* = \frac{dr_*}{dt}, \\ &\text{for } r = \infty \quad p = p_\infty = p_1, \quad v = v_\infty = 0. \end{aligned} \tag{1.176}$$

From (1.169) and (1.172), using (1.176), we find

$$\frac{p - p_1}{\rho_1} = -\frac{1}{2}\frac{r_*^4 (r_*')^2}{r^4} + \frac{(r_*^2 r_*')'}{r}, \qquad v = \frac{r_*^2 r_*'}{r^2}. \tag{1.177}$$

The prime denotes differentiation with respect to time. For $r = r_*$ we have

$$\frac{p_1}{\rho_1} = \frac{1}{2}(r_*')^2 - \frac{(r_*^2 r_*')'}{r^*}. \tag{1.178}$$

The law of motion $r_*(t)$ of the boundary of the cavity formed during the explosion is found from (1.178). If a new variable $y = r'$ is introduced, then (1.178) admits reduction in order, since the time t does not appear explicitly. As a result of integration we find that

$$y^2 = k_2 r_*^{-3} - \frac{2}{3}\frac{p_1}{\rho_1}. \tag{1.179}$$

We determine the constant k_2 from the condition that for $p_1 = 0$ we must obtain a self-similar solution. It turns out that

$$k_2 = \frac{E_0}{2\pi\rho_1}.$$

The maximum radius of the cavity r_{max} is defined as that value of r_* for which $y = dr_*/dt = 0$. From (1.179) we obtain

$$r_{max} = \left(\frac{3}{4\pi}\frac{E_0}{\rho_1}\right)^{1/3}.$$

If we introduce the dimensionless variables

$$l = \frac{r}{r_{max}}, \qquad \tau = \left(\frac{2}{3}\frac{\rho_1}{\rho_1}\right)^{1/2}\frac{t}{r_{max}}, \qquad l_* = \frac{r_*}{r_{max}},$$

then (1.179) takes the form

$$\frac{dl_x}{d\tau} = \pm\frac{(1-l_*^3)^{1/2}}{l_*^{3/2}}. \tag{1.180}$$

From this the dependence $\tau(l_*)$ is determined by quadrature:

$$\tau(l_*) = \pm\int_0^{l_*}\frac{l_*^{3/2}dl}{(1-l_*^3)^{3/2}}. \tag{1.181}$$

Formulas (1.180) and (1.181) show that $l \leqslant 1$, and, since τ increases, the cavity begins to move toward the center after reaching a maximum radius; peculiar oscillatory motions thus arise. In addition the quantity $dl_*/d\tau$ passes through zero and becomes negative, i.e. in (1.181) the plus sign corresponds to motion away from the center and the minus sign to motion toward the center. After finding the law of motion of the boundary of the cavity the pressure and velocity are found from the formulas

$$\frac{p}{p_1} = 1 + \frac{1}{3}\left[2\frac{d}{d\tau}\left(l_*^2\frac{dl_*}{dt}\right) - l_*^4\left(\frac{dl_*}{d\tau}\right)^2 l^{-3}\right]l^{-1}, \qquad \frac{v}{v_*} = \frac{l_*^2}{l^2}. \tag{1.182}$$

Finally, it is possible to define the period of oscillation τ of the cavity:

$$\tau = 2\tau(1) = 2\int_0^1\frac{l_*^{3/2}dl_*}{(1-l_*^3)^{1/2}}.$$

Using the change of variable $l_* = x^{1/3}$, this integral can be transformed to Euler's integral $B(p, q)$ which can be expressed in terms of Euler's Γ function:

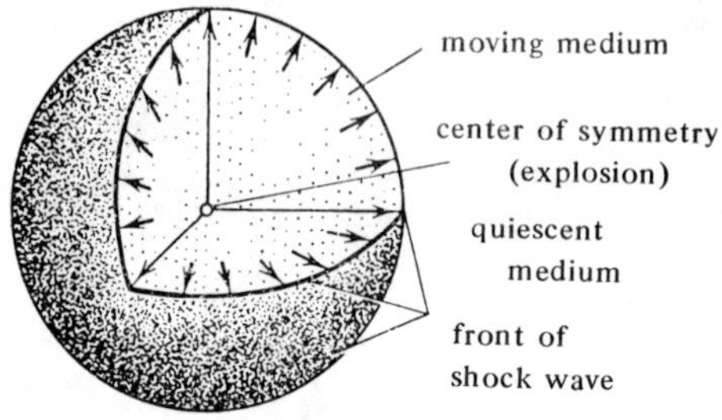

FIGURE 1. Scheme of motion of a compressible medium on explosion of a spherical charge

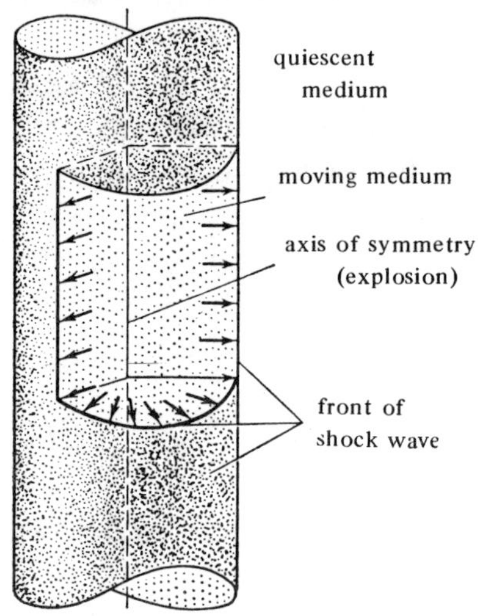

FIGURE 2. Scheme of motion on explosion of a cylindrical charge

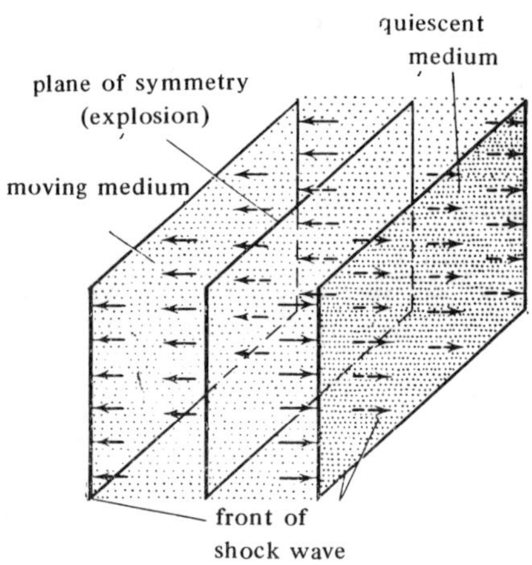

FIGURE 3. Scheme of motion on explosion of a plane charge

$$\tau = \frac{2}{3}\int_0^1 x^{-1/6}(1-x)^{-1/2}\,dx = \frac{2}{3}B\left(\frac{5}{6},\ \frac{1}{2}\right) = 2\sqrt{\pi}\,\frac{\Gamma(5/6)}{\Gamma(1/3)} \approx 1.5.$$

Going over to dimensional coordinates, we obtain the following formula for the period of oscillation:

$$T = 1.14\rho_1^{1/2}E_0^{1/3}p_1^{-5/6}.$$

7.3. *Point explosion in a gas.* Suppose that in an unbounded gas energy E_0 is released at a point, along a line, or over a plane; the energy is reckoned per unit length or per unit area for the linear or plane energy sources respectively. As a result of this, the initial state of the gas receives a strong local perturbation. Suppose that the unperturbed state of the gas does not depend on time, and $v_0 = v_0(r)$, $p_0 = p_0(r)$ and $\rho_0 = \rho_0(r)$ at $t = 0$. If we denote by E the total energy of a unit volume of the gas, by the law of energy conservation for $t \geqslant 0$ we have $E_0 = \int \Delta E\, d\Omega$, where $\Delta E = E - E^0$, $E^0 = (\rho_0 v_0^2/2) + \rho_0 \epsilon_0$ is the initial energy of the gas, and the integration extends over all of space. Thus, as in the preceding examples, we arrive at the singular Cauchy problem

$$\boldsymbol{v}|_{t=0}=\boldsymbol{v}_0, \quad p|_{t=0}=p_0, \quad \rho|_{t=0}=\rho_0, \quad \Delta E|_{t=+0}=\delta(\boldsymbol{r}). \qquad (1.183)$$

Experiment and theory indicate that during an explosion in a gas a shock wave is formed which propagates through the gas from the site of the explosion. Empirically it is known that if the energy of the explosion E_0 is sufficiently high (say, for $\nu = 3$, greater than the initial energy in a unit volume of the gas) and the volume where it is released is considerably less than the volume for which E_0 becomes comparable in magnitude with its initial energy, then a shock wave propagates through the gas because of the abrupt increase in pressure and temperature at the site of the explosion.

From the theoretical point of view it is known, first of all, that even for the Cauchy problem with smooth initial data discontinuities in the solution may arise [37]–[39], and this is a fortiori the case in a problem where the initial data are given in the form of a δ function for the energy source. Secondly, Sedov [62] has shown (see also Chapter 2) that a solution of the problem of a strong explosion in a gas (with zero initial pressure) cannot be constructed in the class of continuous functions. Thus the desired solution will in fact be a generalized solution in which the regions of continuity will be separated by surfaces of discontinuity.

Suppose that the equation of the front of a shock wave corresponding to the leading front of perturbations in the gas is $F(\boldsymbol{r}, t) = 0$. Then on passing through this surface conditions (1.53) and (1.54) must be satisfied. The problem posed above can be formulated as a boundary value problem with conditions (1.53) and (1.54) on the surface $F = 0$, where the surface $F = 0$ contracts to the point, the line, or tends to the plane of the explosion as $t \to 0$. In addition to these conditions there is the integral law of energy conservation, which we now write in the form

$$E_0 = \int\limits_{\Omega(t)} \Delta E \, d\Omega, \qquad (1.184)$$

where $\Omega(t)$ is the volume occupied by the gas perturbed by the explosion inside the front of the shock wave.

We shall consider in more detail the formulation of the basic problem for the case of an explosion in a quiescent perfect gas when $p_1 = \text{const}$, $\rho = \rho(r)$, and the adiabatic condition is satisfied for the motion of the gas behind the shock wave. In this case the motion will be one-dimensional, and the relation (1.184) takes the form

$$E_0 = \sigma_\nu \int\limits_0^{r_2} \left(\frac{\rho v^2}{2} + \frac{p - p_1}{\gamma - 1}\right) r^{\nu-1} dr, \qquad (1.185)$$

where $\sigma_\nu = 2(\nu - 1)\pi + (\nu - 2)(\nu - 3)$, r_2 is the radius of the shock wave, and on the shock wave, i.e. for $r = r_2$, conditions (1.63) obtain.

At the initial moment of time $t = 0$, for $r > 0$ we have $p = p_1 = \text{const}$, $\rho = \rho_1(r)$, $v_1 = 0$ and, in addition, the following conditions: $r_2(0) = 0$ and at the center of symmetry energy E_0 is released. To determine the unknown functions it is necessary to integrate the system of partial differential equations (1.18)–(1.20) with the indicated initial conditions and conditions on the front of the shock wave. If there is no source of mass at the center of symmetry (i.e. for $r = 0$), then from the conditions of the problem it follows by symmetry (cf. [4], [46]) that

$$v(0, t) = 0. \tag{1.186}$$

The problem is formulated similarly when it is described in the Lagrange variables. The scheme of motion of the gas during a point explosion is shown in Figures 1–3.

The PEP in a gas was first formulated and investigated by L. I. Sedov and G. I. Taylor [63], [64]. The papers of K. P. Stanjukovič [65] and J. von Neumann [66] are among the first in the study of this problem. The PEP in a gas as formulated above can be generalized in a natural way to the case of motion of a gas with chemical reactions and for various types of nonadiabatic motions. If the medium in which the explosion takes place is electrically conducting and, in addition, there is an applied external electromagnetic field, then during a point explosion in such a medium interaction between the electromagnetic field and the motion takes place. The general formulation of the basic problem here is analogous to that for the case of an explosion in a gas without considering electromagnetic effects. However, the picture of the motion becomes considerably more complex. The principal objective of the analysis in the following chapters is the study of the problems of gas dynamics mentioned in the present section and also some applications of the results of these investigations to physical phenomena. In the sequel we shall also consider certain generalizations of the basic problem for the cases of nonsymmetric energy release during the explosion.

BIBLIOGRAPHY

1. L. I. Sedov, *Lectures on continuum mechanics.* Part I, Izdat. Moskov. Gos. Univ., Moscow, 1966. (Russian)

2. L. V. Ovsjannikov, *Lectures on the foundations of gas dynamics,* "Nauka", Novosibirsk, 1967. (Russian)

3. James Serrin, *Mathematical principles of classical fluid mechanics,* Handbuch der Physik (herausgegeben von S. Flügge), Band 8.1: Strömungsmechanik I, Springer-Verlag, Berlin, 1959, pp. 125–263. MR **21** #6836b.

4. V. P. Korobeĭnikov, N. S. Mel'nikova and E. V. Rjazanov, *Theory of point explosion,* Fizmatgiz, Moscow, 1961. (Russian)

5. L. D. Landau and E. M. Lifšic, *Mechanics of continuous media,* 2nd ed., GITTL, Moscow, 1953; English transl., Part I: *Fluid mechanics*; Part II: *Theory of elasticity,* Course of Theoretical Physics, vols. 6, 7, Pergamon Press, New York; Addison-Wesley, Reading, Mass., 1959. MR **16**, 417; **21** #6839.

6. L. I. Sedov, *Continuum mechanics.* Vols. I, II, "Nauka", Moscow, 1970. (Russian)

7. N. E. Kočin, *Vector analysis and the elements of tensor analysis*, 8th ed., Izdat. Akad. Nauk SSSR, Moscow, 1961. (Russian) MR **26** #6887.

8. S. L. Sobolev, *Equations of mathematical physics*, 2nd ed., GITTL, Moscow, 1950; English transl., Pergamon Press, London; Addison-Wesley, Reading, Mass., 1964. MR **13**, 42; **31** #2478.

9. A. G. Kulikovskiĭ and G. A. Ljubimov, *Magnetohydrodynamics*, Fizmatgiz, Moscow, 1962; English transl., Adiwes Internat. Ser. Phys., Addison-Wesley, Reading, Mass., 1965.

10. L. D. Landau and E. M. Lifšic, *Electrodynamics of continuous media*, GITTL, Moscow, 1957; English transl., Course of Theoretical Physics, vol. 8, Pergamon Press, Oxford; Addison-Wesley, Reading, Mass,, 1960. MR **20** #6274; **22** #11796.

11. Shih-i Pai, *Magnetogasdynamics and plasma dynamics*, Springer-Verlag, Berlin, 1962.

12. S. L. Braginskiĭ, *Magnetohydrodynamics of weakly conducting liquids*, Ž. Èksper. Teoret. Fiz. **37** (1959), 1417–1430 = Soviet Phys. JETP **10** (1960), 1005–1014.

13. T. G. Cowling, *Magnetohydrodynamics*, Interscience Tracts in Phys. and Astr., no. 4, Interscience, New York, 1957.

14. G. F. Chew, M. L. Goldberger and F. E. Low, *The Boltzmann equation and the one-fluid hydromagnetic equations in the absence of particle collisions*, Proc. Roy. Soc. London Ser. A **236** (1956), 112–118. MR **18**, 92.

15. L. D. Pičahči, *Discontinuities in a rarefied plasma in the Chew-Goldberger-Low approximation*, Ukrain. Fiz. Ž. **5** (1960), 450–457. (Ukrainian; English summary)

16. T. F. Volkov, *Hydrodynamical description of a strongly rarefied plasma*, Problemy Teorii Plazmy, vyp. 4, Atomizdat, Moscow, 1964, pp. 3–19. (Russian)

17. S. Chapman and T. Cowling, *The mathematical theory of nonuniform gases*, Cambridge Univ. Press, New York, 1970.

18. S. I. Braginskiĭ, *Transport phenomena in a plasma*, Problemy Teorii Plazmy, vyp. 1, Atomizdat, Moscow, 1963, pp. 183–272. (Russian)

19. R. K. Jaggi, *Wave motion in a plasma with anisotropic pressure*, Phys. Fluids **5** (1962), 949–962.

20. L. Spitzer, Jr., *Physics of fully ionized gases*, Interscience Tracts on Phys. and Astr., no. 3, Interscience, New York, 1956; MR **23** #B1483.

21. C. L. Longmire, *Elementary plasma physics*, Interscience, New York, 1963; Russian transl., Atomizdat, Moscow, 1966.

22. D. A. Frank-Kameneckiĭ, *Lectures on plasma physics*, Atomizdat, Moscow, 1964. (Russian)

23. Rolf K. M. Landshoff (editor), *Magnetohydrodynamics* (*a symposium*), Stanford Univ. Press, Stanford, Calif., 1957.

24. I. E. Tamm and A. D. Saharov, *Theory of the magnetic thermonuclear reactor.* I, II, III, Plasma Physics and the Problem of Controllable Thermonuclear Reactions, Vol. I, Izdat. Akad. Nauk SSSR, Moscow, 1958, pp. 3–41. (Russian)

25. D. V. Sivuhin, *Drift theory of the motion of a charged particle in electromagnetic fields*, Problemy Teorii Plazmy, vyp. 1, Atomizdat, Moscow, 1963, pp. 7–97. (Russian)

26. I. E. Tamm, *Foundations of the theory of electricity*, 7th ed., GITTL, Moscow, 1956. (Russian)

27. M. N. Kogan, *Shock waves in magneto-gas-dynamics*, Prikl. Mat. Meh. **23** (1959), 557–563 = J. Appl. Math. Mech. **23** (1959), 784–792. MR **22** #5309.

28. R. V. Polovin, *Shock waves in magnetohydrodynamics,* Uspehi Fiz. Nauk **72** (1960), 33–52 = Soviet Phys. Uspekhi **3** (1961), 677–688. MR **22** #12990.

29. S. V. Iordanskiĭ, *Zemplén's theorem in magnetohydrodynamics,* Dokl. Akad. Nauk SSSR **121** (1958), 610–612 = Soviet Phys. Dokl. **3** (1958), 736–738. MR **21** #3194.

30. R. V. Polovin and G. Ja. Ljubarskiĭ, *The impossibility of rarefaction shock waves in magnetohydrodynamics,* Ž. Èksper. Teoret. Fiz. **35** (1958), 510 = Soviet Phys. JETF **8** (1959), 351–352.

31. F. A. Baum, S. A. Kaplan and K. P. Stanjukovič, *Introduction to cosmic gas dynamics,* Fizmatgiz, Moscow, 1958. (Russian) MR **21** #1210.

32. A. G. Kulikovskiĭ and G. A. Ljubimov, *Gas-ionizing magnetohydrodynamic shock waves,* Dokl. Akad. Nauk SSSR **129** (1959), 52–55 = Soviet Phys. Dokl. **4** (1960), 115–118. MR **22** #6333.

33. V. P. Karlikov and V. P. Korobeĭnikov, *Disturbances in an electromagnetic field by shock waves under the influence of a sudden increase in conductivity,* Prikl. Mat. Meh. **25** (1961), 554–556 = J. Appl. Math. Mech. **25** (1961), 819–823. MR **25** #895.

34. A. A. Barmin and A. G. Kulikovskiĭ, *Gas-ionizing shock waves in an arbitrarily ordered magnetic medium,* Problems of Hydrodynamics and Continuum Mechanics, Contributions in Honor of the Sixtieth Birthday of Academician L. I. Sedov, published simultaneously in Russian and English, "Nauka", Moscow; Soc. Indust. Appl. Math., Philadelphia, Pa., 1969, pp. 39–54.

35. R. Z. Sagdeev, *Collective processes and shock waves in a rarefied plasma,* Voprosy Teorii Plazmy, vyp. 4, Atomizdat, Moscow, 1964, pp. 20–80. (Russian)

36. V. I. Karpman, *Hydromagnetic shock waves,* Problems of Magnetohydrodynamics and Cosmic Gas Dynamics (Voprosy Kosmogonii, Tom X), "Nauka", Moscow, 1964, pp. 36–57. (Russian; English summary)

37. R. Courant, *Methods of mathematical physics.* Vol. II: *Partial differential equations,* Interscience, New York, 1962. MR **25** #4216.

38. B. L. Roždestvenskiĭ and N. N. Janenko, *Systems of quasilinear equations and their applications to gas dynamics,* "Nauka", Moscow, 1968; English transl., Transl. Math. Monographs, vol. 30, Amer. Math. Soc., Providence, R. I. (to appear).

39. A. Jeffrey and T. Taniuti, *Non-linear wave propagation, with applications to physics and magnetohydrodynamics,* Academic Press, New York, 1964.

40. R. V. Polovin, *Some questions of the theory of simple waves,* Report presented at the Second Conf. on Theoret. Appl. Magnetohydrodynamics (Riga, 1960); Abstract published in Questions of Magnetohydrodynamics and Plasma Dynamics. II, Izdat. Akad. Nauk Latv. SSR, Riga, 1962, p. 233. (Russian)*

41. V. P. Korobeĭnikov and S. P. Lomnev, *On the motion of charged particles in a plasma in the presence of a magnetohydrodynamic shock wave,* PMTF Ž. Prikl. Meh. i. Tehn. Fiz. **1964**, no. 6, 89–92. (Russian)

42. V. P. Šabanskiĭ, *The radiation belts,* Geomagnetizm i Aeronomija **5** (1965), 969–1002. (Russian)

43. Y. M. Lynn, *Discontinuities in an anisotropic plasma,* Phys. Fluids **10** (1967), 2278–2280.

Editor's note.* See also R. V. Polovin, *Contribution to the theory of simple magnetohydrodynamic waves,* Ž. Èksper. Teoret. Fiz. **39 (1960), 463–470 = Soviet Phys. JETP **12** (1961), 326–330; and *Flow of a plasma into vacuum in the presence of a magnetic field,* ibid. 657–661 = 461–463.

44. S. S. Grigorjan, *Cauchy's problem and the problem of a piston for one-dimensional nonsteady motions of gas (automodel motion)*, Prikl. Mat. Meh. **22** (1958), 179–187 = J. Appl. Math. Mech. **22** (1958), 244–255. MR **20** #6895.

45. K. P. Stanjukovič, *Transient motions of a continuous medium*, GITTL, Moscow, 1955; English transl., Pergamon Press, New York, 1960. MR **18**, 440; **22** #5244.

46. L. I. Sedov, *Similarity and dimensional methods in mechanics*, 4th ed., GITTL, Moscow, 1957; English transl., Academic Press, New York, 1959. MR **19**, 898; **21** #6840.

47. Ja. B. Zel'dovič and A. S. Kompaneec, *Theory of detonations*, GITTL, Moscow, 1955; English transl., Academic Press, New York, 1960.

48. V. P. Korobeĭnikov, *On one-dimensional motions of a gas in a magnetic field accompanied by shock waves*, PMTF Ž. Prikl. Meh. i Tehn. Fiz. **1960**, no. 2, 47–53. (Russian)

49. L. S. Pontrjagin, *Continuous groups*, 2nd ed., GITTL, Moscow, 1954; English transl., Gordon and Breach, New York, 1966. MR **17**, 171; **34** #1439.

50. N. G. Čebotarev, *Theory of Lie groups*, GITTL, Moscow, 1940. (Russian)

51. L. V. Ovsjannikov, *Group properties of differential equations*, Izdat. Sibirsk. Otdel. Akad. Nauk SSSR, Novosibirsk, 1962. (Russian) MR **26** #264.

52. Garrett Birkhoff, *Mathematics for engineers*. III: *Dimensional analysis of partial differential equations*, Electrical Engrg. **67** (1948), 1185–1188.

53. V. V. Lohin and L. I. Sedov, *Nonlinear tensor-valued functions of several tensor arguments*, Prikl. Mat. Meh. **27** (1963), 393–417 = J. Appl. Math. Mech. **27** (1963), 597–629.

54. G. I. Barenblatt, *On limiting self-similar motions in the theory of unsteady filtration of a gas in a porous medium and the theory of the boundary layer*, Prikl. Mat. Meh. **18** (1954), 409–414. (Russian) MR **18**, 91.

55. V. P. Korobeĭnikov, *On invariant solutions of the equations of magnetohydrodynamics*, Magnitnaja Gidrodinamika **1967**, no. 3, 55–58 = Magnetohydrodynamics **3** (1967), no. 3, 30–31.

56. ———, *On integrals of the equations of unsteady adiabatic motion of a gas*, Dokl. Akad. Nauk SSSR **104** (1955), 509–512. (Russian) MR **17**, 1150.

57. M. L. Lidov, *A finite integral of the equations of one-dimensional self-similar adiabatic motions of a gas*, Dokl. Akad. Nauk SSSR **103** (1955), 35–36. (Russian) MR **17**, 207.

58. V. P. Korobeĭnikov, *Unidimensional automodel models of a conducting gas in a magnetic field*, Dokl. Akad. Nauk SSSR **121** (1958), 613–615 = Soviet Phys. Dokl. **3** (1958/59), 739–742. MR **21** #4005.

59. G. E. Šilov, *Mathematical analysis. Second special course*, 2nd ed., Fizmatgiz, Moscow, 1961; English transl., Pergamon Press, New York, 1965. MR **24** #A1180; **32** #2519.

60. V. S. Vladimirov, *Equations of mathematical physics*, "Nauka", Moscow, 1967; English transl., Pure and Appl. Math., vol. 3, Marcel Dekker, New York, 1971. MR **39** #599; **42** #3394.

61. G. M. Položiĭ, *The equations of mathematical physics*, "Vysšaja Škola", Moscow, 1964. (Russian) MR **31** #6404.

62. L. I. Sedov, *Propagation of strong blast waves*, Prikl. Mat. Meh. **10** (1946), 241–250. (Russian; English summary)

63. ———, *Le mouvement d'air en cas d'une forte explosion*, C. R. (Dokl.) Acad. Sci. URSS **52** (1946), 17–20.

64. G. I. Taylor, *The formation of a blast wave by a very intense explosion*. I, II, Proc. Roy. Soc. London Ser. A **201** (1950), 159–174, 175–186.

65. K. P. Stanjukovič, *Application of particular solutions of gas dynamics equations to the study of detonation and shock waves*, C. R. (Dokl.) Acad. Sci. URSS **52** (1946), 589–592.

66. H. Goldstine and J. von Neumann, *Blast wave calculation*, Comm. Pure Appl. Math. **8** (1955), 327–353. MR **17**, 314.

CHAPTER 2

STRONG EXPLOSION IN A GAS

§1. L. I. Sedov's exact solution of the self-similar problem of strong explosion in a perfect gas for constant and variable initial densities

During a point explosion in a gas a strong shock wave is formed, with the pressure behind its front p_2 much greater than the initial pressure p_1. If the initial pressure is neglected and it is assumed that the initial density varies according to the law

$$\rho_1 = Ar^{-\omega}, \tag{2.1}$$

where A is a constant with dimension $[A] = ML^{\omega-3}$, ω a derived constant, then the PEP in a perfect gas will be self-similar [1], [2]. Indeed, among the defining parameters of the problem there are only two constants E_0 ($[E_0] = ML^{\nu-1}T^{-2}$) and A with independent dimensions.

Because of the self-similarity the system of equations of gas dynamics describing the perturbed motions of the gas behind the shock wave is transformed into a system of ordinary differential equations. If we introduce the dimensionless variables

$$f = \frac{v}{D}, \qquad g = \frac{\rho}{\rho_1}, \qquad h = \frac{p}{\rho_1 D^2}, \qquad \lambda = \frac{r}{r_2}, \tag{2.2}$$

then for the new unknown functions $f(\lambda)$, $g(\lambda)$ and $h(\lambda)$ we obtain from the equations of gas dynamics (1.18)–(1.20) a system of three ordinary differential equations

$$\begin{gathered} g(f-\lambda) f' + h' + \frac{\omega-\nu}{2} fg = 0, \\ (f-\lambda) g' + \left(f' + \frac{\nu-1}{\lambda} f - \omega\right) g = 0, \\ (f-\lambda) h' + \gamma\left(f' + \frac{\nu-1}{\lambda} f\right) h - \nu h = 0. \end{gathered} \tag{2.3}$$

The boundary conditions on the front of the strong shock wave take the form

$$f(1) = \frac{2}{\gamma+1}, \qquad g(1) = \frac{\gamma+1}{\gamma-1}, \qquad h(1) = \frac{2}{\gamma+1}. \tag{2.4}$$

L. I. Sedov [1], [2] found the energy integral (1.157) for the system (2.3) which makes it possible to obtain an exact solution of the basic problem for zero counterpressure. In the Euler variables this solution, taking conditions (2.4) into account, can be written in the form [2]–[4]

$$\frac{r}{r_2}=\lambda(\mu)=\mu^{-\delta}\left[\frac{2\gamma}{\gamma-1}\left(\mu-\frac{\gamma+1}{2\gamma}\right)\right]^{\beta_2}$$
$$\times\left\{\frac{2(\nu\gamma-\nu+2)}{3\nu-2-\gamma(\nu-2)-\omega(\gamma+1)}\left[\frac{\gamma+1}{\varepsilon(\nu\gamma-\nu+2)}-\mu\right]\right\}^{-\beta_1}, \tag{2.5}$$

$$\frac{v}{v_2}=\frac{\gamma+1}{2}f=\lambda\mu, \tag{2.6}$$

$$\frac{\rho}{\rho_2}=\frac{\gamma-1}{\gamma+1}g(\mu)=\mu^{\delta\omega}\left[\frac{2\gamma}{\gamma-1}\left(\mu-\frac{\gamma+1}{2\gamma}\right)\right]^{\beta_3-\omega\beta_2}$$
$$\times\left[\frac{2}{\gamma-1}\left(\frac{\gamma+1}{2}-\mu\right)\right]^{-\beta_5}\left\{\frac{2(\nu\gamma-\nu+2)}{3\nu-2-\gamma(\nu-2)-\omega(\gamma+7)}\right.$$
$$\left.\times\left[\frac{\gamma+1}{\delta(\nu\gamma-\nu+2)}-\mu\right]\right\}^{\beta_4-\omega\beta_1}, \tag{2.7}$$

$$\frac{p}{p_2}=\frac{\gamma+1}{2}h=\mu^{\delta\nu}\left[\frac{2}{\gamma-1}\left(\frac{\gamma+1}{2}-\mu\right)\right]^{1-\beta_5}$$
$$\times\left\{\frac{2(\nu\gamma-\nu+2)}{3\nu-2-\gamma(\nu-2)-\omega(\gamma+1)}\left[\frac{\gamma+1}{\delta(\nu\gamma-\nu+2)}-\mu\right]\right\}^{\beta_4-(\omega-2)\beta_1}, \tag{2.8}$$

$$\frac{T}{T_2}=\theta(\mu)=\mu^{\delta(\nu-\omega)}\left[\frac{2\gamma}{\gamma-1}\left(\mu-\frac{\gamma+1}{2\gamma}\right)\right]^{\omega\beta_2-\beta_3}$$
$$\times\left[\frac{2}{\gamma-1}\left(\frac{\gamma+1}{2}-\mu\right)\right]\left\{\frac{2(\nu\gamma-\nu+2)}{3\nu-2-\gamma(\nu-2)-\omega(\gamma+1)}\right.$$
$$\left.\times\left[\frac{(\nu+2-\omega)(\gamma+1)}{2(\nu\gamma-\nu+2)}-\mu\right]\right\}^{-2\beta_1}, \tag{2.9}$$

where

$$\beta_1=\beta_2+\frac{\gamma+1}{\nu\gamma-\nu+2}-\delta,\qquad \beta_2=\frac{\gamma-1}{2\gamma-2+\nu-\omega\gamma},$$
$$\beta_3=\frac{\nu-\omega}{2\gamma-2+\nu-\omega\gamma}=1+(\omega-2)\beta_2,\qquad \delta=\frac{2}{\nu+2-\omega}, \tag{2.10}$$
$$\beta_4=\frac{2(\nu-\omega)\beta_1}{\delta(2\nu-\nu\gamma-\omega)},\qquad \beta_5=\frac{2\nu-\omega(\gamma+1)}{2\nu-\nu\gamma-\omega}.$$

The values of the density, pressure, and temperature immediately behind the front of the shock wave, i.e. for $\lambda=1-0$, are found from the formulas

$$v_2=\frac{2\delta}{\gamma+1}\left(\frac{E_0}{\alpha A}\right)^{1/2}r_2^{\frac{\omega-\nu}{2}}=\frac{2\delta}{\gamma+1}\left(\frac{E_0}{\alpha A}\right)^{\delta/2}t^{\frac{\omega-\nu}{\nu+2-\omega}},$$
$$\rho_2=\frac{\gamma+1}{\gamma-1}Ar_2^{-\omega},\quad p_2=\frac{2E_0}{\alpha}\delta^2\frac{r_2^{-\nu}}{\gamma+1}=\frac{2\delta^2A}{\gamma+1}\left(\frac{E_0}{\alpha A}\right)^{\frac{2-\omega}{\nu+2-\omega}}t^{-\nu\delta}, \tag{2.11}$$
$$RT_2=\frac{p_2}{\rho_2}=\frac{2(\gamma-1)}{(\gamma+1)^2}\delta^2\frac{E_0}{\alpha A}r_2^{\omega-\nu}=\frac{2(\gamma-1)}{(\gamma+1)^2}D^2,\qquad \delta=\frac{2}{\nu+2-\omega}.$$

For the velocity of the shock wave and its coordinates we have

$$D=\frac{dr_2}{dt}=\delta\frac{r_2}{t}=\delta\left(\frac{E_0}{\alpha A}\right)^{1/2}r^{\frac{\omega-\nu}{2}}=\delta\left(\frac{E_0}{\alpha A}\right)^{\delta/2}t^{\frac{\omega-\nu}{\nu+2-\omega}}, \qquad (2.12)$$

$$\frac{dD}{dt}=(\omega-\nu)\frac{D^2}{2r_2},$$

$$r_2=\left(\frac{E_0}{\alpha A}\right)^{\delta/2}t^{\delta}, \quad \delta=\frac{2}{\nu+2-\omega}. \qquad (2.13)$$

The solution (2.1)–(2.5) describes the behavior of the gas-dynamical functions behind the front of a shock wave propagating through a quiescent medium with variable density. From the structure of this solution it follows that it can be written in terms of the parametric variable μ, and $\lambda(1) = 1$. The parameter α entering in the solution of the problem depends on γ, ω and ν, and is defined from the integral law of energy conservation

$$E_0=\sigma_\nu\int_0^{r_2}\left(\frac{p}{\gamma-1}+\frac{\rho v^2}{2}\right)r^{\nu-1}dr. \qquad (2.14)$$

Considering (2.1)–(2.13), from the integral law of energy conservation (2.14) we obtain a formula for determining $\alpha(\nu, \gamma, \omega)$:

$$\alpha(\nu,\gamma,\omega)=\frac{\delta^2\sigma_\nu}{\gamma-1}\int_0^1\left(h+\frac{(\gamma-1)}{2}gf^2\right)\lambda^{\nu-1}d\lambda. \qquad (2.15)$$

It follows that α is a functional of the solution.

If we take for the Lagrange coordinate the initial coordinate ξ of the particle, then, noting that on the front of the shock wave $r_2 = \xi$, from the conditions on a strong shock wave and formulas (2.11)–(2.12) we find for the quantity p/ρ^γ, which is conserved in the particle,

$$\frac{p}{\rho^\gamma}=\frac{4\delta(\gamma-1)^\gamma E_0}{\alpha(\gamma+1)^{\gamma+1}A^\gamma\xi^{\nu-\omega\gamma}}. \qquad (2.16)$$

The solution (2.5)–(2.10) and the relation (2.16) enable us to find the relation between the Lagrange coordinate ξ, the time t, and the parametric variable μ:

$$\xi=r_2\mu^{-\delta}\left[\frac{2\gamma}{\gamma-1}\left(\mu-\frac{\gamma+1}{2\gamma}\right)\right]^{\beta_6}\left[\frac{2}{\gamma-1}\left(\frac{\gamma+1}{2}-\mu\right)\right]^{\beta_8}$$
$$\times\left\{\frac{2(\nu\gamma-\nu+2)}{3\nu-2-\gamma(\nu-2)-\omega(\gamma+1)}\left[\frac{\gamma+1}{\delta(\nu\gamma-\nu+2)}-\mu\right]\right\}^{\beta_7}, \qquad (2.17)$$

where

$$\beta_6=\frac{\gamma}{2\gamma-2+\nu-\omega\gamma}, \quad \beta_7=\frac{\nu\gamma-\nu+2}{2\nu-\nu\gamma-\omega}\beta_1, \quad \beta_8=\frac{\gamma}{\nu\gamma-2\nu+\omega}. \qquad (2.18)$$

After the exact analytic solution has been found the problem arises of studying the dependence of its properties on the parameters it contains and of determining the values of α.

We shall mention here only certain well-known important properties of the solution [2], [3] for constant initial density $\rho_1 = \text{const}$ and with $\gamma = 1.4$. For the case of constant density the parameter μ varies from 1 to $(\gamma + 1)/2\gamma$ if the solution is continuable to the center, while the value $(\gamma + 1)/2\gamma$ corresponds to the center of symmetry.

Study of the distribution of velocities, densities, pressures, and temperatures leads to the following conclusions:

1) Since $v/v_2 = \mu\lambda$, and the parameter μ, taking values in the interval $[(\gamma + 1)/2\gamma, 1]$ for $\gamma = 1.4$, is not greatly different from one, it follows that the velocity of air particles at a fixed moment of time is approximately proportional to their distances r from the center of symmetry.

2) The density ρ decreases very rapidly in the direction from the shock wave to the center of symmetry. Since on the surface of the shock wave $\rho_2 = ((\gamma+1)/(\gamma-1))\rho_1$, i.e. for $\gamma = 1.4$ the density on the front is 6 times greater than the initial density, it follows from this that the main mass of the moving gas is concentrated in a narrow layer behind the front of the shock wave.

3) For $t > 0$ the pressure at the center of symmetry is always finite and amounts to a certain fraction of its value on the shock wave. It is characteristic also that up to rather large distances (about $0.5\ r_2$) from the center of symmetry the pressure is constant with respect to r: $p(r, t) \approx p(0, t)$, $r \leqslant 0.5\ r_2$. This region where for fixed t the pressure is constant is sometimes called the "pressure plateau".

4) The temperature tends to infinity on approach to the center of symmetry. This fact is explained by our neglect of the thermal conductivity, which has a substantial effect on the distribution of temperatures in a neighborhood of the center.

The special features of the distribution of velocity, pressure, and temperature in space mentioned for the case $\gamma = 1.4$ follow from the solution (2.2)–(2.10) and graphs of the basic functions shown in Figures 4–7 (see §2 of this chapter).

§2. Dependence of the exact solution on the parameters

2.1. *Dependence of the solution for constant initial density on the parameters* γ *and* ν. In studying the dependence of the solution (2.2)–(2.10) on γ and ν in the case $\omega = 0$ the following basic facts were previously established [2]. For spherical symmetry ($\nu = 3$) with $\gamma = 7$ the pressure and density become zero at the center of symmetry. For $\gamma \geqslant 7$ near the center a spherical cavity is formed where $p = 0$ and $\rho = 0$, while with increase in γ the solution tends to the solution for an incompressible fluid considered in the preceding chapter. In the cases $\nu = 1, 2$ the pressure near the center is not equal to zero for any finite γ. However, the following questions were not entirely resolved: the detailed behavior of the basic functions for γ varying in the range $1 < \gamma \leqslant 7$ for all ν and the detailed behavior of the solution for $\gamma > 7$.

We have carried out a detailed investigation of the dependence of the solution on γ when $1 < \gamma \leqslant 7$. The results of this investigation are given below. First of all, the continuous dependence of the solution (2.2)–(2.10) on γ was established. Analysis of this solution shows that the functions it contains are continuous with respect to γ everywhere except at the point $\gamma = 2$. At the point $\gamma = 2$ the functions g and h are not defined. If the solution has a finite limit when $\gamma \to 2$, we can complete the definitions of the functions g and h by setting them equal at the point $\gamma = 2$ to their limiting values.

Further, we can define the solution of the problem at $\gamma = 2$ directly from the differential equations and the boundary data. If the solution thus found coincides with the functions obtained by the limiting process, this will imply the continuous dependence of the solution at the point $\gamma = 2$. Since there is continuous dependence of the solution on γ for all other values of $\gamma \in (1, 7]$, the solution will then depend continuously on γ.

We shall consider the limit with respect to γ for the dependence $g(\mu)$. From (2.7) it follows that for $\omega = 0$

$$g(\mu) = \frac{\gamma+1}{\gamma-1}\left[\frac{2\gamma}{\gamma-1}\left(\mu - \frac{\gamma+1}{2\gamma}\right)\right]^{\beta_3}\left[\frac{2}{\gamma-1}\frac{\gamma+1}{2} - \mu\right]^{-\beta_5}$$
$$\times\left\{\frac{2(\nu\gamma-\nu+2)}{3\nu-2-\gamma(\nu-2)}\left[\frac{\gamma+1}{\delta(\nu\gamma-\nu+2)} - \mu\right]\right\}^{\beta_4}. \quad (2.19)$$

After simple transformations this formula can be brought to the form

$$g(\mu) = \frac{\gamma+1}{\gamma-1}\left[\frac{2\gamma}{\gamma-1}\left(\mu - \frac{\gamma+1}{2\gamma}\right)\right]^{\beta_3}\left[\frac{2(\nu\gamma-\nu+2)}{3\nu-2-\gamma(-2)}\times\right.$$
$$\left.\times\left(\frac{\gamma+1}{\delta(\nu\gamma-\nu+2)} - \mu\right)\right]^{\beta_4-\beta_5}\left[\frac{(\gamma-1)(\nu\gamma-\nu+2)}{3\nu-2-\gamma(\nu-2)}\,\frac{\dfrac{(\nu+2)(\gamma+1)}{2(\nu\gamma-\nu+2)} - \mu}{\dfrac{\gamma+1}{2} - \mu}\right]^{\beta_5}.$$

The last factor can be transformed to the following form:

$$\left[1 + \frac{1}{x}\right]^{-\frac{2\nu(\gamma+1)}{\nu+2}\frac{1-\mu}{[(\gamma+1)/2]-\mu}x},$$

where

$$x = \left(\frac{\gamma+1}{2} - \mu\right)(\nu+2)[(1-\mu)\nu(\gamma+1)(2-\gamma)]^{-1}.$$

We thus seek the limit of the following expression:

$$\lim_{\gamma\to 2} g(\mu) = 3\lim_{\gamma\to 2}\left\{\left[\frac{2\gamma}{\gamma-1}\left(\mu - \frac{\gamma+1}{2\gamma}\right)\right]^{\beta_3}\right\}\times$$

$$\times \lim_{\gamma\to 2}\left\{\frac{2(\nu\gamma-\nu+2)}{3\nu-2-\gamma(\nu-2)}\left(\frac{(\nu+2)(\gamma+1)}{2(\nu\gamma-\nu+2)}-\mu\right)^{\beta_4-\beta_5}\right\}$$

$$\times \lim_{\substack{x\to\infty\\ \gamma\to 2}}\left(1+\frac{1}{x}\right)^{-\frac{2\nu(\gamma+1)}{\nu+2}\frac{1-\mu}{[(\gamma+1)/2]-\mu}}.$$

Passing to the limit, we find that

$$g(\mu)=3\left[4\left(\mu-\frac{3}{4}\right)\right]^{\nu/(\nu+2)}\left[2\left(\frac{3}{2}-\mu\right)\right]^{(\nu-2)/(\nu+2)}\exp\left[-\frac{6\nu}{\nu+2}\frac{1-\mu}{\frac{3}{2}-\mu}\right]. \tag{2.20}$$

The limiting relations for $h(\mu)$ and $\lambda(\mu)$ can be found similarly; the limit for $\lambda(\mu)$ is altogether trivial. For $h(\mu)$ we will have an expression of the type (2.20).

If we substitute the value $\gamma = 2$ into the conditions on the shock wave (2.4) and integrate the system (2.3) with conditions (2.4), we obtain the solution

$$\lambda=\mu^{-2/(\nu+2)}\left[2\left(\frac{3}{2}-\mu\right)\right]^{-2/(\nu+2)}\left[4\left(\mu-\frac{3}{4}\right)\right]^{1/(\nu+2)},$$

$$f=\frac{2}{3}\lambda\mu,\qquad g=g(\mu),$$

$$h=\frac{2}{3}\mu^{2\nu/(\nu+2)}\left[2\left(\frac{3}{2}-\mu\right)\right]^{2(\nu-2)/(\nu+2)}\exp\left[-\frac{6\nu}{\nu+2}\frac{1-\mu}{\frac{3}{2}-\mu}\right], \tag{2.21}$$

where the expression for $g(\mu)$ coincides with (2.20). Comparing the solution (2.21) with the solution obtained from the limiting process $\gamma\longrightarrow 2$, it is found that the two solutions coincide. The limiting values of the functional $\alpha(\nu,\gamma)$ will similarly coincide with the corresponding value obtained for the functions (2.21). This implies that the solution (2.5)–(2.13) depends continuously on γ.

The following theorem has thus been proved.

THEOREM 2. *The solution of the basic problem of strong explosion in a gas with constant initial density depends continuously on the adiabatic exponent* γ *for values of* γ *in the interval* $(1, 7)$.

This result was first published in [5].

We also note for the case $\gamma = 2$ the formula for the dimensionless pressure at the center of the explosion, i.e. for $\mu=\mu_0=3/4$:

$$h(\mu_0)=\frac{2}{3}3^{4(\nu-1)/(\nu+2)}4^{(2-3\nu)/(\nu+2)}e^{-2\nu/(\nu+2)}. \tag{2.22}$$

From this formula it follows that the pressure at the center is finite for any value of ν different from zero. The above analysis removes any doubt as to the behavior of the solution written in the form (2.5)–(2.13) for $\gamma\longrightarrow 2$ and gives the complete solution of the problem for $\gamma = 2$ which will be used in the study of problems of magnetohydrodynamics.

An investigation was conducted of the effect of γ on the functions $f(\lambda)$, $g(\lambda)$, $h(\lambda)$ and $\alpha(\nu, \gamma)$. In view of the fact that the exact analytic solution of the problem on strong explosion is given by rather complicated analytic formulas, work was done to compile tables of the self-similar functions [6]. In these tables the values of the functions in question are given for the following values of γ: 1.1, 1.2, 1.3, 1.4, 5/3, 2, 3. Computation of the self-similar functions was also carried out for the values $\gamma = 4, 5, 6$.

Figures 4–7 show the graphs of the functions p/p_2, v/v_2, T/T_2 and ρ/ρ_2 of λ for $\nu = 3$ characterizing the behavior of the solution for a wide range of values of γ. Here it is possible to trace the extremely fine effect of the change in sign of the curvature of the function ρ/ρ_2 as λ varies from zero to one for $4 < \gamma < 7$.

As already mentioned, the solution contains an energy parameter $\alpha(\nu, \gamma)$ which is a functional of the solution (2.5)–(2.10) for $\omega = 0$. A knowledge of the numerical values of α is required for applications of the theory to physical problems. The numerical values of α were earlier found most precisely for $\gamma = 1.4$ [2], [3]. We undertook the task of computing the parameter α with great accuracy for a wide range of values of γ and $\nu = 1, 2, 3$.

In order to get an idea of the numerical value, we shall give upper and lower bounds for this parameter. With this in mind we write the formula for α in the form

$$\alpha(\nu, \gamma) = \frac{8\sigma_\nu}{(\nu+2)^2(\gamma^2-1)} \int_0^1 \left[\frac{p}{p_2} + \frac{\rho}{\rho_2}\left(\frac{v}{v_2}\right)^2\right] \lambda^{\nu-1} d\lambda.$$

Since $p/p_2 \leqslant 1$, $\rho/\rho_2 \leqslant 1$ and $v/v_2 \leqslant 1$, on replacing these quantities by one we obtain an upper bound for the parameter:

$$\alpha < \alpha_1 = \frac{16\sigma_\nu}{\nu(\nu+2)^2(\gamma^2-1)}.$$

If we take p/p_2, ρ/ρ_2 and v/v_2 equal to their minimum values for $\lambda = 0$, we obtain a lower bound for the integral:

$$\alpha > \alpha_2 = \frac{8\sigma_\nu h(0)}{(\nu+2)^2 2\nu(\gamma-1)}.$$

Here it has been noted that $v(0, t) = \rho(0, t) = 0$ for $\gamma \leqslant 7$. We thus have

$$\frac{4\sigma_\nu h(0)}{\nu(\gamma-1)(\nu+2)^2} < \alpha < \frac{16\sigma_\nu}{\nu(\gamma^2-1)(\nu+2)^2}. \tag{2.23}$$

To obtain sufficiently precise values of α the integral in the expression for α was computed numerically. We remark that the calculation of $\alpha(\nu, \gamma)$ was also

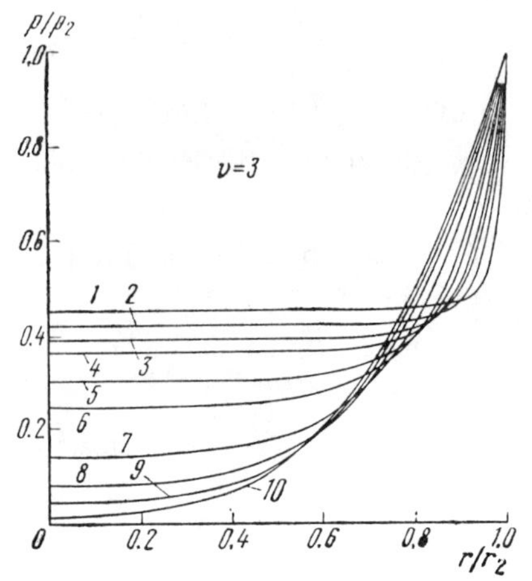

FIGURE 4. Distributions of relative pressures for various γ
1) $\gamma = 1.1$; 2) 1.2; 3) 1.3; 4) 1.4; 5) 5/3; 6) 2; 7) 3; 8) 4; 9) 5; 10) 6

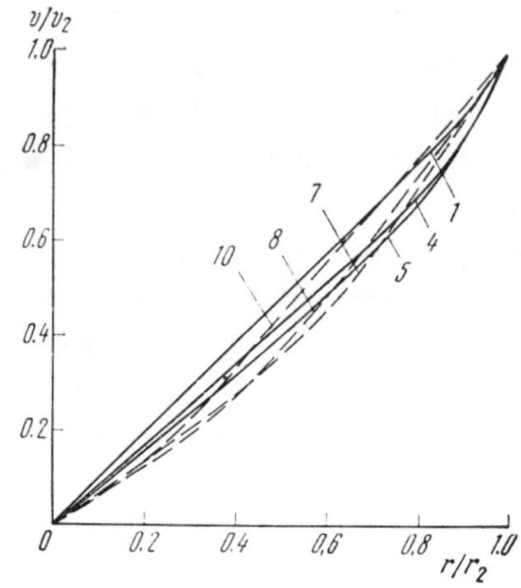

FIGURE 5. Distributions of relative velocities for various γ
For the notation see Figure 4

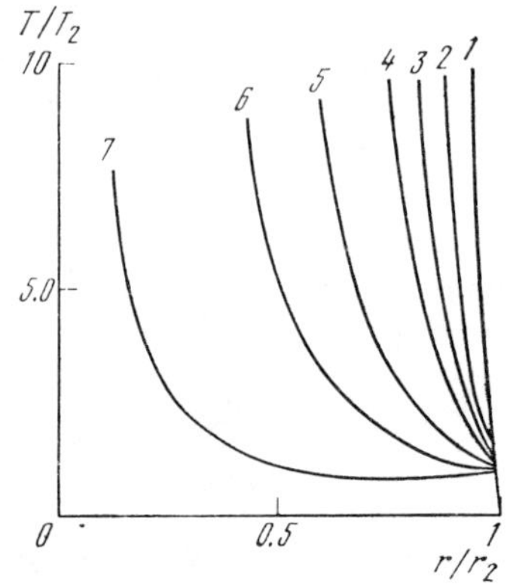

FIGURE 6. Distributions of relative temperatures for various γ
For the notation see Figure 4

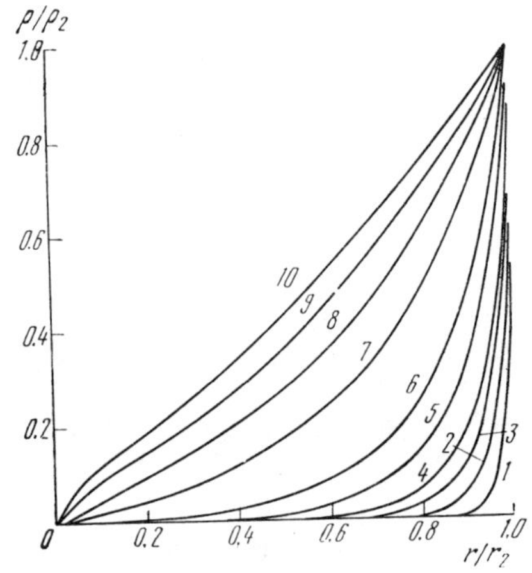

FIGURE 7. Distribution of relative densities for various γ
For the notation see Figure 4

undertaken by the American scientist D. L. Jones [8], but the values of α presented there are given for a narrow range of values of γ and were found to be erroneous (corrections are given in [8]).

The simplest path to the approximate evaluation of the integral in the formula for α for $\gamma < 7$ is the following. We transform the integrand making use of energy integral

$$\alpha = \frac{2\delta^2\sigma_\nu}{(\gamma-1)\gamma}\int_0^1 \frac{h\mu\lambda^{\nu-1}}{u(\mu)}\,d\lambda, \tag{2.24}$$

where

$$u(\mu) = \frac{2}{\gamma-1}\left(\frac{\gamma+1}{2}-\mu\right).$$

The integral in (2.24) is evaluated by the trapezoidal rule with nonuniform subdivision of the segment of integration with respect to λ. When using the computer to evaluate the integral, the points of subdivision for λ were chosen as follows. The segment of integration with respect to μ, $[(\gamma + 1)/2\gamma, 1]$, was divided into n parts, and the integrand was then evaluated at these points in accordance with (2.5)–(2.8), and the points of subdivision of the interval $[0, 1]$ for λ were found. After this the integral was evaluated by the trapezoidal rule. The number n was then doubled and the process was repeated. If when the points of subdivision are increased α does not vary to prescribed accuracy, then that value is taken for α which is obtained in computing the integral with the greatest n. In computations for values of γ in the range $1 < \gamma < 7$ the number n for which good accuracy was achieved was of the order 500. Values of α computed with high accuracy (four significant digits) for $\gamma < 4$ and various ν were published in tables in [6]. We remark that direct evaluation of the integral (2.24) with respect to the variable μ is possible, but here special quadrature formulas must be used, since the integrand will have an (integrable) singularity for $\mu = (\gamma + 1)/2\gamma$. It will follow from what is presented below that for $\gamma = 7$ and $\nu = 3$ the integral in (2.24) can be evaluated exactly.

In calculating α for the values $\nu = 3$ and $\gamma > 7$ it is expedient to transform the integral to the variable μ, since in these cases the solution does not reach the center and the integrand has no singularity. In this case the formula for computing α reduces to the following form:

$$\alpha = \frac{256}{25}\,\frac{\pi\gamma}{(\gamma-1)^2(\gamma-7)}\int_1^{(\gamma+1)/2} z(\mu)\,d\mu, \tag{2.24a}$$

where

$$z(\mu) = \left\{\frac{2(3\gamma-1)}{\gamma-7}\left[\mu-\frac{5(\gamma+1)}{2(3\gamma-1)}\right]\right\}^a\left[\frac{2\gamma}{\gamma-1}\left(\mu-\frac{\gamma+1}{2\gamma}\right)\right]^b$$
$$\times\left[\frac{2}{\gamma-1}\left(\frac{\gamma+1}{2}-\mu\right)\right]^c\left(\mu^2-\frac{\gamma+1}{\gamma}\mu+\frac{\gamma+1}{2\gamma}\right),$$

$$a = \beta_4 - 5\beta_1 - 1, \qquad b = 3\beta_2 - 1, \qquad c = \frac{2}{\gamma-2},$$

$$\beta_2 = \frac{\gamma-1}{2\gamma+1}, \qquad \beta_1 = \beta_2 + \frac{\gamma+1}{3\gamma-1} - \frac{2}{5}.$$

The integral in (2.24a) can be evaluated by any standard method.

Some of the results of the calculation of α by (2.24) and (2.24a) for the values $\nu = 3$ and $\gamma > 3$ are given below:

γ	4	5	6	7	7.1
α	0.076625	0.050894	0.036676	0.027925	0.027248
γ	7.3	8	9	10	11
α	0.025956	0.022161	0.018122	0.015155	0.012899

Since in applications values of α may arise which differ from those considered in [6] and above, we conducted work on approximation of tabulated values by analytic formulas of the form

$$\alpha = k_1 (\gamma - 1)^{k_2 + k_3 \lg (\gamma - 1)} \qquad (\gamma \leqslant 2),$$

$$\alpha = k_1 \gamma^{\kappa} \qquad (\gamma > 2).$$

The following values were obtained [9] as a result of computing k_1, k_2, k_3 and κ by the method of "mean values" and the method of least squares:

$\nu = 1$

$1.1 \leqslant \gamma \leqslant 3$ $\quad \Delta_\alpha = 1.52\%$, $\quad k_1 = 0.36011$, $\quad k_2 = -1.2700$, $k_3 = -0.017912$,

$1.2 \leqslant \gamma \leqslant 2$, $\quad \Delta_\alpha = 0.46\%$, $\quad k_1 = 0.36011$, $\quad k_2 = -1.2537$, $k_3 = -0.18471$;

$\nu = 2$

$1.1 \leqslant \gamma \leqslant 3$, $\quad \Delta_\alpha = 1.55\%$, $\quad k_1 = 0.34649$, $\quad k_2 = -1.19796$, $k_3 = -0.14134$,

$1.2 \leqslant \gamma \leqslant 2$, $\quad \Delta_\alpha = 0.34\%$, $\quad k_1 = 0.56235$, $\quad k_2 = 1.1768$, $k_3 = -0.13945$;

$\nu = 3$, $\qquad \gamma \leqslant 3$

$1.1 \leqslant \gamma \leqslant 3$, $\quad \Delta_\alpha = 1.38\%$, $\quad k_1 = 0.30774$, $\quad k_2 = -1.1598$, $k_3 = -0.11917$,

$1.2 \leqslant \gamma \leqslant 2$, $\quad \Delta_\alpha = 0.30\%$, $\quad k_1 = 0.49210$, $\quad k_2 = -1.1409$, $k_3 = -0.11735$;

$\nu = 3$, $\qquad \gamma > 2$

$2.5 \leqslant \gamma \leqslant 11.5$, $\Delta_\alpha = 0.50\%$, $\quad k_1 = 1.238$, $\kappa = -2.1448 + 0.2325 \lg \gamma$.

Here Δ_α denotes the maximum relative error given by the corresponding formula at points of the tabulated data for the indicated range of variation of the adiabatic exponent γ. These formulas can be used for computations in which it is not necessary to know α with high accuracy.

2.2. *On the analytic properties of the solution in a neighborhood of the center of symmetry for the case of constant initial density.* For the solution of the linearized point explosion problem with consideration of counterpressure which will be presented in Chapter 3 and for studying the behavior of the solution of the general non-self-similar problem (considering counterpressure) near the center of symmetry, it is necessary to know the representation of the solution of the self-similar problem in terms of power series in λ. Such a representation can also be useful for computing the functions f, g and h for small values of λ.

In §1 of this chapter it was shown that the dependence of λ on the parameter μ has the form (2.5), (2.10):

$$\lambda = \mu^{-\frac{2}{\nu+2}}\left[\frac{2(\nu\gamma-\nu+2)}{3\nu-2-\gamma(\nu-2)}\left(\frac{(\nu+2)(\gamma+1)}{2(\nu\gamma-\nu+2)}-\mu\right)\right]^{-\beta_1}\left[\frac{2\gamma}{\gamma-1}\left(\mu-\frac{\gamma+1}{2\gamma}\right)\right]^{\beta_2}.$$

We shall consider the solution for $\gamma < 7$, i.e. the case in which the solution extends to the center of symmetry. At the center of symmetry we have $\lambda = 0$ and $\mu = \mu_0 = (\gamma+1)/2\gamma$. We transform the formula for λ by raising its left and right sides to the power $1/\beta_2$:

$$\lambda^{1/\beta_2} = \frac{2\gamma}{\gamma-1}\mu^{-2/(\nu+2)\beta_2}\left(\mu-\frac{\gamma+1}{2\gamma}\right)\left[\frac{2(\nu\gamma-\nu+2)}{3\nu-2-\gamma(\nu-2)}\left(\frac{(\nu+2)(\gamma+1)}{2(\nu\gamma-\nu+2)}-\mu\right)\right]^{-\beta_1/\beta_2}.$$

Since the dependence of β_2 on γ and ν has the form $\beta_2 = (\gamma-1)/[2(\gamma-1)+\nu]$, we obtain

$$\frac{1}{\beta_2} = \frac{\nu}{\gamma-1} + 2 = s + 2, \quad s = \frac{\nu}{\gamma-1}.$$

We introduce the notation $\lambda^{1/\beta_2} = \lambda^{s-2} = x$. Then the relation for λ^{1/β_2} can be written as follows:

$$\Phi(x, \mu) = x - \frac{2\gamma}{\gamma+1}\mu^{-2/(\nu+2)\beta_2}(\mu-\mu_0)$$
$$\times\left[\frac{2(\nu\gamma-\nu+2)}{3\nu-2-\gamma(\nu-2)}\left(\frac{\nu+2}{2}\,\frac{\gamma+1}{\nu\gamma-\nu+2}-\mu\right)\right]^{-\beta_1/\beta_2} = 0. \qquad (2.25)$$

Equation (2.25) defines μ as an implicit function of x, whereby $\Phi(0, \mu_0) = 0$ and $(\partial\Phi/\partial\mu)(0, \mu_0) \neq 0$.

From the form of the function $\Phi(x, \mu)$ it follows that $\Phi(x, \mu)$ can be expanded in a series of powers of x and $\mu - \mu_0$. From the theory of implicit functions it is known that in this case the function $\mu(x)$ can also be expanded in a convergent (in a neighborhood of $x = 0$) power series in x. Using the theory of series, one can find the explicit dependence of the coefficients of the expansion of the function μ in powers of x on the quantities γ and ν.

Equation (2.25) can be written as follows:

$$\mu = \mu_0 + x\frac{\gamma - 1}{2\gamma}\mu^{2/(\nu+2)\beta_2}\left[\frac{2(\nu\gamma - \nu + 2)}{3\nu - 2 - \gamma(\nu - 2)}\left(\frac{(\nu + 2)(\gamma + 1)}{2(\nu\gamma - \nu + 2)} - \mu\right)\right]^{\beta_1/\beta_2}. \quad (2.25a)$$

In the theory of series it is proved [10] that if a function $y(x)$ is defined by an equation $y = a + x\psi(y)$, where the function $\psi(y)$ can be expanded in a power series in $y - a$, then the expansion of $y(x)$ in a neighborhood of $x = 0$ has the form

$$y = a + x\psi(a) + \frac{x^2}{2}\frac{d}{da}[\psi^2(a)] + \ldots + \frac{x^n}{n!}\frac{d^{n-1}}{da^{n-1}}[\psi^n(a)].$$

This series is a special case of the Lagrange series.

Applying this formula to the implicit function $\mu(x)$ defined by (2.25a), we find the expansion of μ in powers of x in a neighborhood of $x = 0$:

$$\mu = \mu_0 + x\frac{\gamma - 1}{2\gamma}\mu_0^{2/(\nu+2)\beta_2}\left[\frac{2(\nu\gamma - \nu + 2)}{3\nu - 2 - \gamma(\nu - 2)}\left(\frac{(\nu + 2)(\gamma + 1)}{2(\nu\gamma - \nu + 2)} - \mu_0\right)\right]^{\beta_1/\beta_2}$$
$$+ \frac{x^2}{2}\frac{d}{d\mu_0}\left\{\left(\frac{\gamma - 1}{2\gamma}\right)^2 \mu_0^{4/(\nu+2)\beta_2}\left[\frac{2(\nu\gamma - \nu + 2)}{3\nu - 2 - \gamma(\nu - 2)}\right.\right.$$
$$\left.\left.\times\left(\frac{(\nu + 2)(\gamma + 1)}{2(\nu\gamma - \nu + 2)} - \mu_0\right)\right]^{2e\beta_1/\beta_2}\right\} + \ldots. \quad (2.25b)$$

Using the rule given above, we can find any term of this series by differentiation. Since $f = 2\mu\lambda/(\gamma + 1)$ and $x = \lambda^{s+2}$, the formula obtained for μ makes it possible to obtain an expansion of $f(\lambda)$ in a power series in λ. Using (2.25b) and (2.5)–(2.9), we can also find expansions for $g(\lambda)$, $h(\lambda)$, and $\theta(\lambda)$ in series in λ. These expansions have the form

$$\begin{aligned} f(\lambda) &= \frac{2}{\gamma + 1}\lambda(\sigma_0 + \sigma_1\lambda^{s+2} + \sigma_2\lambda^{2(s+2)} + \ldots), \\ g(\lambda) &= \frac{\gamma + 1}{\gamma - 1}\lambda^s(\alpha_0 + \alpha_1\lambda^{s+2} + \alpha_2\lambda^{2(s+2)} + \ldots), \\ h(\lambda) &= \frac{2}{\gamma + 1}(\delta_0 + \delta_1\lambda^{s+2} + \delta_2\lambda^{2(s+2)} + \ldots), \\ \theta(\lambda) &= \lambda^{-s}(\eta_0 + \eta_1\lambda^{s+2} + \eta_2\lambda^{2(s+2)} + \ldots). \end{aligned} \quad (2.26)$$

The dependence of the coefficients $\sigma_0, \alpha_0, \delta_0, \sigma_1, \delta_1$ and α_1 on the quantities γ and ν is given in [3] and will not be reproduced here. Using the formulas presented, it is possible to determine the dependence of $\sigma_i, \alpha_i, \delta_i$ and η_i on γ and ν. The form of the expansions (2.26) and the dependence of the coefficients on γ and ν were first indicated by Sedov in his book [2]. The analysis presented may be interpreted as a justification for the expansions (2.26) and a new approach to the question of determining the coefficients.

2.3. *Dependence of the solution on* ω, ν *and* γ *in the case of variable density.* In the general case of variable density the solution contains three parameters: ω, ν and γ. We shall consider the change in the properties of the solutions depending on the change in these parameters. Suppose that the condition of finite initial mass in any finite volume containing the point $r = 0$ is satisfied. We consider the mass contained in the volume between the origin and the coordinate $r = \xi$:

$$M = \sigma_\nu \int_0^\xi \rho r^{\nu-1} dr = \begin{cases} \sigma_\nu A \dfrac{r}{\nu - \omega}\Big|_0^\xi & (\omega \neq \nu), \\ \sigma_\nu A \ln\ r \,\Big|_0^\xi & (\omega = \nu), \end{cases}$$

where $\sigma_\nu = 2(\nu - 1)\pi + (\nu - 2)(\nu - 3)$. It is obvious that the finite-mass condition inside this volume can be satisfied only under the condition $\omega < \nu$. For $\omega < \nu$ it follows from (2.11) and (2.13) that v_2 and p_2 decrease with increase in time, and the shock wave slows down. Since p/ρ^γ is proportional to e^{s/c_V}, it follows from (2.16) that as the shock wave propagates the entropy of particles behind the front of the shock wave decreases if $\omega < \nu/\gamma$, increases if $\omega > \nu/\gamma$ and remains constant if $\omega = \nu/\gamma$ [2]. It is interesting to note that for values of ω lying the range $\nu/\gamma < \omega < \nu$ the shock wave slows down as it propagates away from the site of the explosion, the pressure behind its front drops, but nevertheless the entropy increases [2], [3]. This effect is explained by the strong decrease in density [2]. [3].

It follows from the solution (2.5)–(2.10) that the parametric variable μ can range over the interval

$$\frac{\gamma+1}{2\gamma} \leqslant \mu \leqslant \frac{\gamma+1}{2}.$$

The value $\mu = \mu_2 = 1$ corresponding to the shock wave $(\lambda = 1)$ lies in this interval; the value $\mu_0 = (\gamma + 1)/2\gamma$ corresponds to the center of symmetry. If $\omega < \nu$, then for

$$\omega < \frac{3\nu - 2 + \gamma(2 - \nu)}{\gamma + 1}$$

the solution extends to the center of symmetry and the variable μ takes values in the interval

$$\frac{\gamma+1}{2\gamma} \leqslant \mu \leqslant 1.$$

If $\nu = 3$ and $(7 - \gamma)/(\gamma + 1) < \omega < 6/(\gamma + 1)$, then about the center of the explosion there is formed a spherical cavity which extends with time, on the boundary of which $\mu = \mu_* = (\gamma + 1)/2$, $\lambda = \lambda_*(\gamma, \omega)$, and the density, pressure, and temperature are equal to zero. The parameter μ varies within the limits $1 \leqslant \mu \leqslant (\gamma + 1)/2$. Asymptotic formulas for finding the basic functions near the center of symmetry and in a neighborhood of the spherical cavity are presented in [2], [3].

E. V. Rjazanov and the author [5] investigated the behavior of the solution in a neighborhood of the singular values of ω and γ, i.e. those values of ω and γ for which either the coefficients on the right sides of (2.5)–(2.10) or the exponents β_j $(j = 1, \cdots, 5)$ become infinite. It was found that there are the following three singular values:

$$\omega_1 = \frac{3\nu - 2 + \gamma(2-\nu)}{\gamma+1}, \qquad \omega_2 = \frac{2(\gamma-1)+\nu}{\gamma}, \qquad \omega_3 = \nu(2-\gamma). \tag{2.27}$$

The solution of the problem for the first singular case where $\omega = \omega_1$ was studied by Sedov [2]. It has the simple form

$$f = \frac{2}{\gamma+1}\lambda, \qquad g(\lambda) = \frac{\gamma+1}{\gamma-1}\lambda^{\nu-2}, \qquad h = \frac{2}{\gamma+1}\lambda^{\nu}. \tag{2.28}$$

The form of the solution in the other two cases, i.e. for $\omega = \omega_2$ and $\omega = \omega_3$, and the continuous dependence of the solution on the parameters γ and ω in a neighborhood of the curves $\omega_2(\gamma)$ and $\omega_3(\gamma)$ are established as in the case $\omega_3 = 0$, $\gamma = 2$. We shall present here only the form of the solution for the case $\omega = \omega_2$. This solution has the form

$$\lambda = \mu^{-\frac{2\gamma}{\nu\gamma-\nu+2}}\left[\frac{2\gamma}{\gamma-1}\left(\mu - \frac{\gamma+1}{2\gamma}\right)\right]^{\frac{\gamma-1}{\nu\gamma-\nu+2}} \exp\left[-\frac{\gamma+1}{\nu\gamma-\nu+2}\,\frac{1-\mu}{\mu-\frac{\gamma+1}{2\gamma}}\right],$$

$$g(\mu) = \frac{\gamma+1}{\gamma-1}\mu^{\frac{2(\nu-2+2\gamma)}{\nu\gamma-\nu+2}}\left[\frac{2}{\gamma-1}\left(\frac{\gamma+1}{2}-\mu\right)\right]^{\frac{\nu-2(\gamma+1)}{\nu\gamma-\nu+2}} \tag{2.29}$$

$$\times\left[\frac{2\gamma}{\gamma-1}\left(\mu-\frac{\gamma+1}{2\gamma}\right)\right]^{\frac{4-\nu-2\gamma}{\nu\gamma-\nu+2}} \exp\left[-\frac{2(\gamma+1)}{\nu\gamma-\nu+2}\,\frac{1-\mu}{\mu-\frac{\gamma+1}{2\gamma}}\right],$$

$$h(\mu) = \mu^{\frac{2\nu\gamma}{\nu\gamma-\nu+2}}\left[\frac{2}{\gamma-1}\left(\frac{\gamma+1}{2}-\mu\right)\right]^{\frac{\gamma(\nu-2)}{\nu\gamma-\nu+2}}\left[\frac{2\gamma}{\gamma-1}\left(\mu-\frac{\gamma+1}{2\gamma}\right)\right]^{-\frac{\nu\gamma}{\nu\gamma-\nu+2}},$$

$$f = \frac{2\lambda}{\gamma+1}\mu.$$

In the third singular case, i.e. for $\omega_3 = \nu(\gamma-2)$, the form of the solution will be analogous to the case $\gamma = 2$ considered above and will not be recorded here. We mention that for the singular case $\omega = \omega_1$ the energy parameter is

$$\alpha = \frac{2\sigma_\nu(\gamma+1)}{\nu(\gamma-1)(\nu\gamma-\nu+2)^2}. \tag{2.30}$$

For arbitrary γ and ω the parameter α can be determined numerically. We remark also that on the basis of Theorem 1 from the solution (2.5)–(2.10) it is possible to obtain the form of solutions in the case of motions which are limits of self-similar motions by limiting processes with respect to δ and transformations of time translation (or translation of the coordinate for $\nu = 1$). N. N. Kočina obtained these solutions by direct integration of the corresponding invariant equations of gas dynamics [11].

§3. The strong explosion problem in a gas with zero temperature gradient

3.1. *The point explosion problem for a nonlinear heat equation.* The solution (2.5)–(2.10) of the strong explosion problem for the case of adiabatic perturbed motions described by (2.3) is characterized by large temperature gradients. For $\lambda = 0$ ($\omega = 0$) the temperature at the center of the explosion is infinite and grows on approach to the center as $r^{-\nu/(\gamma-1)}$. This temperature distribution does not correspond to the real distribution, since processes of heat conduction, primarily heat radiation [12], play a major role in the case of substantial temperature gradients and high temperatures. It is therefore of interest to consider other models for the propagation of perturbations in the case of a point explosion; the simplest such model is that of a body conducting heat nonlinearly. The problem of the propagation of a heat wave for a medium whose coefficient of thermal conductivity depends on the temperature according to a power law has been considered in detail in [12]. The formulation of the problem is analogous to that considered in §7 of Chapter 1 for the ordinary heat equation. It is required to find a solution of equation (1.44) with the initial condition (1.152). The solution of this problem for an arbitrary dependence $\chi(T)$ and nonzero initial energy u ($u = \rho_0\epsilon$) has not been investigated in sufficient detail. Suppose now that $\chi = \kappa_0 u^n$, $u|_{t=0} = 0$ ($r > 0$), where κ_0 is a constant with dimension $[\kappa_0] = L^{n+1}T^{2n}M^{-n}$.

Equation (1.44) takes the form

$$\frac{\partial u}{\partial t} = \frac{1}{r^{\nu-1}} \kappa_0 \frac{\partial}{\partial r}\left(r^{\nu-1} u^n \frac{\partial u}{\partial r}\right).$$

The problem contains two dimensional constants κ_0 and E_0, i.e. it is self-similar.

We remark that we must consider the integral law of energy conservation (1.165) and the condition at infinity $u|_{r=\infty} = 0$. The solution will depend on one dimensionless variable

$$\lambda = \frac{r}{\lambda_0 (\kappa E_0^n t)^{1/(\nu n+2)}}. \tag{2.31}$$

The solution for the cases $\nu = 1$ and $\nu = 3$ has been investigated in [12]. It has the form

$$u = e(1-\lambda)\, u_0 (1-\lambda^2)^{1/n}, \tag{2.32}$$

where $e(1-\lambda)$ is the unit step function and $u_0 = u(0, t)$ is the value of the energy u at the center.

The leading front of the wave of the perturbation, or the front of the heat wave, propagates according to the law

$$r_2 = \lambda_0(\kappa_0 E_0^n t)^{1/(\nu n+2)},$$

where λ_0 is a known constant determined by the integral law of energy conservation.

For $n > 1$ it follows from the solution (2.32) that the energy u is weakly varying in the region near the center, and only near $\lambda = 1$, i.e. $r = r_2$, does it decrease abruptly to zero. If it is assumed that $u \sim T$, then the temperature will behave similarly.

In the approximation of nonlinear heat conductivity we do not consider the motion of the gas. For a strong explosion in a gas this is justified only at the very beginning of the process [12].

If the coefficient κ is extremely large, (1.44) can be approximately replaced by $\partial u/\partial r = 0$ or (since $u = u(T)$) by $\partial T/\partial r = 0$.

We thus arrive at the condition of zero temperature or internal energy gradient. The solution here will be given by

$$u = u(0, t) = u_0 \quad (r < r_2). \tag{2.33}$$

Thus, when there is intense heat exchange among the particles of the medium, the condition of zero temperature gradient can be taken as an approximate condition.

DEFINITION. Processes for which the condition $\partial T/\partial r = 0$ $(T = T(r))$ is satisfied we call *homothermic.* We remark that if $T = \text{const}$, a homothermic process coincides with an isothermic process.

3.2. *Strong explosion with zero temperature gradient.* We now consider the problem of a strong point explosion where in place of the condition of adiabatic flow behind the front of the shock wave it is assumed that intense heat exchange takes place. Because of this we assume that in the region of motion of the gas there is no temperature gradient, i.e. $\partial T/\partial r = 0$. This assumption on the character of the flow corresponds to the initial stage of development of a very powerful explosion (for example, an atomic explosion) when in the region of gas flow there is a high temperature, and due to radiation and thermal conductivity intense heat exchange occurs between particles of the gas. In an atomic explosion this corresponds approximately to that stage of development of the explosion in which the front of the shock wave has not yet torn away from the fireball. Because of the assumptions indicated, the temperature in the flow region depends only on time and not on the distance to the center of the explosion, i.e. $T = T(t)$, and the flow is homothermic.

We shall assume that the initial density ρ_1 of the quiescent gas is constant. The formulation of the basic strong explosion problem for homothermic motions and its solution in the case $\nu = 3$ were first published in our work [13]. A detailed analysis for all ν was published in [3]. The case of spherical symmetry was considered independently in the work of Ryžov and Taganov [14].

The system of partial differential equations describing the one-dimensional transient motion in question has the form

$$\frac{\partial v}{\partial t}+v\frac{\partial v}{\partial r}+\frac{1}{\rho}\frac{\partial p}{\partial r}=0, \quad \frac{\partial T}{\partial r}=0,$$
$$\frac{\partial \rho}{\partial t}+v\frac{\partial \rho}{\partial r}+\rho\left[\frac{\partial v}{\partial r}+(\nu-1)\frac{v}{r}\right]=0. \tag{2.34}$$

Using the equation of state of a perfect gas

$$p=R\rho T \tag{2.35}$$

we eliminate the pressure from the first equation of this system; the system (2.34) then takes the form

$$\frac{\partial v}{\partial t}+v\frac{\partial v}{\partial r}+\frac{RT}{\rho}\frac{\partial \rho}{\partial r}=0, \quad \frac{\partial T}{\partial r}=0,$$
$$\frac{\partial \rho}{\partial r}+v\frac{\partial \rho}{\partial r}+\rho\left[\frac{\partial v}{\partial r}+(\nu-1)\frac{v}{r}\right]=0. \tag{2.36}$$

If the initial pressure p_1 is neglected, the system of defining parameters of this problem has the form $r, t, E_0, \rho_1, \gamma$, where the constant γ is essential only for calculation of the energy balance.

This motion of the gas is self-similar. All dimensionless characteristics of the flow may be considered functions of the following parameters:

$$\lambda=\left(\frac{\rho_1}{E}\right)^{1/(\nu+2)} rt^{-2/(\nu+2)}, \quad \gamma,$$

where E is a constant with the dimension of energy $[E] = ML^{\nu-1}T^{-2}$ related to the energy of the explosion E_0 by the formula $E_0 = \alpha E$, α being a constant.

Since for the shock wave the coordinate r_2 is a function of time t, it follows that $r_2 = \lambda_2(E/\rho_1)^{1/(\nu+2)}t^{2/(\nu+2)}$. We determine the constant α from the condition that $\lambda_2 = 1$ on the shock wave. For λ we then have the formula $\lambda = r/r_2$. For the propagation speed of the shock wave we obtain

$$D=\frac{2}{\nu+2}\frac{r_2}{t}=\frac{2}{\nu+2}\left(\frac{E}{\rho_1}\right)r_2^{-\nu/2}=\frac{2}{\nu+2}\left(\frac{E}{\rho_1}\right)^{1/(\nu+2)}t^{-\nu/(\nu+2)}.$$

In place of the variable λ we introduce a new independent variable Λ by the formula

$$\Lambda=\frac{1}{\theta_2^{1/2}}\lambda, \tag{2.37}$$

where θ_2 is a constant. On the shock wave $\Lambda_2 = 1/\theta_2^{1/2}$, since $\lambda_2 = 1$.

For the velocity, density and temperature we can write

$$v=\theta_2^{1/2}Df(\lambda) \quad \rho=\rho_1 g(\lambda), \quad T=\frac{\theta_2}{R}D^2, \tag{2.38}$$

where $f(\lambda)$ and $g(\lambda)$ are the dimensionless velocity and density.

Using the relation (2.38) between the dimensional and dimensionless variables and the formulas $\partial/\partial r = (1/\sqrt{\theta_2})r_2(\partial/\partial\Lambda)$, $\partial/\partial t = -(D\Lambda/r_2)(\partial/\partial\Lambda)$, the system (2.36) can be transformed into an equivalent system of ordinary differential equations

$$\frac{g'}{g} = (\Lambda - f)f' + \frac{\nu}{2}f, \ f, \quad f' = (\Lambda - f)\frac{g'}{g} + (\nu - 2)\frac{f}{\Lambda}. \tag{2.39}$$

The prime denotes differentiation with respect to Λ. If we introduce the dimensionless pressure h by the formula $p = \rho_1 (r_2^2/t^2) h(\Lambda)$, then from (2.35) and (2.38) we obtain

$$h(\Lambda) = \left(\frac{2}{\nu + 2}\right)^2 \theta_2 g(\Lambda). \tag{2.40}$$

This equation makes it possible to find the pressure $h(\Lambda)$ if the function $g(\Lambda)$ and constant θ_2 are known.

We remark that it is possible to obtain an equation for finding $f(\Lambda)$ from (2.39) by eliminating g'/g:

$$f' = \frac{\frac{\nu}{2}\Lambda f(\Lambda - f) - (\nu - 1) f}{\Lambda[1 - (\Lambda - f)^2]}. \tag{2.41}$$

From the first equation of (2.39) we find that

$$\frac{g'}{g} = -\left(\frac{f^2}{2}\right)' + (\Lambda f)' + \frac{\nu - 2}{2} f.$$

Integrating this equation from some Λ to Λ_2, we obtain

$$\ln\frac{g}{g_2} = \frac{1}{2}(f_1^2 - f^2) + \Lambda f - \Lambda_2 f_2 + \frac{\nu - 2}{2}\int_{\Lambda_2}^{\Lambda} f d\Lambda, \tag{2.42}$$

where g_2 and f_2 are the values of g and f on the shock wave.

For $\nu = 2$ the relation (2.42) gives a first integral([1]) of the system (2.39):

$$\ln\frac{g}{g_2} = \frac{1}{2}(f_2^2 - f^2) + \Lambda f - \Lambda_2 f_2. \tag{2.43}$$

From what has been said it follows that the solution of the strong explosion problem with zero temperature gradient reduces essentially to finding the function $f(\Lambda)$ from the differential equation (2.41). If $f(\Lambda)$ has been found, $g(\Lambda)$ is determined from (2.42) or (2.43). In this connection the functions f and g must satisfy particular boundary conditions. We shall derive these boundary conditions.

From the law of conservation of momentum and the law of mass conservation on passage through the front of a shock wave we have

$$\rho_1 D^2 = \rho_2 (v_2 - D)^2 + R\rho_2 T_2, \qquad -\rho_1 D = \rho_2 (v_2 - D). \tag{2.44}$$

Going over in (2.44) to dimensionless variables according to (2.37) and (2.38), we obtain boundary conditions for $f(\Lambda)$ and $g(\Lambda)$ at $\Lambda = \Lambda_2$:

([1]) This integral was first found in another manner by M. L. Lidov and was communicated to the author.

$$f_2(\Lambda_2) = \frac{1}{2}\left(\Lambda_2 + \sqrt{\Lambda_2^2 - 4}\right), \tag{2.45}$$

$$g_2(\Lambda_2) = \frac{\Lambda_2}{\Lambda_2 - f_2}. \tag{2.46}$$

There is, moreover, still another boundary condition for $f(\Lambda)$ from the condition that the velocity be zero at the center of symmetry:

$$f(0) = 0. \tag{2.47}$$

From the boundary conditions on the shock wave it follows that a solution of (2.41) satisfying condition (2.45) depends on the quantity $\Lambda_2 = 1/\theta_2^{1/2}$ as a parameter. The quantity Λ_2 should be chosen such that the integral curve $f(\Lambda)$ satisfies the boundary condition (2.47). Moreover, from (2.45) it follows that the condition $\Lambda_2 \geqslant 2$ must be satisfied.

It must be pointed out that in the problem being considered the dimensionless functions f and g do not depend on γ, as follows from the system of equations (2.39) and the boundary conditions (2.45)–(2.47). The effect of γ appears only in reckoning the energy balance.

We shall express the constant E entering in the formulas for the characteristics of the motion in terms of the energy of the explosion E_0 (which is equal in our formulation to the total energy of the perturbed gas). For the total energy we have

$$E_0 = \sigma_\nu \int_0^{r_2} \frac{\rho v^2}{2} r^{\nu-1} dr + \sigma_\nu \int_0^{r_2} \frac{R\rho T}{\gamma - 1} r^{\nu-1} dr.$$

Here the first term corresponds to the kinetic energy and the second to the thermal energy of the gas. Going over to dimensionless variables, we obtain

$$\frac{E_0}{E} = \alpha = \left(\frac{2}{\nu+2}\right)^2 \frac{\sigma_\nu}{\Lambda_2^{\nu+2}} \left[\int_0^{\Lambda_2} \frac{gf^2}{2} \Lambda^{\nu-2} d\Lambda + \frac{1}{\gamma-1} \int_0^{\Lambda_2} g \Lambda^{\nu-1} d\Lambda \right]. \tag{2.48}$$

To evaluate the second term in square brackets it is possible to use the relations which follow from the condition of mass conservation in the motion of the gas:

$$\sigma_\nu \int_0^{r_2} \rho r^{\nu-1} dr = \frac{\sigma_\nu}{\nu} r_2^\nu. \tag{2.49}$$

In dimensionless variables (2.49) takes the form

$$\frac{\sigma_\nu}{\Lambda_2^\nu} \int_0^{\Lambda_2} g \Lambda^{\nu-1} d\Lambda = \frac{\sigma_\nu}{\nu}. \tag{2.50}$$

Using (2.50), formula (2.48) can be written as follows:

$$\frac{E_0}{E}=\left(\frac{2}{\nu+2}\right)^2\frac{\sigma_\nu}{\Lambda^2}\left[\frac{e_\nu}{\Lambda_2^\nu}+\frac{1}{\nu(\gamma-1)}\right], \tag{2.51}$$

where

$$e_\nu=\int_0^{\Lambda_2}\frac{gf^2}{2}\Lambda^{\nu-1}d\Lambda. \tag{2.52}$$

Formulas (2.51) and (2.52) give the expression for the constant E in terms of the energy of the explosion E_0 and the constant γ.

We proceed to consider concrete special cases of this problem. We shall investigate solutions of (2.41), which we write as

$$f'=\frac{\frac{\nu}{2}(\Lambda-f)\Lambda-(\nu-1)}{1-(\Lambda-f)^2}\frac{f}{\Lambda}. \tag{2.53}$$

In order to study the behavior of the solution of (2.53) we consider the field of integral curves of this equation for the various values of ν. This will enable us to find from the set of integral curves that unique integral curve which satisfies the boundary conditions (2.45) and (2.46).

a) *The case of spherical symmetry.* For $\nu = 3$ equation (2.53) in the region of the Λ, f plane of interest to us has the following singular points:

$$O(0,\ 0),\qquad A(1,\ 0),\qquad B=\left(\frac{4}{3},\ \frac{1}{3}\right).$$

The singular point O is a saddle point. Two integral curves converge at this point: the line $\Lambda = 0$ and the line $f = 0$. The singular point A is a node; the integral curve $f = 0$ enters at the point A. At the singular point B there is a saddle; through it there pass two integral curves with slope of tangents $N_1 = 1.3624$ and $N_2 = 0.4624$. The boundary condition on the shock wave

$$f_2(\Lambda_2)=\frac{1}{2}\left(\Lambda_2+\sqrt{\Lambda_2^2-4}\right)$$

gives in the Λ, f plane a curve which must intersect the desired integral curve.

A complete picture of the field of integral curves is given in Figure 8. From consideration of the field of integral curves it follows that the unique curve which can satisfy all the boundary conditions is the curve $OABC$ beginning at the point $O(0, 0)$, passing through the singular points A and B, and intersecting the curve $f_2(\Lambda_2)$ at the point C.

b) *The case of cylindrical symmetry.* For $\nu = 2$ we have from (2.53)

$$f'=\frac{f}{\Lambda}\frac{\Lambda(\Lambda-f)-1}{1-(\Lambda-f)^2}. \tag{2.54}$$

For finite values of Λ this equation has the two singular points $O(0, 0)$ and $A(1, 0)$.

The singular point O is a saddle point. Into it come the lines $\Lambda = 0, f = 0$. The point A is a singular point of rational character [15]. The critical directions, i.e. the directions along which the curves come into the singular point, are $N_1 = 0, N_2 = 1$. One integral curve comes into the singular point along each of the critical directions. The field of integral curves of (2.54) is shown in Figure 9.

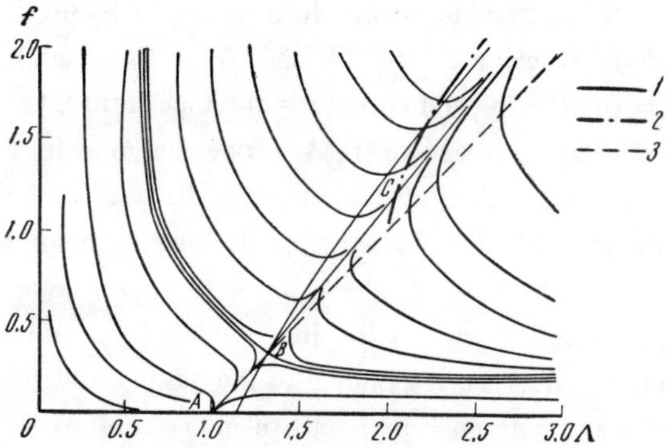

FIGURE 8. The field of integral curves of $\nu = 3$
1) integral curves and lines of zero derivatives; 2) dependence of f_2 on Λ_2; 3) lines of infinite derivatives

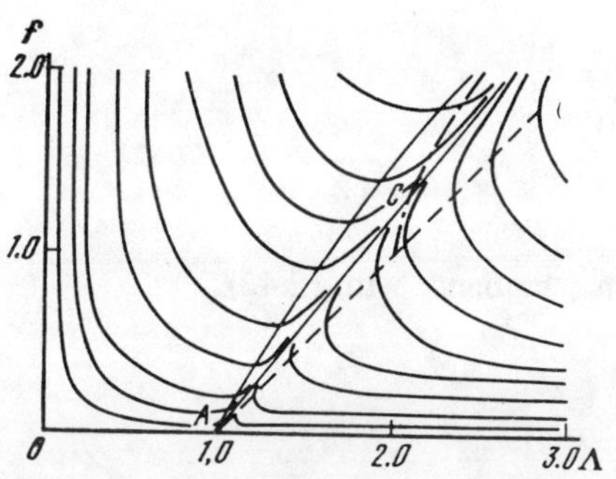

FIGURE 9. The field of integral curves for $\nu = 2$
For the notation see Figure 8

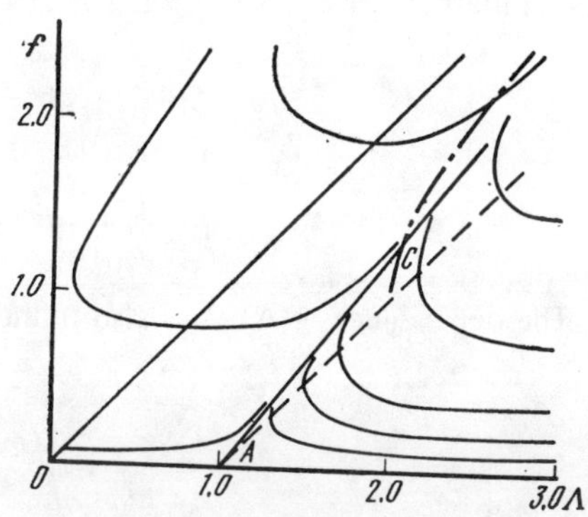

FIGURE 10. The field of integral curves for $\nu = 1$
For the notation see Figure 8

The integral curve OAC corresponds to the solution of the problem. This curve issues from the point A with slope $N_2 = 1$ and intersects the boundary curve $f_2(\Lambda_2)$ at the point C.

The distribution of the integral curves in the f, Λ plane for the cylindrical case differs from that in the spherical case in that the singular point B (Figure 8) is absent for the problem with cylindrical symmetry. This is also the situation for $\nu = 1$, as will be shown below.

c) *The case of plane symmetry.* In the case of plane symmetry we have

$$f' = \frac{f(\Lambda - f)}{2[1 - (f - \Lambda^2)]}. \tag{2.55}$$

The picture of the integral curves of (2.55) is given in Figure 10. For finite Λ and f (2.55) has the unique singular point $A(1, 0)$, a saddle point. Into this point there come two integral curves with slope of the tangents $N_1 = 5/4$ and $N_2 = 0$. In addition, (2.55) has the trivial solution $f = 0$. The integral curve $f = 0$ comes into the point A. Thus from the center of symmetry to the curve $f_2(\Lambda_2)$ one can go as follows: from the point O to the point A along the integral curve $f = 0$, and from the point A to the point C along the integral curve issuing from the singular point A with slope $N_1 = 5/4$.

The study of the fields of integral curves shows that for the complete solution of the problem one must find the dependence $f(\Lambda)$ corresponding to the desired integral curve passing through the singular points and determine its points of intersection with the curve $f_2(\Lambda_2)$. This problem has been solved by numerical integration of (2.53). The computations gave the following values of the constants Λ_2 and θ_2:

$$\Lambda_2 = 2.024, \qquad \theta_2 = 0.244 \qquad (\nu = 3),$$
$$\Lambda_2 = 2.040, \qquad \theta_2 = 0.240 \qquad (\nu = 2),$$
$$\Lambda_2 = 2.076, \qquad \theta_2 = 0.232 \qquad (\nu = 1).$$

The dependence $g(\Lambda)$ was also found in accordance with (2.42).

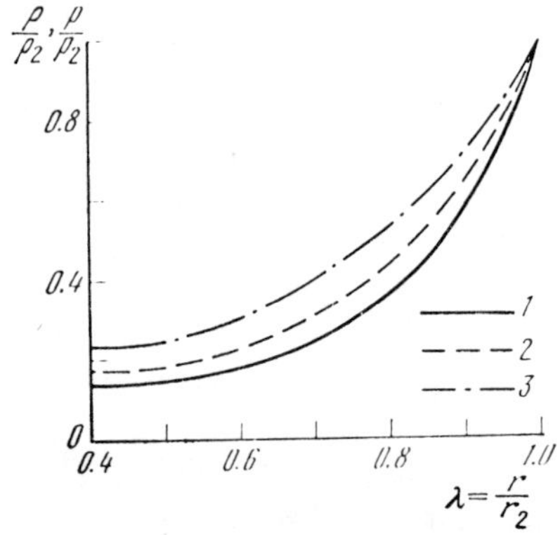

FIGURE 11. Dependence of the dimensionless pressure and density on λ
1) $\nu = 3$; 2) $\nu = 2$; 3) $\nu = 1$

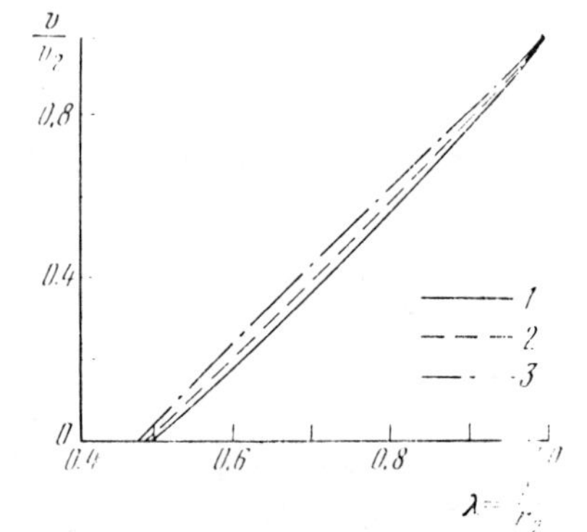

FIGURE 12. Dependence of the dimensionless velocity on λ
1) $\nu = 3$; 2) $\nu = 2$; 3) $\nu = 1$

The results of the solution of the problem for all three symmetry cases are represented in the graphs of $(p/p_2)(\lambda)$, $(\rho/\rho_2)(\lambda)$ and $(v/v_2)(\lambda)$ (Figures 11 and 12). The solution obtained is characterized first of all by the fact that near the center up to the value $r_* = \theta_2^{1/2} r_2$ there is a region of quiescent gas. This region expands as time goes on. In the f, Λ plane this region corresponds to the curve $f = 0$. For $r > r_*$ the velocity of the gas is directed away from the center and varies with increasing r according to a law which is close to being linear. In the central quiescent region the density of the gas is constant; the pressure and temperature vary in time as $t^{-2\nu/(2+\nu)}$. For values of $r > \theta_2^{1/2} r_2$ the density and pressure increase with increasing r, reaching maximum values on the front of the shock wave. The gradients of the pressure and density increase on approach to the shock wave. The values of the parameter e_ν lie in the interval $1 < e_\nu < 2.5$; thus for $\nu = 3$ it was found that $e_\nu = 1.088$. If the density is assumed to be variable, changing according to the law (2.1), the problem in question remains self-similar. The behavior of the solution of this problem for various ω has been investigated by E. V. Rjazanov [3]. In addition to the PEP for the homothermic model, other problems have also been studied (cf., for example, [16]).

§4. Consideration of high-temperature effects in the problem of a strong explosion

4.1. *Thermodynamic properties of gases at high temperatures.* During the propagation of strong shock waves in a gas high pressures and temperatures are created. At high temperatures effects of dissocation of molecules and ionization of atoms and also effects related to radiative processes have an effect on the motion of the gas.

We shall assume that the processes of dissociation and ionization are equilibrium processes, and we shall not consider radiation for the time being. We consider as an example the case of single ionization of a diatomic gas, i.e. when before the front of the shock wave it is possible to take $\gamma = 1.4$. For the thermal equation of state of the mixture of gas particles we can write

$$p = \sum k n_j T, \tag{2.56}$$

where $j = 1, 2, 3, 4$; n_1 through n_4 are the numerical densities of the neutral atoms, molecules, positive ions, and electrons respectively, where because of quasi-neutrality $n_4 = n_3$; k is the Boltzmann constant.

For the process $A_2 \rightleftarrows n_1 A_1 + n_2 A_2 + n_3 A_3 + n_4 A_4$ in chemical equilibrium the law of mass action [17] holds:

$$m_2 = 2m_1, \quad m_1 = m_3 + m_4, \quad d_1 = 2^{3/2}\left(\frac{2h}{m_1}\right)^{-3}, \quad d_2 = \left(\frac{h}{m_4}\right)^{-3},$$

$$\frac{n_1^2}{n_2} = d_1 \left(\frac{2\pi}{m_1} kT\right)^{3/2} \frac{\left[\sum_j \exp(-\varepsilon_{ij}/kT)\right]^2}{\left[\sum_j \exp(-\varepsilon_{2j}/kT)\right] \exp(-W_2/kT)}, \tag{2.57}$$

$$\frac{n_3 n_4}{n_1} = \left(\frac{m_1}{m_3}\right)^{3/2} d_2 \left(\frac{2\pi kT}{m_4}\right)^{3/2} \frac{\left[\sum_j \exp(-\varepsilon_{3j}/kT)\right] \exp(-W_1/kT)}{\sum_j \exp(-\varepsilon_{ij}/kT)}, \tag{2.58}$$

where the m_j are the masses of the particles, W_1 is the energy of ionization, i.e. the energy required for the ionization of atom A_1 at $T = 0°\text{K}$, W_2 is the energy of dissociation, i.e. the energy necessary for the dissociation of a molecule at $T = 0°\text{K}$, h is Planck's constant, and ϵ_{ij} is the energy of particle A_i in the various states. For the mass density of the gas we have

$$\rho = \sum_j m_j n_j, \tag{2.59}$$

where $m_4/m_3 \ll 1$ and $m_2 \approx m_3$.

For known values of p and ρ formulas (2.56)–(2.59) make it possible to determine the unknown quantities n_1, n_2, n_4 and T. In the equations of motion of the gas the thermodynamic properties of the gas enter through the relation between the internal energy, the density, and the pressure. For the internal energy there is the relation [12], [17], [18]

$$\rho\varepsilon = \frac{3}{2} nkT + \sum_i n_i (u_i + w_i), \tag{2.60}$$

where

$$u_i = kT^2 \frac{\partial}{\partial T} \ln z_i, \qquad z_i = \sum_s \exp \frac{\varepsilon_{is}}{kT}, \qquad z_4 = 1,$$
$$w_1 = \frac{1}{2} W_2, \qquad w_2 = w_3 = 0, \quad w_4 = W_1. \tag{2.61}$$

A simple example of a gas for which it is meaningful to solve the explosion problem taking into account high-temperature effects is hydrogen. For the case of hydrogen formulas (2.56)–(2.61) give the thermodynamic functions right up to complete ionization. It is true that Coulomb interaction and the appearance of negative ions are not considered here, but in a number of cases it is possible to neglect these effects.

For a more complex mixture of gases and multiple ionization of the atoms the formulas for calculating the thermodynamic functions become much more complicated [12]. At the present time the calculation of the thermodynamic functions of air over a broad range of temperatures and pressures has been completed by N. M. Kuznecov [21]. To obtain rough estimates and an approximate calculation of the influence of ionization and dissociation processes on variations of temperature and density one can resort to the following schemes.

1) We assume the gas to be perfect with some mean "effective" adiabatic exponent γ. Estimates carried out by various authors indicate that in the stage of a strong shock wave the effective adiabatic exponent of gases lies in the interval $1.2 \leqslant \gamma \leqslant 1.3$. The thermodynamic functions of the gas are hereby reduced to those for a perfect gas

with constant heat capacities. For rough estimates of temperatures it can be assumed that the variation of the thermodynamic properties of the gas at high temperatures has a weak effect on the pressures and densities in the gas flow. Then, solving the problem for the effective γ and assuming p and ρ to be known, by using equations of the type (2.56)–(2.61) or tables of thermodynamic functions we can determine T and the concentrations of the components of the gas mixture. Such estimates of temperatures for single ionization of air have been carried out in [22].

2) For a more accurate estimate of properties of gases it is possible to approximate the dependence of the internal energy ϵ on p and ρ or on ρ and T by analytic formulas and then apply approximation methods to solve the differential equations. This method is widely used in various problems of gas dynamics. In application to a strong explosion in air this approach has been used in [23], [24].

Examples of approximation of the internal energy by simple analytic relations are also given in [12]. As an approximate formula here a formula of the following form is used:

$$\varepsilon = \varepsilon_0 T^{\alpha} \rho^{\beta}, \qquad \varepsilon_0 = \text{const.} \tag{2.62}$$

Thus, if the density of air varies between $10\rho_0$ and $10^{-3}\rho_0$ (ρ_0 is the normal density) and the temperature between 10^4°K and 2.5×10^5°K, formula (2.62) can be taken in the form

$$\varepsilon = \varepsilon_0 \left(\frac{T}{10^4}\right)^{1.5} \left(\frac{\rho_0}{\rho}\right)^{0.12}, \tag{2.63}$$

where ϵ_0 is the value of the internal energy at $T = 10^4$°K and $\rho = \rho_0$.

We conducted work on approximation of the relations on a strong shock wave and of the internal energy of air. Tables [21] were used for the basic data on the thermodynamic functions. The approximation of the tabulated data for relations on the shock wave were carried out by two methods. In the first method the dependence of the enthalpy on the speed of the shock wave D was approximated, and in the second method the dependence of the pressure behind the shock wave p_2 on the velocity behind the shock wave v_2 was approximated.

We shall consider first the first method of approximation. As was noted in [12], the relation between the enthalpy i and D^2 depends weakly on the thermodynamic properties of air at high temperatures. Indeed, from the jump conditions it follows that

$$i_2 = \frac{D^2}{2}\left[1 - \left(\frac{\rho_1}{\rho_2}\right)^2\right]$$

and for strong compression the ratio i_2/D^2 varies weakly. We introduce the dimensionless variables

$$\mathcal{J}=\frac{\rho_1 i_2}{p_1},\qquad \bar{D}=\frac{D}{\sqrt{p_1/\rho_1}}$$

and approximate the dependence $\mathcal{J}(\bar{D})$ according to the formula

$$\mathcal{J}=\frac{k_j}{2}\bar{D}^2+j\mathcal{J}_0,\qquad j=0,\ 1. \tag{2.64}$$

We shall use (2.64) both in the case of the rougher approximation $j = 0$ and in the more accurate case when $j = 1$. Concrete calculations were carried out only for the case of the initial parameters of the standard atmosphere at sea level: $p_1 = 1.01325 \times 10^6$ dyne/cm^2, $\rho_1 = 1.224 \times 10^{-3}$ g/cm^3.

For $j = 0$ the approximation of the tables gives the value $k_0 = 0.992$, i.e.

$$\mathcal{J}=0.4959\bar{D}^2. \tag{2.65}$$

Comparison with tabulated data shows that in the range of temperatures behind the wave front from 1300°K to 2×10^{6}°K the errors in determining i by formula (2.65) do not exceed 3%. It is natural that for temperatures below 1000°K the errors increase rapidly. More accurate results are obtained by approximating with formula (2.64) when $j = 1$. Here the values $k_1 = 2 \times 0.49446$ and $\mathcal{J}_0 = 0.32$ were obtained, i.e. (2.64) has the form

$$\mathcal{J}=0.49446\bar{D}^2+0.32. \tag{2.66}$$

The dependence (2.66) has higher accuracy than (2.65). With variation of T from 1×10^{3}°K to 3×10^{6}°K the error of (2.66) does not exceed 3%.

Thus these formulas together with the dynamical conditions on the shock wave can replace altogether the tables of [21] for determining p_2 and ρ_2. Formulas (2.64) are to be taken here in place of the law of energy conservation on passage through the jumps. Temperatures T_2 are to be found from the tables once p_2, i_2 or p_2, ρ_2 are known from the approximate formulas.

The second method of approximation of the data involves finding the approximate relation between $P_2 = p_2/p_1$ and $v_2/v_1 = V_2$. For normal initial density and pressure at sea level we have

$$\gamma_{\text{eff}}=-0.2875\,(T_2\cdot 10^{-3})+1.4155$$

(T_2 is the temperature in °K)

$$\theta_2=a+\sqrt{\frac{T_2}{T_1}+a_1^2},\qquad a=\frac{1}{2}\left(\frac{\gamma_{\text{eff}}+1}{\gamma_{\text{eff}}-1}+6\frac{T_1}{T_2}\right),$$
$$\frac{p_2}{p_1}=\theta_2\frac{T_2}{T_1}\left(1\leqslant\frac{p_2}{p_1}\leqslant 88.47\right), \tag{2.67}$$

$$P_2=(1.65V_2^2+1)-[0.0243\,(\ln V_2)^{6.791}]\,(88.47<P_2\leqslant 77410), \tag{2.68}$$

$$P_2 = 1.65 V_2^2 + 1 \; (77410 < P_2 \leqslant 567400), \tag{2.69}$$

$$P_2 = 1.65 V_2^2 + 1 + 222.56 V_2 - 137750 \; (567400 < P_2 \leqslant 936600). \tag{2.70}$$

Formulas (2.67)–(2.70) correspond to a range of T_2 from 288.16 to 3×10^{6}°K and have an error not exceeding 3% with respect to tabulated data.

For the dependencies of the thermodynamic functions on the density and pressure or on the density and temperature no simple formulas of high accuracy were found. We shall mention here only one result.

We seek an approximation of the internal energy ϵ in the form

$$\varepsilon = \frac{p}{\rho} \varphi\left(\rho, \frac{p}{\rho}\right). \tag{2.71}$$

The function φ is to be chosen from tables. If φ depends only on ρ, we can obtain only a very rough approximation of $\epsilon(p, \rho)$. A more accurate approximation is given, for example, by formulas of the form

$$\varphi = b \coth\left(\frac{g + \delta_0}{a}\right), \quad g = \frac{\rho}{\rho_0}, \tag{2.72}$$

where b, a_0 and δ_0 depend on the ratio p/ρ and change little with change in p/ρ; ρ_0 is a characteristic density ($\rho_0 \approx \rho_1$) and the pressure refers to p_1. Thus the approximation given by the following relations is not bad:

$$b = \frac{1}{2.013}\left(\frac{p}{g}\right)^{-0.06687}, \qquad a = 2.8996\left(\frac{p}{g}\right)^{0.0342}, \qquad \delta_0 = 5.606\left(\frac{p}{g}\right)^{-0.641}. \tag{2.73}$$

Naturally, (2.72) and (2.73) can be generalized and made more precise.

At high temperatures it is sometimes important to consider the kinetics of chemical reactions between various components of the gas and in the processes of dissociation and ionization, i.e. attention is not restricted to equilibrium approximations. The effect of nonequilibrium states and rates of chemical reactions on the composition of the gas in a strong explosion in air is discussed in [12]. We shall not consider these questions here. Questions closely related to these will be discussed in Chapter 6 in studying the chemical reactions of rapid combustion.

4.2. *On the effect of heat radiation on the motion of the gas.* At temperatures of several thousand degrees and higher heat radiation will have an essential effect on the motion of the gas. Without going into the general questions of the theory of heat transfer by radiation, we restrict ourselves to only a very simple model of the theory of heat transfer by radiation and specifically assume: 1) the existence of local thermodynamic equilibrium between matter and the photon gas-radiation; 2) the process of heat transfer by radiation takes place by radiant conduction, i.e. the approximation of radiant conduction is used.

In the approximation of radiant heat conduction the flow of radiant energy under conditions of local equilibrium is proportional to the temperature gradient, i.e. the transfer of radiation has the character of heat conduction, where the coefficient of thermal conductivity depends on the temperature (and density). For the model of radiant heat conduction the equations of hydrodynamics retain their form if for the pressure p in the momentum equation we understand the sum $p^* = p_g + p_r$, where p_g is the pressure of the gas and $p_r = (4/3)(\sigma/c)T^4$ is the radiation pressure (σ is the Stefan-Boltzmann constant and c is the speed of light), to the internal energy $\rho\epsilon$ we add the density of radiant energy $\epsilon_r = 4\sigma T^4/c$, and for the vector of heat flow we take

$$\boldsymbol{q} = -[\lambda_\mu + \kappa]\nabla T,$$

where λ_μ is the coefficient of molecular heat conductivity, κ is the coefficient of radiant conductivity

$$\kappa = \frac{16}{3}\sigma c l_R T^3, \tag{2.74}$$

and l_R is the mean free path of radiation for all frequencies.

In gases at high temperatures ($T > 2000°K$) we have $\lambda_\mu \ll \kappa$, while to estimate l_R one may apply the formula [12]

$$l_R = A_r T^{\alpha_1} n^{-\alpha_2}, \tag{2.75}$$

where A_r is a constant, n is the numerical density, and α_1 and α_2 are constants. For air α_1 has a value of about 2 and α_2 about 1.7 [12]. The values of l_R for air are presented in tables in [12].

The equations for one-dimensional motions of the gas for the model of radiant heat conduction have the form

$$\begin{gathered} \rho\frac{dv}{dt} = -\frac{\partial p^*}{\partial r}, \qquad \frac{d\rho}{dt} + \rho\left(\frac{\partial v}{\partial r} + (\nu - 1)\frac{v}{r}\right) = 0, \qquad \kappa = \kappa(\rho, T), \\ \frac{\partial}{\partial t} r^{\nu-1}\left(\rho\varepsilon + \rho\frac{v^2}{2} + \varepsilon_{\text{и}}\right) + \frac{\partial}{\partial r} r^{\nu-1} v\left(\rho\varepsilon + \varepsilon_{\text{и}} + \rho\frac{v^2}{2} + p^*\right) - \frac{\partial}{\partial r} r^{\nu-1}\kappa\frac{\partial T}{\partial r} = 0. \end{gathered} \tag{2.76}$$

Questions of the effect of radiation on the motion of the gas in a strong explosion have been considered in [12], [25].

4.3. *Formulations of problems of point explosion in a gas with inclusion of high-temperature effects.* From the background material presented concerning processes in gases at high temperatures it follows that in order to account for them in detail we must complicate the original system of equations of gas dynamics considerably. We shall note some possible formulations of problems which follow from the conclusions drawn above. We consider here only one-dimensional motions.

Suppose that radiation is neglected entirely. Then, assuming thermodynamic equilibrium, together with the original system (1.18)–(1.20) we must solve equations of the type (2.56)–(2.61) to determine the composition of the mixture of gases and the internal energy. The boundary conditions on the shock wave must correspondingly be more general than for the usual perfect gas.

The simplest case of a problem of this type is the problem of a strong explosion in atomic hydrogen or one involving only single ionization of more complex atoms. Indeed, in this case the dependence of the internal energy on the temperature and pressure is given by an expression of the type (2.60):

$$\varepsilon = \frac{3}{2}(1+\alpha) R_0 T + \alpha \frac{W_1}{m_0}, \tag{2.77}$$

$$\alpha = \left[\frac{h^3}{(2\pi m_e)^{3/2}} \frac{p}{(kT)^{5/2}} \frac{g_0}{2g_1} \exp\left(\frac{W_1}{kT}\right) + 1\right]^{-1/2}, \tag{2.78}$$

where α is the degree of ionization, m_e is the electron mass, m_0 is the mass of the atom, W_1 is the energy of ionization, R_0 is the gas constant per unit mass, and g_0 and g_1 are statistical weights of the basic level of the neutral atom and ion.

The thermal equation of state assumes the form

$$p = \rho(1+\alpha) R_0 T. \tag{2.79}$$

In the system of defining parameters of the problem there are added two dimensional constants, W_1/m_0 and the factor in front of the exponential function in (2.78). This implies that the problem is not longer self-similar. In Sedov's book [2] classes of caloric equations of state of a gas are considered for which the problem of strong explosion in a medium with constant density remains self-similar. For self-similarity the dependence of the internal energy on p and ρ must have the form

$$\varepsilon = \frac{p}{\rho_0} \varphi\left(\frac{\rho}{\rho_0}\right), \tag{2.80}$$

where ρ_0 is some constant with the dimension of density. If the initial density is variable, the function (2.80) reduces simply to the form of a perfect gas

$$\varepsilon = \beta \frac{p}{\rho}, \tag{2.81}$$

where β is a derived constant.

Thus the solution of the strong explosion problem for an equation of state of the form (2.77) does not reduce to the integration of ordinary differential equations, and requires the use of numerical methods.

If this problem is considered in the approximation of the effective adiabatic exponent γ, it reduces to the solution (2.5)–(2.13) considered earlier. The considerations with regard to the approximation of tables of thermodynamic functions of air indicate that the strong explosion problem will be self-similar only for rough approximation of these tables. We remark that strong explosion problems including dissociation and ionization which are not self-similar have at the present time been studied rather little.

The problem becomes still more complicated if radiation is considered, even in the approximation of local thermodynamic equilibrium. Of course, we considered the effect of radiation indirectly in the problem of a homothermic strong explosion, but this approximation is valid only in the initial stage of a strong explosion. If we adopt the model of radiant heat conduction, then for this model it is necessary to investigate the solution of the system of equations (2.76) taking into account the usual initial conditions on the shock wave for the heat-conducting mixture of gas-radiation. These conditions on the shock wave for one-dimensional motions can be written as follows:

$$[\rho(D-v)]=0, \quad [\rho v(D-v)-p]=0,$$
$$\left[\rho(D-v)\left(\frac{v^2}{2}+\varepsilon^*\right)-p^*v+\kappa(\rho, T)\frac{\partial T}{\partial r}\right]=0, \quad [T]=0, \tag{2.82}$$

where

$$\varepsilon^*=\varepsilon+\frac{1}{\gamma_1-1}\frac{p_r}{\rho}, \quad \gamma_1=\frac{4}{3}. \tag{2.83}$$

Since a thermal wave can propagate in advance of the shock wave, it is necessary to give conditions at infinity:

$$T(\infty, t)=v(\infty, t)=p(\infty, t)=0, \quad \rho(\infty, t)=\rho_1. \tag{2.84}$$

For the complete solution of the problem posed we must consider the relations (2.74), (2.75) and the effect of dissociation and ionization on the form of the function $\epsilon(p, \rho)$. The problem formulated is, in general, not self-similar and requires application of numerical methods for its solution.

We shall consider only certain special cases of this complex problem. First of all, we shall assume that the gas is perfect with constant (effective) adiabatic exponent γ. Secondly, we consider two cases separately: 1) $\kappa = 0$, $p_r \neq 0$; 2) $\kappa = \kappa(\rho, T) \neq 0$, $p_r = 0$.

In the first case the PEP can be solved exactly using a method proposed by E. Lariš and I. Šehtman [26] (cf. also §3). Here the problem is solved for a perfect gas where $p = p^*$ and $\epsilon = \epsilon^*$, and then the temperatures are found from the thermal equation of state for the gas-radiation mixture. We shall not consider this in detail but rather only point out that for this model the temperatures at the center of the explosion are finite and vary in time as

$$T_r(0, t)=\left[\frac{3}{a}\frac{8\rho_1}{(\nu+2)^2(\gamma+1)}\frac{p}{p_2}(0)\right]^{1/4}\left(\frac{E_0}{\alpha\rho_1}\right)^{1/2(\nu+2)}t^{-\nu/2(\nu+2)}. \tag{2.85}$$

In a neighborhood of the center of the explosion the temperature depends weakly on the coordinate. For small duration times of the process this solution of the problem is qualitatively close to the solution where a zero temperature gradient is assumed.

In the second case, where the pressure of the equilibrium radiation (and its energy) is not considered, the problem can be studied in the self-similar formulation. Suppose that the initial density is constant and that the coefficient of thermal conductivity depends on density and temperature as follows:

$$\kappa = \kappa_1 \rho^{a_1} T^{a_2}. \tag{2.86}$$

Then, as in the case $a_1 = 0$ considered by Bam-Zelikovič [28], the problem will be self-similar for any a_1 if $a_2 = (\nu - 2)/2\nu$ and the gas is perfect.

We have obtained [3], [27] conditions of self-similarity for the case of an explosion in a medium with variable density $\rho_1 = Ar^{-\omega}$ when $a_1 = 0$. Here dimensional analysis of the constants entering in the problem shows that the strong explosion problem is self-similar if the following relation holds between the parameters ν, a_2, and ω:

$$\omega + \nu - 2 - 2a_2(\nu - \omega) = 0. \tag{2.87}$$

For the case $a_1 = 0, \nu = 3$ and constant density the self-similar strong explosion problem has been studied by the authors of [27], [29], [30].

We shall consider briefly the results of the investigation of this problem. The author studied this problem [27], [29] in the approximate formulation where the thermal wave propagating through the cold gas in front of the shock wave was not considered. The following results were obtained. The gas set in motion by the shock wave moves in a direction away from the center and is contained in a sphere whose radius increases with time according to the law

$$r_2 = \left(\frac{E_0}{\alpha\rho_1}\right) t^{2/5}. \tag{2.88}$$

The temperature at the center of the explosion is finite and decreases monotonically with increasing r, having a maximum at the center. The velocities of particles near the center are small. For $r < 0.5r_2$ the particles of the gas are nearly in a state of rest. The distribution of pressures remains qualitatively the same as in the case of the adiabatic and homothermic flow models. Near the center of the explosion we have the asymptotic formulas

$$v = \frac{1}{\kappa_1} b_1(t) r^3 + o(r^3), \qquad \rho = \rho(0, t) + \frac{1}{\kappa_1} b_2(t) r^2 + o(r^2),$$
$$T = T_0(0, t) + \frac{b_3}{\kappa_1} r^2 + o(r^2). \tag{2.89}$$

If the thermal conductivity is small, it has a substantial effect on the temperature only in some small neighborhood of the center. Analysis of the solution of the problem indicated that with increasing κ_1 the zone of weak variation of temperature expands, i.e. the flow becomes approximately homothermic.

V. E. Neuvažaev [30] carried out detailed calculations of the problem taking into account the thermal wave and the gas flow in front of the shock wave caused by it. We remark that for small κ_1 the author proposed [29] an approximate method for solving the problem based on "splicing" the solution of the heat-conducting problem near $r = 0$ with the solution of the adiabatic problem. An analogous approach was subsequently developed in [31].

The case of a strong cylindrical explosion when $\kappa = \text{const}$ was studied by I. O. Bežaev [32]. In studying the homothermic model the variations of density found in the shock wave are considerably lower than in the adiabatic theory, and the density actually changes by little more than twice. This situation obtains for large but finite values of κ.

We now consider a rough scheme of reckoning with the thermal wave in advance of the shock wave with homothermic motion inside the latter. We assume that the explosion is spherical, the density is constant, and the problem is self-similar. Then the problem has an energy integral [2], [3], [28]

$$r^2\left[\left(\frac{2}{5}\frac{r}{t}-v\right)\left(\rho\frac{v^2}{2}+\rho\frac{\tilde{T}}{\gamma-1}\right)-\rho\tilde{T}v+\frac{\kappa}{R}\frac{\partial\tilde{T}}{\partial r}\right]=C. \tag{2.90}$$

Here C is an arbitrary constant, and $\widetilde{T} = RT$. Assuming for sufficiently large r that $v = T = \kappa(T) = 0$, we set the constant C equal to zero. Going over to dimensionless variables $f = v/D$, $g = \rho/\rho_1$, $\theta = \widetilde{T}/D^2$ and $\lambda = r/r_2$, we find

$$B\theta^{1/6}\frac{d\theta}{d\lambda}-g(\lambda-f)((\gamma-1)f^2+\theta)=0, \tag{2.91}$$

where B is a dimensionless constant proportional to the coefficient κ_1 [3].

We assume that the temperature does not suffer a jump on passage through the discontinuity (this is satisfied when heat conduction is considered); then for (2.91) we have the boundary condition

$$\theta(1)=\theta_2. \tag{2.92}$$

We now neglect variations in the velocity and density caused by the thermal wave, i.e. we assume that $g = \text{const} = g_0$ and $f = 0$ for $\lambda > 1$. It is now elementary to integrate (2.91). Using the boundary condition (2.92), we find that

$$\theta=\left[\theta_2^{1/6}+\frac{g_0}{12B}(1-\lambda^6)\right]^6. \tag{2.93}$$

Suppose that θ becomes zero for $\lambda = \lambda_T$; then for the coordinate of the leading front of the thermal wave λ_T we have

$$\lambda_T=\sqrt[6]{1+\frac{12B}{g_0}\theta_2^{1/6}}, \tag{2.94}$$

where the values of θ_2 are given by (2.37) and $g_0 = 1$. The coordinate λ_T is large for large B $(B > 1)$. For $B \to 0$ the boundary of the thermal wave approaches the front of the shock wave. If the effective value of κ_1 is not large, then $B < 1$ and λ_T is close to one. In this case the temperature drops abruptly to zero in a narrow zone in front of the wave and for $\lambda = \lambda_T$ becomes zero together with five of its derivatives with respect to λ. The neglecting of the velocities v and the variation of the density ρ for $\lambda > 1$ will here be justified.

For the dimensional temperature T we have the following distribution in space: $T = T_2(t)$ for $r \leqslant r_2$; for $r > r_2$ the temperature decreases with increasing r, becoming zero for $r = \lambda_T r_2$. Moreover, since $\theta \neq 0$ in the region $1 \leqslant \lambda \leqslant \lambda_T$, some fraction of the energy E_0 will be contained here. It then follows from the condition that the integral law of energy conservation (1.184) be satisfied that the constant $\alpha = E_0/E$ introduced above receives a certain increment $\Delta\alpha$ as compared with the value $\alpha = \alpha_{hT}$ computed according to (2.48). Denoting by α^* the new value of E_0/E, we have $\alpha^* = \alpha_{hT} + \Delta\alpha$. In the present approximation we can write for $\Delta\alpha$

$$\Delta\alpha = \frac{8}{25}\,\frac{\sigma_\nu}{(\gamma^2 - 1)\, g_2\theta_2}\int_1^{\lambda_T} \theta\lambda^2 d\lambda,$$

where the quantities g_2 and θ_2 are known from the solution of the homothermic problem (cf. §3).

§5. Some applications of the theory of strong explosion

5.1. *The initial stage of gas motion in an atomic explosion.* The application to atomic explosion was a major stimulus to the development of the theory of strong explosion. We shall not consider this question in detail here, since it is treated in [2], [3], [12], [33]–[36], etc.

We shall mention only the following:

1) Experimental data on the dependence of the radius of the shock wave on time are well approximated by a formula of the type

$$\frac{5}{2}\lg r_2 - \lg t = C_1, \tag{2.95}$$

where C_1 is a constant.

According to data obtained during the explosion of an atomic bomb in New Mexico in 1945 and published by G. I. Taylor [33], for the constant C_1 one should take the value $C_1 = 11.915$.

The theoretical formula $r_2 = (E_0/\alpha\rho_1)^{1/5}t^{2/5}$ agrees with (2.95) if we take

$$0.5\lg\frac{E_0}{\alpha\rho_1} = 11.915. \tag{2.96}$$

If the values of ρ_1 and α are known, this formula makes it possible to estimate the fraction of the energy E_0 which went to the motion of the gas.

The numerical value of α (and hence also the theoretical estimate of E_0) depends on the flow model adopted.

Thus for the adiabatic model of a perfect gas with constant γ for $\nu = 3$ we have (cf. §2)

$$\alpha_{\mathrm{ad}} = 0.4921\,(\gamma - 1)^{-(1.1469+0.11735\,\lg\,(\gamma-1))} \qquad (1.2 \leqslant \gamma \leqslant 2).$$

For the homothermic flow model with $\nu = 3$ we have $\alpha_{\mathrm{h}T} = 0.0643 + 0.163/(\gamma - 1)$. For the same γ the inequality $\alpha_{\mathrm{ad}} > \alpha_{\mathrm{h}T}$ holds. This implies that estimates of the energy E_0 going to motion of the gas (obtained using α_{ad}) will be high if processes of heat conduction are not considered. Since the energy of dissociation and ionization is accounted for by the choice of the effective γ in only a very approximate way, a precise evaluation of these processes may also change the magnitude of E_0 [37].

2) The distribution of temperatures in the region between the center and the leading front of the perturbation is closer to the real distribution when effects of heat conduction are taken into account. During the first moments after explosion (when the temperature is higher than $3 \times 10^5\,^\circ$K) the velocity of the thermal wave is greater than the velocity of the shock wave [12], and the temperature distribution is close to that given by (2.32). Further, the solution of the homothermic problem may be used approximately up to the moment when the fireball breaks away [14]. In this model it is essentially assumed that $\kappa = \infty$ behind the front of the shock wave and $\kappa = 0$ in advance of the front of the shock wave. The energy of explosion radiated forward is not considered. The model of radiant conduction can be used for later moments of time with a density close to normal near the earth's surface. As the temperature drops further in the stage of a strong shock wave the flow is sufficiently well described by the adiabatic model.

3) As we have already noted, effects of heat conduction have a substantial influence not only on the distribution of temperatures but on the distribution of velocities, densities, and pressures as well. As an illustration, in Figures 13–15 there is shown a comparison of the distribution of velocities, densities, and pressures for the adiabatic model (curve 1) and homothermic model (curve 2) of the flow.

5.2. *Electrical explosion of conductors and discharges in a gas.* The conclusions of the theory of strong explosion can also be used to consider processes of perturbation of a gas caused by electrical discharges through conductors formed of fine metallic wires and plates [38]–[45]. In view of the relatively low temperatures, the adiabatic model with single ionization is often used here.

Experiments conducted afford the following conclusions. For fine straight conductors (wire thickness less than 1 mm) energies of explosion are of the order 3×10^8 erg/cm, the dependence of the coordinate of the basic shock wave (as a rule, secondary shock waves are formed) is nearly self-similar, and $r_2 = (E_0/\alpha\rho_1)^{1/4} t^{1/2}$ if $r_2 > 10 r_0$, where r_0 is the radius of the conductor and t is the time after termination of the discharge. This is corroborated also by the calculations of Rouse [43] using a theoretical model of an exploding wire in air taking dissociation and ionization into account. In particular, is is shown here that the ratio of the energy parameters for air α_a, considering equilibrium ionization and dissociation, for the case of a perfect gas with $\gamma = 1.4$ is approximately equal to 2.18, i.e. $\alpha_a = 2.18\alpha(\gamma)$. This corresponds to an effective exponent γ for air γ_{eff} close to 1.2.

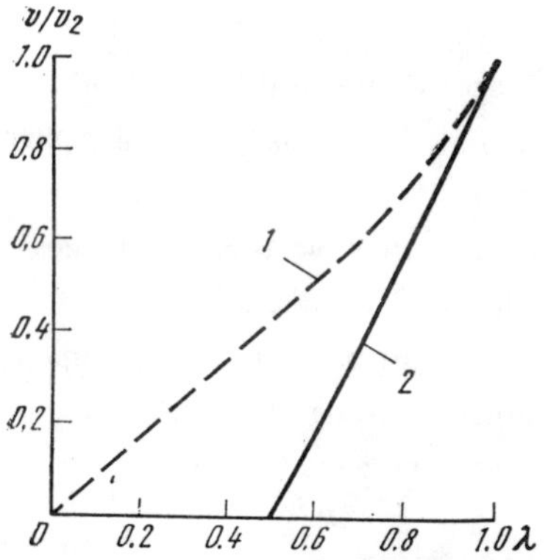

FIGURE 13. Comparison of the distributions of dimensionless velocities

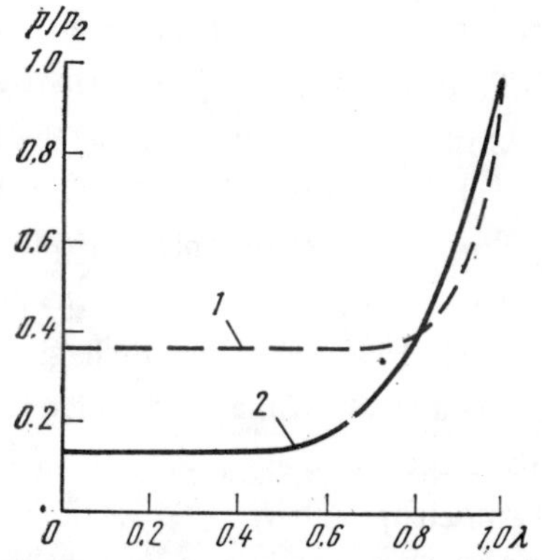

FIGURE 14. Comparison of the distributions of dimensionless pressures

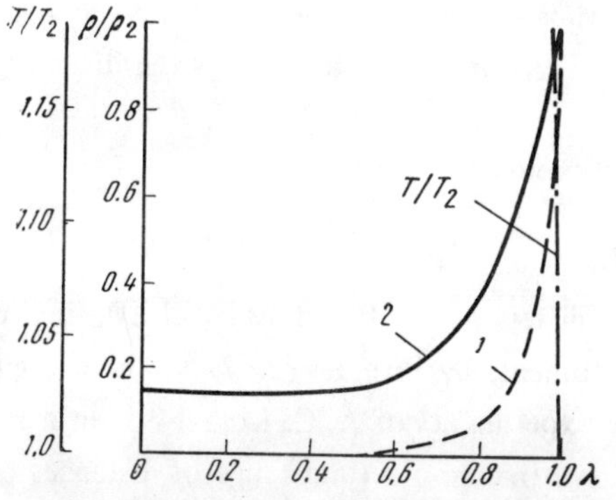

FIGURE 15. Comparison of the distributions of dimensionless densities and temperatures ($T/T_2 = 1$ for case 2)

The secondary shock waves arising in the explosion of wires are caused by expansion of metal vapors. This phenomenon cannot be described within the framework of the usual theory of a point explosion without the supply of mass. Similar conclusions can be drawn also for the case of explosion of fine metallic plates if this process is equated with the phenomenon of a strong planar point explosion.

In addition to the explosion of wires and plates, in order to obtain high temperatures discharges are induced in a gas between flat or cylindrical electrodes. The general qualitative picture of the process here is as follows. In the spark gap a fine electrically conducting channel is formed as the result of ionization by the electric field. Due to Joule heating, the air in this channel is heated to temperatures of the order of several tens of thousands of degrees and is ionized. As a result of the rapid increase in temperature, the pressure rises abruptly and a strong cylindrical shock wave propagates through the gas. Phenomena of spark discharge have been studied by S. L. Mandel′štam and his coworkers (see the bibliography in [12]). Without considering this phenomenon in detail, we mention only that measurements of the density distribution behind the front of the shock wave [46] have shown that the process of gas motion is close to that of a cylindrical strong explosion, while the average density in the region of the channel of the spark is approximately 1/1000 of the initial air density. Similar phenomena occur in the atmosphere during a thunderstorm.

Processes close to a planar explosion are observed during electrical discharges along flat electrically conducting channels [47] or during sharp discharges in T-shaped tubes [48], [49]. Here the discharge is produced in a short portion of a T-shaped tube. As a result of the abrupt increase in temperature, a strong plane shock wave begins to propagate along the part of the T-shaped tube perpendicular to the direction of the discharge tube.

In the experiments of Kash [48] the theory of a strong plane explosion is not used in interpreting results for discharges in air. However, it is indicated here that the dependence between the time of arrival of the shock wave and its position is well approximated by the formula $t = ax^{8/5}$, where x is the distance from the discharge and a is a constant.

For a planar strong explosion we have

$$t = (\alpha\rho_1/E_0)^{1/2} r_2^{3/2}. \tag{2.97}$$

Since the exponent 8/5 differs from 3/2 only by 1/10, it is clear that interpretation of the results of the experiments by formula (2.97) would also give good agreement with experiment. In the experiments of A. C. Kolb [49] on discharges in hydrogen and deuterium it was shown that in a certain range of distances from the site of the discharge satisfactory agreement is found between the theory of strong explosion and experiments measuring the velocities of shock waves along the length of the tube. Using Kolb's data one can determine the specific energy E_0 expended in motion of the gas behind the front of the wave.

Indeed, in [49] a table is given according to which $D = 5.6 \times 10^6$ cm/sec for $r_2 = 9$ cm. Since all the energy passes to one side of the explosion site, it follows that $2E_0 = \alpha\rho_1(3/2D)^2 r_2$. Taking for γ the value 1.3 and for the initial density $\rho_1 = 1.8 \times 10^{-7}$ g/cm^3 (the initial density was 0.7 mm Hg), we find that $E_0 \sim 10^8$ erg/cm^2. We note also that from the data presented in the table just mentioned it follows that the product $D^2 r_2$ is slowly varying in the range of distances from 3 to 11 cm along the tube (the total length of the tube was 12 cm).

The analysis above gives some idea of the applicability of the theory of strong explosion to still another circle of problems of practical importance.

5.3. *Local heating of a gas by a laser beam.* In recent years the theory of point explosions has found an important new application to processes related to the heating of a gas by focusing a laser pulse. The laser beam affords the possibility of obtaining in the focusing region a very high temperature of the gas (around a million degrees) in a volume of the order of 10^{-3}cm^3. This gives rise to a strong shock wave which ionizes the gas and propagates with a velocity of about 10^8cm/sec. The time of energy supply is very small and is of the order of 10^{-7}sec. (Here we mean not the time of energy supply by the laser, which is of the order of 5×10^{-9}sec, but the time of formation of the shock wave.) The laser pulse can radiate energy of the order of 100 J and higher. Two basic methods are used to obtain high temperatures in gases: breakdown (or "spark") in the gas caused by the laser or focusing on a target placed in a sufficiently rarefied gas. In the first method the evolution of energy in the gas, as a rule, is not isotropic in a neighborhood of the focusing site (cf., for example, [50], [51]). Here even a long spark can be formed with a length of about 1 m or more [52], [53] with variable specific energy introduced in the gas, i.e. we have here a case of evolution of energy along a line when $E_0 = E(z)$, z being the coordinate along the line. If end effects are neglected, then for times greater than the time of formation of the shock wave there is a cylindrical explosion along the line when $E_0 = E(z)$. We shall also consider problems of this type in Chapter 5.

When a target is used, as a rule, a strong spherical shock wave forms at the focusing site [53], [55]. We shall estimate the fraction of energy going to the shock wave and gas-dynamical motion in the explosion stage. It follows from [54] that during breakdown in air (at atmospheric pressure) at the moment of time $t = 1.5 \times 10^{-6}$sec after breakdown the coordinate of the shock wave measured in the experiment in the direction of the axis of the laser beam is approximately equal to 1 cm. Using (2.13), we have

$$E_0 = \alpha \rho_1 r_2^5 / t^2.$$

Taking an approximate value of α equal to 1.5, which corresponds to $\gamma = 1.25$, and $\rho_1 = 1.25 \times 10^{-3}\text{g/cm}^3$, we find from this formula that $E_0 \approx 7$ J. The estimate obtained agrees well with the experimental value of the pulse energy of 10 J. If we consider that the energy of the explosion did not pass through the complete solid angle of 4π but through a smaller part of space (this follows from figures of the wave form), we will obtain a somewhat smaller value of E_0, i.e. approximately 70% of the energy goes to the formation of the shock wave. The remaining energy is expended in organization of the breakdown process and leaves as luminous radiation and also in other losses.

The data presented bear witness to the appearance of new applications of the theory of point explosions and the necessity of new formulations of problems in this theory. We shall consider the formulation of one such problem. Suppose that we have a spherical volume of radius r_0 filled with quiescent plasma with parameters

ρ_1, p_1 and T_1, and that outside this volume the gas has parameters ρ_0, p_0 and T_0 $(\rho_0 < \rho_1)$.

At the moment $t = 0$ energy $E_0 4\pi r_0^2$ is released in a narrow spherical layer of thickness $\Delta r = r_0 - r_1 \ll r_0$. It is required to determine the resulting motion. This problem can arise in the spherical focusing of a laser beam on a plasma target if the absorption length of the laser radiation is small and the time of energy release is considerably less than the quantity $r_0(\rho_1/\gamma p_1)^{1/2}$. For $\Delta r \to 0$ we have the problem of a surface explosion on the boundary separating two media with specific energy E_0 (in this connection see also Chapter 5). Here at the beginning there arise a contact surface and shock waves proceeding toward and away from the center; in the process of motion of the plasma it is possible that a zone with extremely high densities and temperatures is formed.

BIBLIOGRAPHY

1. L. I. Sedov, *Le mouvement d'air en cas d'une forte explosion,* C. R. (Dokl.) Acad. Sci. URSS **52** (1946), 17–20.

2. ———, *Similarity and dimensional methods in mechanics,* 4th ed., GITTL, Moscow, 1957; English transl., Academic Press, New York, 1959. MR **19**, 898; **21** #6840.

3. V. P. Korobeĭnikov, N. S. Mel'nikova and E. V. Rjazanov, *Theory of point explosions,* Fizmatgiz, Moscow, 1961. (Russian) MR **29** #5494.

4. N. S. Burnova (Mel'nikova), *Investigation of the point explosion problem,* Candidate's Dissertation, Moscow State Univ., Moscow, 1953. (Russian)

5. V. P. Korobeĭnikov and E. V. Rjazanov, *Solutions of singular cases of point explosion in a gas,* Prikl. Mat. Meh. **23** (1959), 384–387 = J. Appl. Math. Mech. **23** (1959), 539–544. MR **22** #6299.

6. V. P. Korobeĭnikov, P. I. Čuškin and K. V. Šarovatova, *Tables of gas-dynamical functions for the initial stages of a point explosion,* Soobšč. po Vyčisl. Mat., vyp. 2, Izdat. Vycisl. Centr Akad. Nauk SSSR, Moscow, 1963. (Russian) MR **32** #3408.

7. D. L. Jones, *Strong blast waves in spherical, cylindrical, and plane shocks,* Phys. Fluids **4** (1961), 1183–1184; errata, ibid. **5** (1962), 637.

8. ———, *The energy parameter B for strong blast waves,* Tech. Note No. 155, U. S. Nat. Bur. Standards, Washington, D. C., 1962.

9. K. A. Semendjaev, *Empirical formulas,* ONTI, Moscow, 1933. (Russian)

10. G. M. Fihtengol'c, *A course in differential and integral calculus.* Vol. 2, 4th ed., Fizmatgiz, Moscow, 1958; German transl., VEB Deutscher Verlag, Berlin, 1966. MR **39** #1b.

11. N. N. Kočina, *Some exact solutions of equations of one-dimensional unsteady motion of a perfect gas,* Prikl. Mat. Meh. **21** (1957), 449–458. (Russian) MR **19**, 1121.

12. Ja. B. Zel'dovič and Ju. P. Raĭzer, *The physics of shock waves and high-temperature hydrodynamic phenomena,* 2nd ed., "Nauka", Moscow, 1966; English transl., vols. 1, 2, Academic Press, New York, 1966, 1967.

13. V. P. Korobeĭnikov, *The problem of a strong point explosion in a gas when the temperature gradient is zero,* Dokl. Akad. Nauk SSSR **109** (1956), 271–274. (Russian)

14. O. S. Ryžov and G. I. Taganov, *A second limit case of the point explosion problem,* Prikl. Mat. Meh. **20** (1956), 545–548. (Russian)

15. M. Frommer, *Die Integralkurven einer gewöhnlichen Differentialgleichungen erster Ordnung in der Umgebung rationaler Unbestimmtheitsstellen*, Math. Ann. **99** (1928), 222–272; Russian transl., Uspehi Mat. Nauk **9** (1941), 212–253.

16. N. S. Mel'nikova, *On transient motions of a gas displaced by a piston under zero temperature gradient*, Bul. Inst. Politehn. Iaşi **10** (1964), 289–302. (Russian; Romanian summary)

17. G. Ludwig and M. Heil, *Boundary-layer theory with dissociation and ionization*, Advances in Appl. Mech., vol. **6**, Academic Press, New York, 1960, pp. 39–118; Russian transl., Problemy Meh., vyp. 4, 16, Moscow, 1963. MR **23** #B744.

18. F. A. Baum, S. A. Kaplan and K. P. Stanjukovič, *Introduction to cosmic gas dynamics*, Fizmatgiz, Moscow, 1958. (Russian) MR **21** #1210.

19. R. A. Gross and C. L. Eisen, *Some properties of a hydrogen plasma*, Dynamics of Conducting Gases (Proc. Third Biennial Gas Dynamics Sympos., 1959), Northwestern Univ. Press, Evanston, Ill., 1960, pp. 15–24.

20. Gilda M. Harris, *Equilibrium properties and equation of state of a hydrogen plasma*, Phys. Rev. A**133** (1964), A427–A435.

21. N. M. Kuznecov, *Thermodynamic functions and shock adiabatics of air at high temperatures*, "Mašinostroenie", Moscow, 1965. (Russian)

22. R. A. Gross, *Continuum radiation behind a blast wave*, Phys. Fluids **7** (1964), 1078–1080.

23. T. Ginsburg, *Ausbreitung und Reflexion von starken Explosionswellen*, Sympos. Numerical Treatment of Partial Differential Equations with Real Characteristics (Rome, 1959), Liberia Eredi Virgilio Veschi, Rome, 1959, pp. 145–156. MR **22** #449.

24. H. L. Brode, *Blast wave from a spherical charge*, Phys. Fluids **2** (1959), 217–229.

25. *Effects of atomic weapons*, Los Alamos Scientific Laboratory, N. M., 1950.

26. E. Larish and I. Schechtman, *On the introduction of radiation in a problem of gas dynamics*, Dokl. Akad. Nauk SSSR **113** (1957), 1010–1012. (Russian) MR **19**, 1005.

27. V. P. Korobeĭnikov, *Investigation of certain problems of transient one-dimensional motions of a gas*, Candidate's Dissertation, Steklov Inst. Math., Moscow, 1956. (Russian)

28. G. M. Bam-Zelikovič, *Propagation of strong blast waves*, Teoret. Gidromeh., No. 4, Oborongiz, Moscow, 1949. (Russian)

29. V. P. Korobeĭnikov, *On propagation of a strong spherical blast wave in a gas with heat conduction*, Dokl. Akad. Nauk SSSR **113** (1957), 1006–1009. (Russian) MR **19**, 915.

30. V. E. Neuvažaev, *The propagation of a spherical blast wave in a heat-conducting gas*, Prikl. Mat. Meh. **26** (1962), 1094–1099 = J. Appl. Math. Mech. 26 (1962), 1657–1665.

31. V. V. Syčev, *On the theory of strong explosion in a heat-conducting gas*, Prikl. Mat. Meh. **29** (1965), 997–1003 = J. Appl. Math. Mech. **29** (1965), 1176–1184.

32. I. O. Bežaev, *On the effect of viscosity and thermal conductivity of a gas on the propagation of an explosion*, Teoret. Gidromeh., No. 11, Oborongiz, Moscow, 1953. (Russian)

33. G. I. Taylor, *The formation of a blast wave by a very intense explosion*. I, II, Proc. Roy. Soc. London Ser. A**201** (1950), 159–174, 175–186.

34. O. I. Leĭpunskiĭ, *Gamma radiation of an atomic explosion*, Atomizdat, Moscow, 1959. (Russian)

35. P. A. Jampol'skiĭ, *Neutrons of an atomic explosion*, Atomizdat, Moscow, 1961. (Russian)

36. V. N. Lavrenčik, *Global fallout of the products of nuclear explosions*, Atomizdat, Moscow, 1965. (Russian)

37. V. V. Lunev, *The method of effective energy in the problem of strong explosion in a real gas*, PMTF Ž. Prikl. Meh. i Tehn. Fiz. **1968**, no. 5, 50–52 = J. Appl. Mech. Tech. Phys. 9 (1968), 560–561.

38. N. N. Sobolev, *Investigation of the electrical explosion of fine wires,* Ž. Èksper. Teoret. Fiz. **17** (1947), 986–997. (Russian)

39. S. V. Levedev, *Explosion of a metal by an electric current,* Ž. Èksper. Teoret. Fiz. **32** (1957), 199–207 = Soviet Phys. JETP **5** (1957), 243–252.

40. M. N. Plooster, *Shock waves from line sources. Numerical solutions and experimental measurements,* Phys. Fluids **13** (1970), 2664–2675.

41. I. F. Kvarchava et al., *Electrical explosion of wires in a vacuum,* Ž. Èksper. Teoret. Fiz. 31 (1956), 737–744 = Soviet Phys. JETP **4** (1957), 637–644.

42. I. F. Kvarchava et al., *Electrical explosion of metal wires,* Ž. Èksper. Teoret. Fiz. 30 (1956), 42–53 = Soviet Phys. JETP **3** (1956), 40–51.

43. W. G. Chace and H. K. Moore (editors), *Exploding wires.* Vol. 1, Plenum Press, New York; Chapman and Hall, London, 1959; Russian transl., IL, Moscow, 1963.

44. ———, *Exploding wires.* Vol. 2, Plenum Press, New York, 1962; Russian transl., "Mir", Moscow, 1965.

45. ———, *Exploding wires.* Vol. 3, Plenum Press, New York, 1964.

46. G. G. Dolgov and S. L. Mandel'štam, *Density and temperature of a gas in spark discharge,* Ž. Èksper. Teoret. Fiz. **24** (1953), 691–700. (Russian)

47. A. E. Voĭtenko and I. Š. Model', *Production of strong shock waves during electrical discharges in gaps,* Ž. Èksper. Teoret. Fiz. 44 (1963), 1760–1764 = Soviet Phys. JETP 17 (1963), 1180–1182.

48. Rolf K. M. Landshoff (editor), *Magnetohydrodynamics (a symposium),* Stanford Univ. Press, Stanford, Calif., 1957.

49. A. C. Kolb, *Production of high-energy plasmas by magnetically driven shock waves,* Phys. Rev. **107** (1957), 345–350.

50. V. V. Korobkin et al., *Investigation of the air "spark" produced by focussed laser radiation.* III, Ž. Èksper. Teoret. Fiz. **53** (1967), 116–123 = Soviet Phys. JETP **26** (1968), 79–85.

51. E. Panarella and P. Savic, *Blast waves from a laser-induced spark in air,* Canad. J. Phys. **46** (1968), 183–192.

52. N. G. Basov et al., *Formation of a long spark in air by weakly focussed laser radiation,* Dokl. Akad. Nauk SSSR 173 (1967), 538–541 = Soviet Phys. Dokl. **12** (1967/68), 248–251.

53. N. G. Basov, *Lasers in physical investigations,* Priroda **1967**, no. 10, 17–27. (Russian)

54. J. W. Daiber and H. M. Thompson, *Laser-driven detonation waves in gases,* Phys. Fluids **10** (1967), 1162–1169.

55. *The action of laser radiation,* Collection of papers, "Mir", Moscow, 1968. (Russian)

CHAPTER 3

LINEARIZED ONE-DIMENSIONAL PROBLEMS WHICH ARE NOT SELF-SIMILAR

§ 1. Linearization of the equations of gas dynamics

In the preceding chapter we considered the exact solution of a self-similar problem of strong point explosion in a quiescent gas. It was assumed there that the counterpressure p_1 could be neglected. This assumption is valid only during the initial stage of the explosion, i.e. for small times. As the shock wave progresses, the effect of the initial pressure becomes substantial. An additional dimensional parameter p_1 therefore appears in the initial conditions of the problem; because of this the problem ceases to be self-similar, and all dimensionless characteristics of the flow will depend not on one but on two variables. A non-self-similar problem can be solved by numerical integration of the system of equations of gas dynamics. These questions will be considered in Chapter 4. It is, however, possible to set forth a simpler path [1]–[4] to solving the problem for values $(p_2 - p_1)/p_1 < 1$ which is based on linearization of the original equations about the self-similar solution. For greater generality we shall assume [5], [6] that the initial pressure p_1 is variable, with

$$\rho_1 = Ar^{-\omega}, \tag{3.1}$$

$$p_1 = Cr^{-\omega_*}. \tag{3.2}$$

We relinquish also the condition that the initial velocity of the gas v_1 be zero and assume [5], [6]

$$v_1 = Br^{\omega-\nu+1}. \tag{3.3}$$

In the sequel (Chapter 6) we shall deal with the case in which there is evolution of heat on the front of the shock wave.

The question of the realization of the initial states (3.1)–(3.3) will not be discussed here. We note only that the density ρ_1 and velocity v_1 satisfy the stationary equation of continuity. The pressures and densities in isothermic atmospheres of stars can be approximated (for $\omega = \omega_*$) by the distributions (3.1), (3.2). If the state (3.1)–(3.3) is not strictly an equilibrium state, we shall neglect the nonstationary flow which may appear in advance of the blast wave. Naturally, for $B = 0$ and $\omega_* = 0$ the initial state can be assumed to be an equilibrium state. Since for arbitrary ω and ω_* the dimensions of the constants C and B are not expressed in terms of the dimensions of E_0 and A, the point explosion problem (PEP) in a gas with initial state (3.1)–(3.3) is not self-similar; self-similarity is lost if even one of the constants C or B is nonzero. The general formulation of the problem on linearization of the equations of gas dynamics has been considered in §1 of Chapter 1. We shall consider this question in more detail for the case in which the basic motion corresponds to the exact solution (2.5)–(2.12).

We label the functions of this exact solution by the index 0. We shall seek a solution in the form

$$\begin{aligned} v &= v_0(r, t) + \varepsilon v_{(1)}(r, t), \\ p &= p_0(r, t) + \varepsilon p_{(1)}(r, t), \\ \rho &= \rho_0(r, t) + \varepsilon \rho_{(1)}(r, t), \end{aligned} \tag{3.4}$$

where ϵ is a small parameter which may be taken to be the constant C or B, and the new unknown functions (additions to the base solution or perturbations) are labeled by the subscript (1). Carrying out the standard linearization procedure as indicated in §1 of Chapter 1, we arrive at a system of linear partial differential equations with coefficients which depend on v_0, p_0 and ρ_0. In obtaining the linearized equations from a system of the form (1.31) we assume that the functions $v_{(1)}, p_{(1)}$ and $\rho_{(1)}$ and their derivatives have finite limits for $\epsilon \to 0$. If this assumption is valid, the linearization is justified. We shall therefore try to find conditions on solutions of the linearized system under which the conditions that $v_{(1)}, p_{(1)}$ and $\rho_{(1)}$ be finite are satisfied. In some problems these conditions are not satisfied, as a rule, even locally (for example, in a neighborhood of some singular points). In such cases the linearization is, strictly speaking, without meaning, but in a number of problems outside neighborhoods of certain singular points the solution of the linearized equations may provide a correct qualitative and quantitative description of the motion in question. From this point of view solutions which are unbounded at isolated points may still be useful. In the sequel we shall make use of the formalism of reducing the equations to dimensionless form and introducing small dimensionless parameters proportional to the constants C or B.

The original system of equations of gas dynamics for adiabatically perturbed motions of a perfect gas is taken in the form

$$\begin{aligned} &\frac{\partial v}{\partial t} + v\frac{\partial v}{\partial r} + \frac{1}{\rho}\frac{\partial p}{\partial r} = 0, \\ &\frac{\partial \rho}{\partial t} + v\frac{\partial \rho}{\partial r} + \rho\left(\frac{\partial v}{\partial r} + (\nu - 1)\frac{v}{r}\right) = 0, \\ &\frac{\partial p}{\partial t} + v\frac{\partial p}{\partial r} + \gamma p\left(\frac{\partial v}{\partial r} + (\nu - 1)\frac{v}{r}\right) = 0. \end{aligned} \tag{3.5}$$

We denote by $r_2(t)$ the leading front of the perturbation of the gas which we shall assume is an ordinary shock wave. Then, in addition to the initial conditions (3.1)–(3.3), we have, as before, the condition $r_2(0) = 0$ and the condition of energy release E_0 at $r = 0$. According to (1.60), the conditions on the shock wave for $r = r_2$ have the form

$$\begin{aligned}
&\rho_1 (D - v_1) = \rho_2 (D - v_2),\\
&\rho_1 v_1 (D - v_1) - p_1 = \rho_2 v_2 (D - v_2) - p_2, \qquad (3.6)\\
&\rho_1 (D - v_1)\left(\frac{v_1^2}{2} + \frac{p_1}{\rho_1}\frac{1}{\gamma - 1}\right) - p_1 v_1 = \rho_2 (D - v_2)\left(\frac{v_2^2}{2} + \frac{p_2}{\rho_2(\gamma - 1)}\right) - p_2 v_2.
\end{aligned}$$

In addition to (3.6), it is possible to give conditions at the center of symmetry of the flow. Thus, if at the center there is no constantly moving mass source,

$$v(0, t) = 0. \qquad (3.7)$$

We shall consider separately the two cases: a) $B \neq 0, C = 0$; b) $C \neq 0, B = 0$. We introduce the system of dimensionless variables

$$f = \frac{v}{D}, \quad g = \frac{\rho}{\rho_1}, \quad h = \frac{p}{\rho_1 D^2}. \qquad (3.8)$$

We made use of analogous variables in Chapter 2. Because of the absence of self-similarity the functions (3.8) will now depend on two dimensionless variables.

We introduce two new dimensionless variables:

$$y = \frac{v_1}{D}, \quad q = \frac{\gamma p_1}{\rho_1 D^2}. \qquad (3.9)$$

For the case $\omega \geqslant \nu - 1$ the parameter y is small for finite B and small t. By property 3 for shock waves (§2 of Chapter 1) we have $D - v_1 \geqslant a_1$ and $D \geqslant a_1 + v_1$, i.e. we always have $D > v_1$ or $y < 1$. The parameter q is the ratio of the square of the sound speed in the unperturbed medium to the square of the speed of the shock wave. This parameter is also small in the initial stage of development of an explosion.

Problems taking into account the velocity v_1 and pressure p_1 during propagation of the blast wave can be considered separately. In place of r and t in the case of problem a) we introduce the new independent variables

$$\lambda = \frac{r}{r_2}, \quad y \qquad (3.10)$$

and assume that $f = f(\lambda, y)$, $h = h(\lambda, y)$ and $g = g(\lambda, y)$. When considering only counterpressure (problem b)), we introduce the independent variables

$$\lambda = \frac{r}{r_2}, \quad q, \qquad (3.11)$$

where $f = (\lambda, q)$, $h = h(\lambda, q)$ and $g = g(\lambda, q)$. In order to write down the dimensionless equations for both systems of variables (3.8), (3.10) and (3.8), (3.11) at once, we introduce the notation $z_1 = y$ and $z_2 = q$.

For simplicity we assume for the time being that $\omega = \omega_*$. Using the relations

$$\frac{\partial}{\partial r} = \frac{1}{r_2}\frac{\partial}{\partial \lambda}, \qquad \frac{\partial}{\partial t} = \frac{D}{r_2}\left(r_2 \frac{dz_i}{dr_2}\frac{\partial}{\partial z_i} - \lambda\frac{\partial}{\partial \lambda}\right)$$

for the derivatives, the system of equations of gas dynamics (3.5) in the new variables appears as follows:

$$(f_i - \lambda)\frac{\partial f_i}{\partial \lambda} + \eta_i\left(-\frac{f_i}{iz_i} + \frac{\partial f_i}{\partial z_i}\right) + \theta_i f_i + \frac{1}{g_i}\frac{\partial h_i}{\partial \lambda} = 0, \tag{3.12}$$

$$(f_i - \lambda)\frac{\partial g_i}{\partial \lambda} - \omega g_i + \eta_i \frac{\partial g_i}{\partial z_i} + g_i\left(\frac{\partial f_i}{\partial \lambda} + (\nu - 1)\frac{f_i}{\lambda}\right) = 0, \tag{3.13}$$

$$(f_i - \lambda)\frac{\partial h_i}{\partial \lambda} + \varkappa_i h_i + \eta_i\left(\frac{\partial h_i}{\partial z_i} - \frac{2h_i}{iz_i}\right) + \gamma h_i\left(\frac{\partial f_i}{\partial \lambda} + (\nu - 1)\frac{f_i}{\lambda}\right) = 0, \tag{3.14}$$

where

$$\eta_i = r_2\frac{dz_i}{dr_2},\ \theta_1 = \omega - \nu + 1,\ \theta_2 = 0,\ \varkappa_1 = \omega - 2\nu + 2,\ \varkappa_2 = -\omega,\ i = 1, 2. \tag{3.15}$$

In order to study the effect of the velocity v_1 or pressure p_1, we seek a solution of the system (3.12)–(3.15) in the form

$$\begin{aligned} f_i(\lambda,\ z_i) &= f_0(\lambda) + z_i f_{i1}(\lambda) + o(z_i), \\ g_i(\lambda,\ z_i) &= g_0(\lambda) + z_i g_{i1}(\lambda) + o(z_i), \\ h_i(\lambda,\ z_i) &= h_0(\lambda) + z_i h_{i1}(\lambda) + o(z_i), \end{aligned} \tag{3.16}$$

$$\frac{dz_i}{d\ln r_2} = \frac{b_i z_i}{1 + A_{i1} z_i + o(z_i)}. \tag{3.17}$$

Here f_0, g_0 and h_0 are the known self-similar functions, $b_1 = (\nu + \omega - 4)/2$, $b_2 = \nu - \omega'$, $o(z_i)$ denotes a quantity of higher order of smallness than z_i as $z_i \to 0$, and A_{i1} are constants which remain to be determined. If we substitute the solution (3.16), (3.17) into the system (3.12)–(3.14) and neglect all quantities of order higher than z_i, we obtain the following system of linear ordinary differential equations for the functions f_{i1}, g_{i1} and h_{i1}:

$$g_0[mf'_{i1} + (f'_0 + b_{i1})f_{i1} + s_i A_{i1} f_0] + h'_{i1} + \left(m_0 f'_0 - b_2\frac{f_0}{2}\right)g_{i1} = 0, \tag{3.18}$$

$$mg'_{i1} + g_0\left(f'_{i1} + \frac{\nu - 1}{\lambda}f_{i1}\right) + g'_0 f_{i1} + \left(k_i + f'_0 + \frac{(\nu - 1)f_0}{\lambda}\right)g_{i1} = 0, \tag{3.19}$$

$$\begin{aligned} mh'_{i1} + \gamma h_0\left(f'_{i1} + (\nu - 1)\frac{f_{i1}}{\lambda}\right) + h'_0 f_{i1} + \left(m_{i1} + \gamma f'_0 + (\nu - 1)\gamma\frac{f_0}{\lambda}\right)h_{i1} + \\ + m_{i0} A_{i1} h_0 = 0, \end{aligned} \tag{3.20}$$

where

$$m = f_0 - \lambda, \quad b_{11} = \omega - \nu + 1, \quad b_{21} = \frac{b_2}{2}, \quad s_i = \frac{b_i}{i},\ k_i = b_i - \omega,$$
$$m_{11} = b_1 - \nu, \quad b_{21} = -\omega, \quad m_{i0} = \frac{2b_i}{i}, \quad f'_0 = \frac{df_0}{d\lambda}, \quad f'_{i1} = \frac{df_{i1}}{d\lambda}.$$

In considering linearized solutions we must also subject the boundary conditions of the problem to linearization.

We write the conditions on the shock wave (for $\lambda = 1 - 0$) in dimensionless form:

$$(1-y)=g_2(1-f_2), \quad y(1-y)-\frac{g}{\gamma}=g_2f_2(1-f_2)-h_2,$$
$$(1-y)\left(\frac{y^2}{2}+\frac{q}{\gamma(\gamma-1)}\right)-\frac{yq}{\gamma}=(1-f_2)\left(g_2\frac{f_2^2}{2}+\frac{h_2}{\gamma-1}\right)-\frac{f_2h_2}{\gamma}. \tag{3.21}$$

Linearizing with respect to the parameter y (for $q = 0$) or with respect to the parameter q (for $y = 0$), we obtain conditions for f_{i1}, g_{i1} and h_{i1}:

$$f_{11}(1)=\frac{\gamma-1}{\gamma+1}, \quad g_{11}(1)=0, \quad h_{11}(1)=-\frac{4}{\gamma+1}, \tag{3.22}$$

$$f_{21}(1)=-\frac{2}{\gamma+1}, \quad g_{21}=-\frac{2(\gamma+1)}{(\gamma-1)^2}, \quad h_{21}=-\frac{\gamma-1}{\gamma(\gamma+1)}. \tag{3.23}$$

The symmetry condition takes the form

$$f_{i1}(0)=0. \tag{3.24}$$

Thus the question of finding the additions to the self-similar solution reduces to solving the systems (3.18)–(3.20) with boundary conditions (3.22)–(3.24) for $i = 1$ or (3.23), (3.24) for $i = 2$.

In the general case of arbitrary γ, ω and ν the system (3.18)–(3.20) has no exact analytic solution, and numerical methods are used to study the linearized problem. Since this system contains the unknown constant A_{i1}, the problem is not a Cauchy problem, and condition (3.24) or equivalent conditions are necessary to determine A_{i1}. Hence this problem can be considered a two-point boundary value problem for the system (3.18)–(3.20). In place of condition (3.24), it is possible to use the conservation laws to determine the constant:

$$\frac{E_0}{\sigma_\nu}=\int_0^{r_2}\left(\frac{\rho v^2}{2}+\frac{p}{\gamma-1}\right)r^{\nu-1}dr-\int_0^{r_2}\left(\frac{\rho_1v_1^2}{2}+\frac{p_1}{\gamma-1}\right)r^{\nu-1}dr \tag{3.25}$$

–the integral law of energy conservation, and

$$\int_0^{r_2}(\rho-\rho_1)\,r^{\nu-1}dr=\int_0^{t}\rho_1v_1r_2^{\nu-1}dt \tag{3.26}$$

–the integral law of mass conservation.

On writing these laws in dimensionless form and linearizing them, we obtain certain integral relations containing the functions f_{i1}, g_{i1}, h_{i1} and the constant A_{i1}. If the method of determining A_{i1} is based on (3.24), the integral relations for mass and energy can be utilized for control of computational precision or for approximate methods of constructing solutions.

We shall consider in more detail numerical methods of solving the linearized problems by means of the example of an explosion problem with counterpressure and with constant density, i.e. for the case $\omega = 0$ and $v_1 = 0$ (cf. §5). In the case of variable density we shall be mainly interested in the case of spherical symmetry ($\nu = 3$). In place of the dimensionless variables (3.8) the functions

$$\bar{f} = \frac{v}{v_2}, \quad \bar{g} = \frac{\rho}{\rho_2}, \quad \bar{h} = \frac{p}{p_2} \tag{3.27}$$

are sometimes used, where v_2, ρ_2 and p_2 are the values of the corresponding functions ($v_1 = 0$) on the shock wave. We shall indicate the relation between f_0, g_0, h_0 and $\bar{f}_0, \bar{g}_0, \bar{h}_0$ and also between f_{21}, g_{21}, h_{21} and $\bar{f}_1, \bar{g}_1, \bar{h}_1$, i.e. taking into account only the counterpressure. Simple computations give

$$\bar{f}_0 = \frac{\gamma+1}{2} f_0, \quad \bar{f}_1 = \frac{\gamma+1}{2}(f_0 + 2f_{21}), \quad \bar{g}_0 = g_0 \frac{\gamma-1}{\gamma+1}, \quad \bar{g}_1 = \frac{2g_0}{\gamma+1} + \frac{\gamma-1}{\gamma+1} g_{21},$$
$$\bar{h}_0 = \frac{\gamma+1}{2} h_0, \quad \bar{h}_1 = \frac{\gamma+1}{2}\left(\frac{\gamma-1}{2\gamma} h_0 + h_{21}\right). \tag{3.28}$$

We adopt the following system of dimensionless variables in studying linear problems: for $v_1 \neq 0$ and $p_1 = 0$ we take the variables λ, y, f, g and h; for $v_1 = 0$, $p_1 = Cr^{-\omega}$ and $\rho_1 = Ar^{-\omega}$ we take the variables λ, q, f, g and h. If $v_1 = 0$, $p_1 = \text{const}$ and $\rho_1 = Ar^{-\omega}$, we use the variables $\lambda, q, \bar{g}$ and $\bar{h}$.

The corresponding systems of linear equations can be obtained from (3.18)–(3.20) using (3.28), or directly from the equations of gas dynamics. We mention certain general properties of the system (3.18)–(3.20).

1) This system (and the one equivalent to it for $\omega \neq \omega_*$) has an adiabatic integral, the existence of which was proved by M. L. Lidov [7].

2) For the singular self-similar solution (2.28) with

$$\omega = \omega_1 = \frac{3\nu - 2 + \gamma(2-\nu)}{\gamma+1}, \quad f_0 = \frac{2}{\gamma+1}\lambda, \quad g_0 = \frac{\gamma+1}{\gamma-1}\lambda^{\nu-1}, \quad h_0 = \frac{2}{\gamma+1}\lambda^{\nu} \tag{3.29}$$

the system (3.18)–(3.20) and those equivalent to it have an exact analytic solution.

3) To find the relation between the coordinate of the shock wave and time the expression for the η_i and the formula $dr_2/dt = D$ are used.

We proceed to a more detailed consideration of concrete special cases.

§2. An explosion with variable initial density and constant counterpressure

2.1. *The system of linearized equations.* From the system of defining parameters $\gamma, \omega, A, E_0, p_1, r, t$ of this problem it follows that the desired dimensionless functions

$f, \bar{g}$, and $\bar{h}$ will depend on two dimensionless parameters $\lambda = r/r_2$ and $q = a_1^2/D^2$ and the constants γ and ω. We shall henceforth omit the bar in the notation for the dimensionless density and pressure.

We shall consider in detail the behavior of the solution when p_1 is assumed to be nonzero but small as compared with p_2 [4], [8]. It is required to find a solution of the system (3.5) with boundary conditions (3.6) for $v_1 = 0$. Going over to the dimensionless variables above, the system (3.5) can be written in the form [4]

$$\begin{gathered}
(f-\lambda)\frac{\partial f}{\partial \lambda} + \frac{(\gamma-1+2q)\,[2\gamma-(\gamma-1)\,q]}{\gamma\,(\gamma+1)^2}\,\frac{1}{g}\,\frac{\partial h}{\partial \lambda} + \left(\frac{\partial f}{\partial q} - \frac{f}{2q}\right)\eta + \frac{\omega f}{2} = 0, \\
(f-\lambda)\frac{\partial \ln g}{\partial \lambda} + \frac{\partial f}{\partial \lambda} + \frac{\nu-1}{\lambda} f + \left(\frac{\partial \ln q}{\partial q} - \frac{2}{\gamma-1+2q}\right)\eta - \omega = 0, \\
(f-\lambda)\frac{\partial \ln h}{\partial \lambda} + \gamma\left(\frac{\partial f}{\partial \lambda} + \frac{\nu-1}{\lambda} f\right) + \left(\frac{\partial \ln h}{\partial q} - \frac{2\gamma}{[2\gamma-(\gamma-1)\,q]\,q}\right)\eta = 0, \\
\eta = r_2\frac{dq}{dr_2}.
\end{gathered} \tag{3.30}$$

We define the dimensionless radius R_2 of the shock wave by the formulas

$$R_2 = \frac{r_2}{r^0}, \qquad r^0 = \left(\frac{E_0}{p_1}\right)^{1/\nu},$$

where r^0 is a dynamic characteristic length.

In order to obtain the complete solution of the problem in these variables, it is necessary to determine the dependencies $f(\lambda, q)$, $g(\lambda, q)$, $h(\lambda, q)$ and also $R_2(q)$. For this it is necessary to find a solution of the system (3.30) in the λ, q plane inside the square $0 \leqslant \lambda \leqslant 1$, $0 \leqslant q \leqslant 1$ which satisfies the boundary conditions on the jump and at the center of symmetry

$$f(1, q) = \frac{2}{\gamma+1}(1-q), \qquad g(1, q) = h(1, q) = 1, \qquad f(0, q) = 0 \tag{3.31}$$

and the initial conditions

$$f(\lambda, 0) = f_0(\lambda), \qquad g(\lambda, 0) = g_0(\lambda), \qquad h(\lambda, 0) = h(\lambda), \tag{3.32}$$

where f_0, g_0 and h_0 are known functions corresponding to the self-similar problem. They satisfy the system

$$\begin{gathered}
mf_0' + \frac{2(\gamma-1)}{(\gamma+1)^2}\frac{h_0'}{g_0} + \frac{\omega-\nu}{2} f_0 = 0, \\
m\frac{g_0'}{g_0} + f_0' + \frac{\nu-1}{\lambda} f_0 - \omega = 0, \\
m\frac{h_0'}{h_0} + \gamma\left(f_0' + \frac{\nu-1}{\lambda} f_0\right) - \nu = 0,
\end{gathered} \tag{3.33}$$

and

$$f_0(1) = \frac{2}{\gamma+1}, \qquad g_0(1) = h_0(1) = 1, \qquad f_0(0) = 0. \tag{3.34}$$

Here and everywhere below the primes denote derivatives with respect to λ. We remark that the system (3.33) can be obtained from (3.30) by passing to the limit as $q \to 0$. For small q, i.e. for times when the explosion is still sufficiently strong, the solution of the problem posed can be sought in the form (3.16), (3.17), assuming that $b_2 = \nu$.

Since the linearized problem is to be solved, from (3.30), using the system (3.33) and neglecting terms of order higher than q, we obtain a system of linear equations (the analogue of the system (3.18)–(3.20) for determining the functions $f_1(\lambda), g_1(\lambda)$, $h_1(\lambda)$ and the constant A_1

$$mg_0 f_1' + \frac{2(\gamma-1)}{(\gamma+1)^2} h_1' + \left(f_0' + \frac{\omega+\nu}{2}\right) g_0 f_1 + \left[mf_0' + \frac{\omega-\nu}{2} f_0\right] g_1 + \frac{4\gamma-(\gamma-1)^2}{\gamma(\gamma+1)^2} h_0' + \frac{\nu}{2} f_0 g_0 A_1 = 0,$$

$$g_0 f_1' + mg_1' + \left(\frac{\nu-1}{\lambda} g_0 + g_0'\right) f_1 + \left(\frac{\nu-1}{\lambda} f_0 + f_0' + \nu - \omega\right) g_1 - \frac{2\nu}{\gamma-1} g_0 = 0 \quad (m = f_0 - \lambda), \tag{3.35}$$

$$\gamma h_0 f_1' + mh_1' + \left(h_0' + \gamma\frac{\nu-1}{\lambda} h_0\right) f_1 + \gamma\left(f_0' + \frac{\nu-1}{\lambda} f_0\right) h_1 - \nu\left(\frac{\gamma-1}{2\gamma} - A_1\right) h_0 = 0.$$

Noting (3.34), from conditions (3.31) we obtain the boundary conditions for the desired functions $f_1(\lambda), g_1(\lambda)$ and $h_1(\lambda)$:

$$f_1(1) = -\frac{2}{\gamma+1}, \quad g_1(1) = h_1(1) = f_1(0) = 0. \tag{3.36}$$

The system (3.35) can be transformed to a form more convenient for subsequent investigations. To this end we introduce new unknown functions $F(\lambda), G(\lambda)$ and $H(\lambda)$ related to the functions $f_1(\lambda), g_1(\lambda)$ and $h_1(\lambda)$ by

$$f_1(\lambda) = mF, \quad g_1(\lambda) = g_0 G, \quad h_1(\lambda) = h_0 H. \tag{3.37}$$

After transformations the system (3.35) can be written

$$m^2 F' + \frac{2(\gamma-1)}{(\gamma+1)^2}\frac{h_0'}{g_0} H' + m\left(2f_0' + \frac{\omega+\nu-2}{2}\right) F + \frac{2(\gamma-1)}{(\gamma+1)^2}\frac{h_0'}{g_0} H + \left[mf_0' + \frac{\omega-\nu}{2} f_0\right] G + \frac{4\gamma-(\gamma-1)^2}{\gamma(\gamma+1)^2}\frac{h_0'}{g_0} + \frac{\nu}{2} f_0 A_1 = 0, \tag{3.38}$$

$$mF' + mG' + (f_0' - 1) F + \left(\frac{\nu-1}{\lambda} + \frac{g_0'}{g_0}\right) mF + \left(\frac{\nu-1}{\lambda} f_0 + f_0' + \nu - \omega\right) G + m\frac{g_0'}{g_0} G - \frac{2\nu}{\gamma-1} = 0, \tag{3.39}$$

$$\gamma mF' + mH' + (f_0' - 1)\gamma F + m\left(\frac{h_0'}{h_0} + \gamma\frac{\nu-1}{\lambda}\right) F + m\frac{h_0'}{h_0} H + \gamma\left(f_0' + \frac{\nu-1}{\lambda} f_0\right) H - \nu\left(\frac{\gamma-1}{2\gamma} - A_1\right) = 0. \tag{3.40}$$

From the last two equations of this system it is possible to obtain a first integral. We shall demonstrate this.

2.2. *An integral of the system* (3.35) *and the law of motion of the shock wave.* If we eliminate the quantities g_0'/g_0 and h_0'/h_0 from (3.39) and (3.40), we have

$$m(F' + G') + (\omega - \nu)F + \nu G - \frac{2\nu}{\gamma - 1} = 0,$$

$$m(\gamma F' + H') - \nu(\gamma - 1)F + \nu H - \nu\left(\frac{\gamma - 1}{2\gamma} - A_1\right) = 0.$$

Changing the independent variable according to the formula $\varphi = \int_\lambda^1 m^{-1} d\lambda$, we obtain

$$\frac{d}{d\varphi}(F + G) + (\omega - \nu)F + \nu G - \frac{2\nu}{\gamma - 1} = 0,$$

$$\frac{d}{d\varphi}(\gamma F + H) - \nu(\gamma - 1)F + \nu H - \nu\left(\frac{\gamma - 1}{2\gamma} - A_1\right) = 0.$$

Multiplying the first equation by $\nu(2\gamma - 1)/(\omega - 2\nu)$, adding to this the second equation, and integrating the result, we find

$$\left[\frac{\nu(2\gamma - 1)}{\omega - 2\nu} + \gamma\right]F + \frac{\nu(2\gamma - 1)}{\omega - 2\nu}G + H = C_1 \exp(-\nu\varphi) + \frac{2\nu}{\gamma - 1}\frac{2\gamma - 1}{\omega - 2\nu} + \frac{\gamma - 1}{2\gamma} - A_1.$$

From the boundary conditions $F(1) = 2/(\gamma - 1)$, $G(1) = 0$, $H(1) = 0$ and $\varphi(1) = 0$ we find the constant of integration:

$$C_1 = \frac{(3\gamma - 1)(\gamma + 1)}{2\gamma(\gamma - 1)} + A_1.$$

If we use the adiabatic equation for self-similar motions

$$m\left(\frac{h_0'}{h_0} - \gamma\frac{g_0'}{g_0}\right) + \gamma\omega - \nu = 0$$

and note that $\varphi = 0$, $g_0 = 1$ and $h_0 = 1$ for $\lambda = 1$, we obtain

$$e^{-\varphi} = \left(\frac{h_0}{g_0^\gamma}\right)^{1/(\gamma\omega - \nu)}.$$

Thus a first integral of the system (3.35) has been found which satisfies the boundary conditions on the shock wave,

$$\left[\frac{\nu(2\gamma - 1)}{\omega - 2\nu} + \gamma\right]F + \frac{\nu(2\gamma - 1)}{\omega - 2\nu}G + H$$
$$= \left[\frac{(3\gamma - 1)(\gamma + 1)}{2\gamma(\gamma - 1)}A_1\right]\left(\frac{h_0}{g_0^\gamma}\right)^{\nu/(\gamma\omega - \nu)} + \frac{2\nu}{\gamma - 1}\frac{2\gamma - 1}{\omega - 2\nu} + \frac{\gamma - 1}{2\gamma} - A_1. \quad (3.41)$$

Using the integral (3.41), the problem reduces to solving a system of two linear equations, which for arbitrary values of ω and γ can be accomplished by numerical integration.

After the quantity A_1 has been found, it is possible to find $R_2(q)$ and $\tau(q)$; here $R_2 = r_2/r^0$ and $\tau = t/t^0$, where r^0 is the dynamical length introduced earlier and t^0 is the dynamical time, defined by

$$t^0 = E_0^{1/\nu} A^{1/2} r^{0-\omega/2} p_1^{-(\nu+2)/2\nu}.$$

For finding the self-similar solution we have the relations

$$r_2(t) = \left(\frac{E_0}{\alpha A}\right)^{\delta/2} t^\delta, \qquad D^2 = \delta^2 r_2^2 t^{-2}, \qquad \delta = \frac{2}{\nu + 2 - \omega}, \tag{3.42}$$

where $\alpha(\gamma, \omega)$ is a known quantity (cf. Chapter 2).

If (3.42) is transformed to the dimensionless variables above, it is possible to find the relations $R_2(q)$ and $\tau(q)$ in the self-similar problem

$$R_2^\nu(q) = \frac{\delta 2}{\gamma\alpha} q, \qquad \tau(q) = \left(\frac{\delta^2}{\alpha\gamma}\right)^{(2-\nu\delta)/2\nu\delta} (\nu\gamma^{1/2})^{-1} q^{1/\nu\delta}. \tag{3.43}$$

From (3.17) by integrating and using (3.43) we obtain for the linearized problem

$$R_2^\nu(q) = \frac{\delta 2}{\alpha\gamma} q \exp(A_1 q). \tag{3.44}$$

We shall now find $\tau(q)$. Using the definitions of q, τ and R_2, it is easy to show that

$$\frac{dR_2}{d\tau} = \left(\frac{\gamma R_2^\omega}{q}\right)^{1/2}.$$

Since $d\tau/dq = (dR_2/dq)(d\tau/dR_2)$, using (3.44) we find that

$$\frac{d\tau}{dq} = \left(\frac{\delta 2}{\gamma\alpha}\right)^{(2-\omega)/2\nu} (\nu\gamma^{1/2})^{-1} (1 + A_1 q) q^{(2-\nu-\omega)/2\nu} \exp\left(\frac{2-\omega}{2\nu} A_1 q\right).$$

For small values of q we may write

$$(1 + A_1 q) \exp\left(\frac{2-\omega}{2\nu} A_1 q\right) = 1 + \frac{2\nu + 2 - \omega}{2\nu} A_1 q + 0(q).$$

Thus to determine $\tau(q)$ we obtain a differential equation

$$\frac{d\tau}{dq} = \left(\frac{\delta 2}{\gamma\alpha}\right)^{(2-\omega)/2\nu} (\gamma^{1/2\nu})^{-1} \left[1 + \frac{2\nu + 2 - \omega}{2\nu} A_1 q\right] q^{(2-\nu-\omega)/2\nu}.$$

Integrating this equation and determining the constant of integration from the condition $\tau(0) = 0$, we find

$$\tau(q) = \left(\frac{\delta 2}{\gamma\alpha}\right)^{(2-\nu\delta)/2\nu\delta} (\nu\gamma^{1/2})^{-1} q^{1/\nu\delta} \left[\nu\delta + \frac{\nu\delta + 2}{2(\nu\delta + 1)} A_1 q\right]. \tag{3.45}$$

The relations (3.44) and (3.45) give the law of motion of the shock wave in parametric form, i.e. the function $R_2(\tau)$. Using (3.44), (3.45) and the conditions on the shock wave, we can determine the dependence of all the characteristics of the front of the shock wave on its radius and time.

2.3. *Exact solution of the problem for* $\omega = \omega_1$. It was stated above that the solution of the linearized problem of explosion in a medium with variable density can be obtained by numerical integration. However, in the case where $\omega = \omega_1$ the solution of this problem can be given by closed formulas. This is explained by the fact that for this value the self-similar solution has the simple form (cf. the solution (2.28))

$$f_0 = \frac{2}{\gamma+1}\lambda, \qquad g_0 = \lambda^{\nu-2}, \qquad h_0(\lambda) = \lambda^{\nu}.$$

Substituting $f_0(\lambda)$, $g_0(\lambda)$ and $h_0(\lambda)$ from this solution into the coefficients of (3.38)–(3.40), we obtain a system of three ordinary nonhomogeneous equations with coefficients depending on the parameters γ and λ:

$$\begin{gathered}\lambda F' + \frac{2\gamma}{\gamma-1}H' - \frac{2(\nu+1)}{\gamma-1}F - \frac{2\nu}{\gamma-1}G + \frac{2\nu}{\gamma-1}H \\ + \frac{\nu}{(\gamma-1)^2}\left[\frac{4\gamma-(\gamma-1)^2}{\gamma} + A_1(\gamma+1)\right] = 0, \\ \lambda F' + \lambda G' + 2(\nu-1)F + \frac{2\nu(\gamma+1)}{(\gamma-1)^2} - \frac{\nu(\gamma+1)}{\gamma-1}G = 0, \\ \lambda\gamma F' + \lambda H' + \nu(\gamma+1)F - \frac{\nu(\gamma+1)}{\gamma-1}H + \frac{\gamma+1}{\gamma-1}\nu\left(\frac{\gamma-1}{2\gamma} - A_1\right) = 0.\end{gathered} \tag{3.46}$$

This system is a system of equations with constant coefficients if $l = \ln\lambda$ is taken as the independent variable.

In solving the system (3.46), it would be possible to use the integral (3.4) for $\omega = \omega_1$. However, since the complete system (3.46) is easily integrated, this integral need not be used. It may nevertheless be useful for calculating the dependence of the desired functions on λ as well as for control of computations.

We shall find the general solution of the homogeneous system of equations (3.46). The characteristic equation of this system can be written

$$\begin{vmatrix} n - \frac{2(\nu+1)}{\gamma-1} & -\frac{2\nu}{\gamma-1} & \frac{2}{\gamma-1}(n+\nu) \\ n + 2(\nu-1) & n - \frac{(\gamma+1)\nu}{\gamma-1} & 0 \\ \gamma n + \nu(\gamma+1) & 0 & n - \frac{\nu(\gamma+1)}{\gamma-1} \end{vmatrix} = 0. \tag{3.47}$$

To each root n_i of (3.47) there corresponds a solution $\exp(n_i l) = \lambda^{n_i}$ of the homogeneous system. The first root of (3.47) is

$$n_1 = \nu(\gamma+1)/(\gamma-1). \tag{3.48}$$

The second and third roots of (3.47) satisfy the quadratic equation

$$n^2+\frac{5\nu\gamma+3\nu+2}{\gamma+1}\,n-\frac{2\nu(\gamma\nu+1)(3-\gamma)}{\gamma^2-1}=0 \tag{3.49}$$

and are given by

$$n_2=-\frac{\tilde{b}_1}{2}+\sqrt{\frac{b_1^2}{4}+b_2},\qquad n_3=-\frac{\tilde{b}_1}{2}-\sqrt{\frac{b_1^2}{4}+b_2},$$

$$\tilde{b}_1=\frac{5\nu\gamma+3\nu+2}{\gamma+1},\qquad \tilde{b}_2=\frac{2\nu(\nu\gamma+1)(3-\gamma)}{\gamma^2-1}. \tag{3.50}$$

It is evident from (3.50) that $n_2=0$ for $\gamma=3$. The dependence of the roots n_1, n_2 and n_3 on ν and γ is given in Table 1.

TABLE 1

Values of the constants for various γ

γ	ω	n_1	n_2	$-n_3$	$-C_1$	C_2	A_1
1.2	2.6364	33	5.9131	19.095	11.923	2.9058	1.9778
1.4	2.3333	18	3.1540	16.487	6.5665	1.7540	1.8755
5/3	2.0000	12	1.7684	15.268	4.2717	1.1789	1.8211
3.0	1.0000	6	0.0000	14.000	1.6667	0.4667	1.7778
7.0	0.0000	4	−0.8031	13.697	0.6098	1.1699	1.7979

The general solution of the homogeneous system corresponding to the system (3.46) can be written in the form

$$F(\lambda)=C_2\lambda^{n_2}+C_3\lambda^{n_3},$$

$$G(\lambda)=C_1\lambda^{n_1}+\frac{(n_2+2\nu-2)(\gamma-1)}{(\gamma+1)\nu-(\gamma-1)n_2}C_2\lambda^{n_2}+\frac{(n_3+2\nu-2)(\gamma-1)}{(\gamma+1)\nu-(\gamma-1)n_3}C_3\lambda^{n_3}, \tag{3.51}$$

$$H(\lambda)=\frac{\gamma-1}{2\gamma}C_1\lambda^{n_1}+\frac{(\gamma n_2+\nu\gamma+\nu)(\gamma-1)}{(\gamma+1)\nu-(\gamma-1)n_2}C_2\lambda^{n_2}+\frac{(\gamma n_3+\nu\gamma+\nu)(\gamma-1)}{(\gamma+1)\nu-(\gamma-1)n_3}C_3\lambda^{n_3}.$$

A particular solution of the inhomogeneous system of equations (3.46) is the following:

$$F=\alpha_1,\quad G=\alpha_2,\quad H=\alpha_3, \tag{3.52}$$

where α_1, α_2 and α_3 are expressed in terms of A_1, ν and γ by the formulas

$$\alpha_1=\frac{\nu(\gamma+1)A_1}{2(\gamma-1)(\nu\gamma+1)},\qquad \alpha_2=\frac{(\nu-1)A_1}{\nu\gamma+1}+\frac{2}{\gamma-1},$$

$$\alpha_3=\frac{1}{2}\left[\frac{\gamma-1}{\gamma}-\frac{\nu\gamma-\nu+2}{\nu\gamma+1}A_1\right]. \tag{3.53}$$

Using (3.51) and (3.52), we obtain a solution of the system (3.35) for $\omega = \omega_1$ in the following form:

$$f_1(\lambda) = \frac{1-\gamma}{\gamma+1}\lambda(\alpha_1 + C_2\lambda^{n_2} + C_3\lambda^{n_3}),$$

$$g_1(\lambda) = \lambda^{\nu-2}\Big[\alpha_2 + C_1\lambda^{n_1} + \frac{(n_2+2\nu-2)(\gamma-1)}{\nu(\gamma+1)-(\gamma-1)n_2}C_2\lambda^{n_2} + \frac{(n_3+2\nu-2)(\gamma-1)}{\nu(\gamma+1)-(\gamma-1)n_3}C_3\lambda^{n_3}\Big], \tag{3.54}$$

$$h_1(\lambda) = \lambda^{\nu}\Big[\alpha_3 + \frac{\gamma-1}{2\gamma}C_1\lambda^{n_1} + \frac{(\gamma n_2+\nu\gamma+\nu)(\gamma-1)}{\nu(\gamma+1)-(\gamma-1)n_2}C_2\lambda^{n_2} + \frac{(\gamma n_3+\nu\gamma+\nu)(\gamma-1)}{\nu(\gamma+1)-(\gamma-1)n_3}C_3\lambda^{n_3}\Big].$$

We determine the constants C_1, C_2, C_3 and A_1 so that the functions $f_1(\lambda)$, $g_1(\lambda)$ and $h_1(\lambda)$ satisfy the boundary conditions (3.36). Thus from the last condition of (3.36) (the velocity at the center is equal to zero) we find $C_3 = 0$, because n_3 for any γ is a negative quantity and is greater than one in modulus; from the other conditions of (3.36) we obtain a system of inhomogeneous linear equations with coefficients depending on γ and ν:

$$\alpha_1 + C_2 + \frac{2}{\gamma-1} = 0, \tag{3.55}$$

$$\alpha_2 + C_1 + \frac{(n_2+2\nu-2)(\gamma-1)}{\nu(\gamma+1)-(\gamma-1)n_2}C_2 = 0, \tag{3.56}$$

$$\alpha_3 + \frac{\gamma-1}{2\gamma}C_1 + \frac{(\gamma n_2+\nu\gamma+\nu)(\gamma-1)}{\nu(\gamma+1)-(\gamma-1)n_2}C_2 = 0. \tag{3.57}$$

If in (3.57) we substitute C_1 from (3.56) and then use (3.55) and (3.53) we obtain a relation determining A_1. The quantity A_1 will depend on γ and ν. Having found A_1 from the system of equations (3.53), (3.55), (3.56), we find the dependence of the quantities $\alpha_1, \alpha_2, \alpha_3, C_1$ and C_2 on γ and ν.

The results of calculations for $\nu = 3$ are given in Table 1.

Using the constants calculated and formulas (3.54) (with $C_3 = 0$), the functions $f_1(\lambda), g_1(\lambda)$ and $h_1(\lambda)$ were constructed (Figure 16, a, b, c respectively) for various values of γ. By means of the functions $f_1(\lambda), g_1(\lambda)$ and $h_1(\lambda)$, knowing the constant A_1, it is possible to compute the characteristics of the motion for small q.

The dependence of $R_2(q)$ and $\tau(q)$ is found from (3.44) and (3.45) in which it is necessary to set $\delta = \delta_1$, where

$$\delta_1 = \delta(\omega_1) = \frac{\gamma+1}{3\gamma-1}, \qquad \omega_1 = \frac{7-\gamma}{\gamma+1}.$$

Figure 17 shows the distribution of the dimensionless pressure $h = h_0 + qh_1$ in

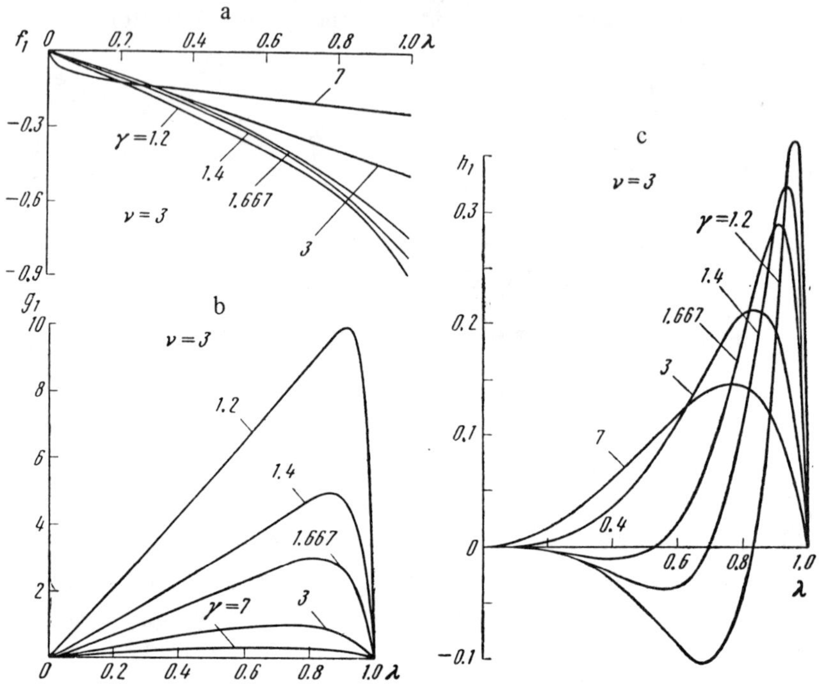

FIGURE 16. Graphs of the functions $f_1(\lambda)$ (a), $g_1(\lambda)$ (b), and $h_1(\lambda)$ (c) characterizing the linear perturbations

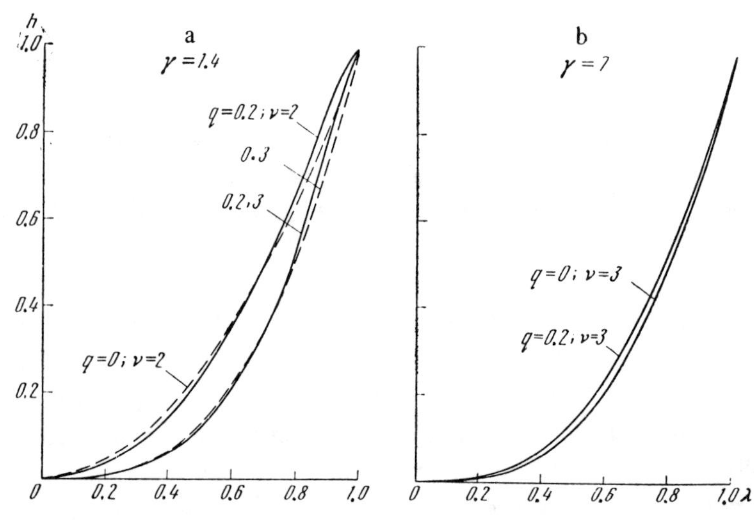

FIGURE 17. Graphs of the dimensionless pressure for $\gamma = 1.4$ and $\gamma = 7$.

the spherical case for $\gamma = 1.4$ and $\gamma = 7$ and the values $q = 0$ and $q = 0.2$. These graphs illustrate the effect of the counterpressure on the development of the explosion in the initial stage.

The analysis conducted affords the conclusion that for $\omega = \omega_1$ and $\gamma \leqslant 3$ the functions f_1, g_1, h_1 and their derivative with respect to λ are bounded everywhere on the segment $[0, 1]$. We thus conclude that a bounded solution of the linearized problem exists and is unique.

Since the functions f_1, g_1 and h_1 may be interpreted as the values of the derivatives with respect to q for $q = 0$, we can conclude that in the solution of the PEP for a medium with variable density in the case $\omega = \omega_1, \gamma \leqslant 7$ the functions v/D, p/p_2 and ρ/ρ_2 have bounded derivatives with respect to q. Analogous conclusions can be drawn also for the case of cylindrical symmetry considered by E. V. Rjazanov [4], [8] (for $\nu = 2$ the graphs of $h(\lambda, q)$ are given in Figure 17a for comparison).

It follows from Table 1 that the differential properties of the solution become worse for $\gamma > 3$, since the exponent n_2 becomes negative.

§3. Accounting for counterpressure in the case of an isothermic atmosphere of variable density

As in §1, we take for dimensionless variables the quantities λ, q, f, g and h, and assume that $\nu = 3$. We seek a solution in the form (3.16), (3.18) for $i = 2$ (problem b), §1). We shall investigate this problem in analogy with the preceding case [5], [6].

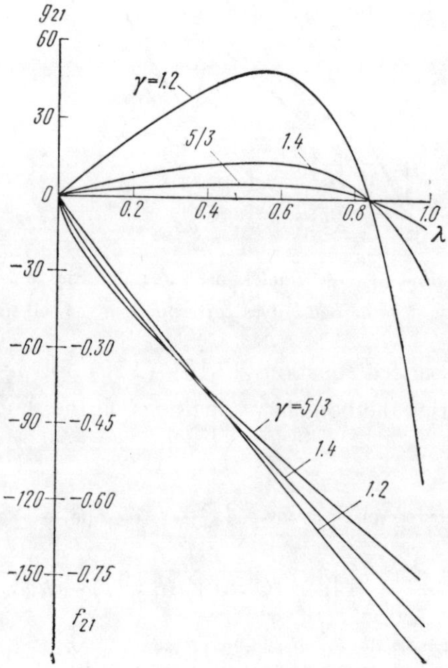

FIGURE 18. Graphs of the functions $g_{21}(\lambda), f_{21}(\lambda)$ for various γ

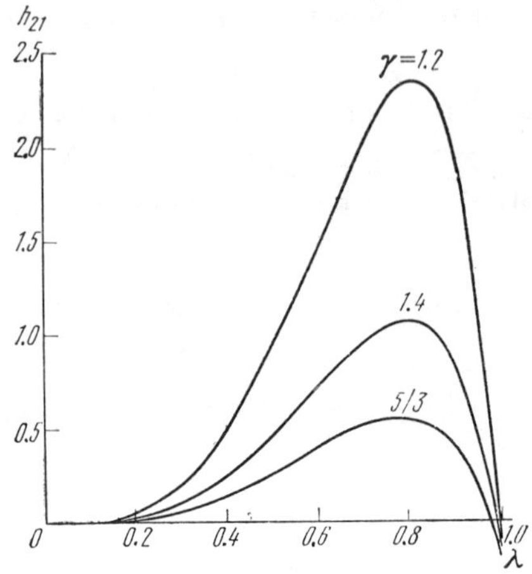

FIGURE 19. Graphs of the functions $h_{21}(\lambda)$ for various γ

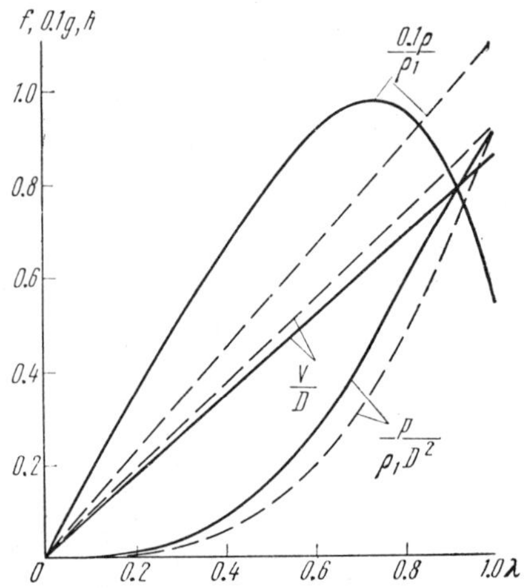

FIGURE 20. Distributions of dimensionless pressures, densities and velocities ($q = 0.05$, $\gamma = 1.2$). The dashed curves correspond to the self-simolar solution.

The system of linearized equations (3.18)–(3.20) for $i = 2$ has an adiabatic integral which, considering the boundary conditions on the shock wave (3.23), can be written in the form

$$H_{(2)} + kG_{(2)} - (\gamma + k) F_{(2)} = -A_{21} + \tilde{C} \exp(\omega - \nu) \varphi, \qquad \varphi = \int_1^\lambda \frac{d\lambda}{m},$$

$$k = -\left[\frac{\gamma}{2} + \frac{\nu}{2} \frac{\gamma - 1}{\nu - \omega}\right], \qquad \tilde{C} = A_{21} H_{(2)}(1) - kG_{(2)}(1) - (\gamma + k) F_{(2)}(1),$$

$$(\lambda - f_0) F_{(2)} = f_2(\lambda), \qquad g_0 G_{(2)} = g_{21}(\lambda), \qquad h_0 H_{(2)} = h_{21}(\lambda). \tag{3.58}$$

Here the function φ can be expressed explicitly in terms of g_0, h_0 and λ as in the analogous integral considered in §2.

The boundary value problem for the system (3.18)–(3.20) with $i = 2$ can be solved numerically. For $\omega = \omega_1$ it is possible to obtain an analytic solution. This solution, which satisfies the condition $f_{21}(0) = 0$, will have the form

$$\begin{aligned} f_{21} &= \frac{\gamma-1}{\gamma+1}\lambda\left[F_0 + C_2\lambda^{-n_2[(\gamma+1)/(\gamma-1)]}\right],\\ g_{21} &= \frac{\gamma+1}{\gamma-1}\lambda\left[C_1\lambda^4 + \frac{n_2-3+\omega_1}{n_2+3-\omega_1}C_2\lambda^{-n_2[(\gamma+1)/(\gamma-1)]} + G_0\right],\\ h_{21} &= \frac{2}{\gamma+1}\lambda^3\left[\frac{3}{7}C_1\lambda^4 + \frac{\gamma n_2 - 3(\gamma-1)}{n_2+3-\omega_1}C_2\lambda^{-n_2[(\gamma+1)/(\gamma-1)]} + H_0\right], \end{aligned} \tag{3.59}$$

where F_0, G_0 and H_0 are given by

$$\begin{aligned} G_0 &= -F_0, \quad H_0 = \frac{1}{3-\omega_1}\left[(3-\omega_1)A_{21} + 3(\gamma-1)F_0\right],\\ F_0 &= A_{21}\left(\frac{6}{\gamma+1} - \frac{3-\omega_1}{\gamma-1}\right)\left[1 - \frac{6}{\gamma+1}\left(1 - \frac{3(\gamma-1)}{3-\omega_1}\right)\right]^{-1}, \end{aligned} \tag{3.60}$$

and n_2 is the smallest root of the quadratic equation

$$n_2^2 - \frac{\gamma-1}{(\gamma+1)^2}(17\gamma-3)\,n - \left(\frac{\gamma-1}{\gamma+1}\right)^2\left(\frac{10-14\gamma}{\gamma+1}\right) = 0.$$

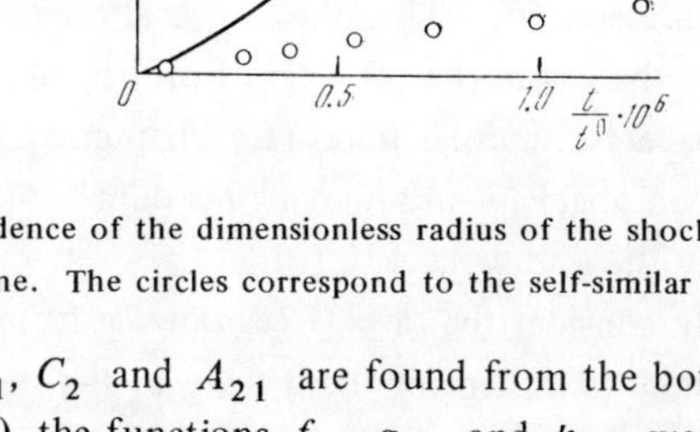

FIGURE 21. Dependence of the dimensionless radius of the shock wave on dimensionless time. The circles correspond to the self-similar solution.

The constants C_1, C_2 and A_{21} are found from the boundary conditions (3.23). For the solution (3.59) the functions f_{21}, g_{21} and h_{21} were calculated for various γ and $\omega = \omega_1$. The values of the constants A_{21}, C_1, C_2, F_0, H_0 and n_2 are given in Table 2. Figures 18 and 19 show the graphs of the functions f_{21}, g_{21} and h_{21} for various γ, and Figure 20 shows the graphs of f, g and h for $q = 0.05$ with $\gamma = 1.2$.

If we introduce the dimensionless radius $R = r_2/r^0$, $r^0 = (E_0/c)^{1/(3-\omega)}$, and

TABLE 2
Values of the constants in (3.59)

γ	A_{21}	C_1	C_2	F_0	H_0	n_2
1.2	7.8750	−19.863	−0.73533	−9.2647	7.412	0.037482
1.4	4.5411	− 9.7302	−1.2158	−3.7842	2.271	0.083213
5/3	3.2711	− 5.7455	−1.0374	−1.9626	0.654	0.13981

dimensionless time $\tau_0 = t/t_0$, $t_0 = r^0 (A/C)^{1/2}$, then the following relations can be used to find the law of motion of the shock wave $R(\tau_0)$:

$$R^{3-\omega} = \frac{\delta^2}{\alpha\gamma} q e^{A_{21}q}, \qquad \delta = \frac{2}{5-\omega}, \qquad z = \frac{1}{3-\omega} - \frac{1}{2},$$

$$\tau_0 = \gamma^{-1/2}\left(\frac{\delta^2}{\alpha\gamma}\right)^{1/(3-\omega)} \frac{q^{z+1}}{(z+1)(3-\omega)}\left[1 + \frac{z+1}{z+2}\left(1 + \frac{1}{3-\omega}\right)A_{21}q\right]. \qquad (3.61)$$

We note that for the self-similar explosion we have

$$R = \alpha^{-1/(5-\omega)}\tau_0^{\delta}, \qquad R^{3-\omega} = \frac{\delta^2}{\alpha\gamma} q.$$

The function $R(\tau_0)$ given in (3.61) is shown in Figure 21 for $\gamma = 1.2$ and $\omega = \omega_1$. The differential properties of the solution become worse in a neighborhood of $\lambda = 0$ as compared with the case $p_1 = \text{const}$. The functions $f_{21}(\lambda)$ and $g_{21}(\lambda)$, which are bounded on the segment $[0, 1]$, have unbounded derivatives at the point $\lambda = 0$.

§4. Strong explosion in a moving gas

In a number of cases it may be of interest to investigate the question of an explosion in a moving perfect gas as it flows from a vessel with high pressure p into surrounding space with low pressure p_1. This question is also encountered when considering the initial motion of the gas in the problem of propagation of disturbances through interplanetary space during solar flares (see Chapter 8). Formulations of one-dimensional point explosion problems in a moving medium taking into account the initial velocity were given by the author in [5], [6].

Here we shall briefly consider the case of gas flowing from a vessel. If we use the hydraulic approximation [9], then for the density of the gas flowing out through a conical ($\nu = 3$) or wedge-shaped ($\nu = 2$) duct we have $\rho_1 = M/v_1 r^{\nu-1}$, where M denotes the gas output. When p_1 is small, at a sufficiently great distance from the site of effluence the velocity v_1 is close to maximal, and we have the case of an initial state in which the density ρ_1 is variable (for the duct shape under consideration $\omega = \nu - 1$) and the initial velocity is nonzero. For the case of an explosion in the narrow end of this duct we arrive at the problem formulated above in a medium with variable density where $v_1 = \text{const}$, $p_1 = \text{const}$ and $\omega = \nu - 1$. We shall now consider the solution of this problem in the linearized form with v_1 and ρ_1 given by (3.1) and (3.3) and the initial pressure equal to zero ($p_1 = 0$).

As we noted in §1, by linearizing the system (3.12)–(3.15) we arrive at the system of linear ordinary differential equations (3.18)–(3.20) with $i = 1$. The boundary conditions on the shock wave for this system have the form (3.22). At the center of the explosion it is possible either to require that the velocity be zero for any ω or to postulate the condition of reestablishment of a source with velocity proportional or equal to v_1. Of course, the motion may be such that this latter condition starts to be satisfied only for sufficiently large t.

Suppose that we require that the condition $v(0, t) = 0$ be satisfied. From the formula for v_1 it follows that $v_1(0) = 0$ if $\omega > 2$. For $\omega = 2$ we have the case of a constant initial velocity v_1 = const; the case $\omega < 2$ corresponds to an infinite velocity v_1 at the center and will not be considered here. We remark also that the initial kinetic energy of the gas inside some volume of radius r

$$E_{1k} = 4\pi \int_0^r \frac{\rho_1 v_1^2}{2} r^2 dr = 4\pi A B^2 \int_0^r r^{\omega-2} dr \tag{3.62}$$

will be unbounded if $\omega < 1$. All this indicates the appearance of new singularities in the solution which may lead to its differential properties becoming worse for decreasing ω. Furthermore, for finiteness of the mass we must require that $\omega < 3$. Thus the most interesting range of ω lies between the limits

$$2 \leqslant \omega < 3. \tag{3.63}$$

We shall suppose condition (3.63) is satisfied. As in the preceding cases, the system of linearized equations (3.18)–(3.20) has an adiabatic integral (i.e. an integral of the latter two equations of this system). The law of mass conservation (3.26) can be used for integral control of computations. Performing the linearization, from (3.26) we find

$$\int_0^1 g_{11} \lambda^2 d\lambda = -\frac{2}{5-\omega}. \tag{3.64}$$

In the particular case $\omega = \omega_1$ for the system (3.18)–(3.20) with $i = 1$ it is possible to find an exact solution. Following the course set in §2, we write this solution as follows:

$$\begin{gathered} f_{11} = \frac{\gamma-1}{\gamma+1} \lambda (C_2 \lambda^{n_2} + F_0), \quad n_1 = \frac{3-\gamma}{\gamma-1}, \\ g_{11} = \frac{\gamma+1}{\gamma-1} \lambda \left[C_1 \lambda^{n_1} + \frac{n_2+4}{n_2-n_1} C_2 \lambda^{n_2} + G_0 \right], \\ h_{11} = \frac{2\lambda^3}{\gamma+1} \left[\frac{3}{n+3} C_1 \lambda^{n_1} + \frac{\gamma n_2 + 3(\gamma-1)}{n_2 - n_1} C_2 \lambda^{n_2} + H_0 \right]. \end{gathered} \tag{3.65}$$

The solution (3.65) contains only two particular solutions of the system (3.18)–(3.20). The third solution corresponds to the negative characteristic root n_3 of largest modulus obtained from the equation

$$n^2+\frac{13\gamma+1}{\gamma+1}n+\frac{20\gamma-12}{\gamma+1}=0.$$

From the condition of least growth of the function f_{11} near $\lambda=0$ we have set the constant C_3 equal to zero in the solutions (3.65) and thus have not considered this particular solution.

For the constants F_0, G_0 and H_0 we have

$$G_0=\frac{2(\omega_1-3)}{\omega_1-1}F_0,\qquad H_0=-(3\gamma-3)\frac{2}{\omega_1-1}F_0-2A_{11},$$

$$F_0=0{,}5\left(12-\frac{\gamma+1}{\gamma-1}(\omega_1-1)\right)\left[\gamma-1-\frac{6(\omega_1-6+3\gamma)}{\omega_1-1}\right]^{-1}A_{11}.$$

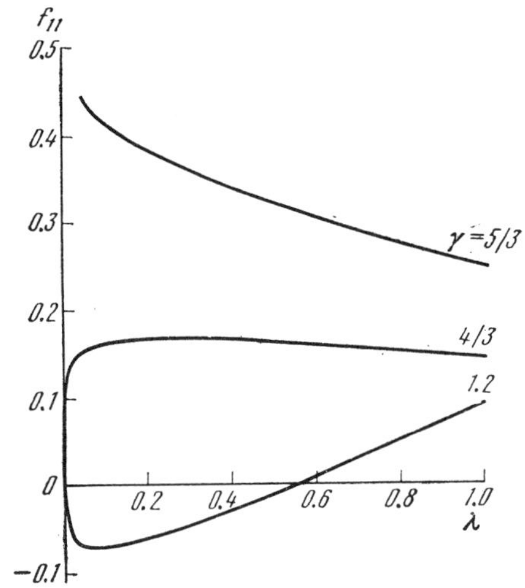

FIGURE 22. Graphs of the functions $f_{11}(\lambda)$ for various γ

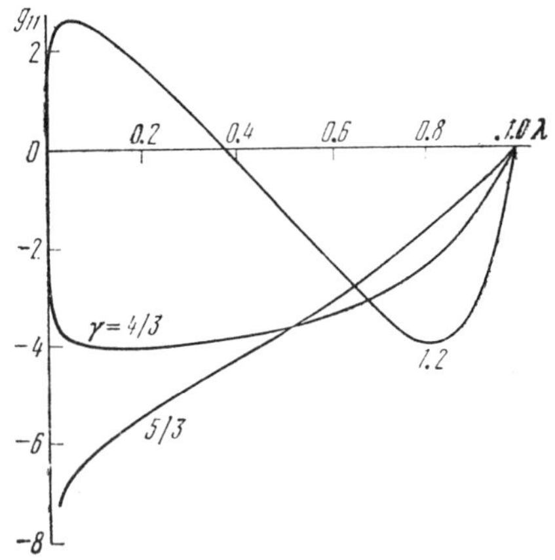

FIGURE 23. Graphs of the functions $g_{11}(\lambda)$ for various γ

The constants A_{11}, C_1 and C_2 are determined from the boundary conditions (3.22). Results of the calculations are given in Table 3.

The computations which have been carried out indicate that for $\gamma>1.2$ there is a conspicuous region near $\lambda=0$ where the above solution becomes meaningless, since the curves of density and pressure pass through zero and become negative. This

TABLE 3

Values of the constants in (3.65)

γ	A_{11}	C_1	C_2	n_2
1.2	0.58341	0.63825	−1.6254	−0.80980
4/3	0.68072	0.39647	1.46413	−0.90401
1.4	0.69335	0.35282	1.6933	−0.94495
5/3	0.56286	0.50591	1.4221	−1.0779

is explained by the singular character of the base solution f_0, g_0, h_0 for $\omega = \omega_1$, the behavior of the solution, and the shortcomings of the method of linearization. Here it is also possible to consider a variant of the solution by introducing a spherical cavity with zero pressure in a neighborhood of the point $\lambda = 0$. It can be hoped, however, that such essentially integral characteristics of the motion as the constant A_{11} will by way of improving the law of motion of the shock wave give correct values and not be strongly dependent on the local properties of the solution near the center of symmetry.

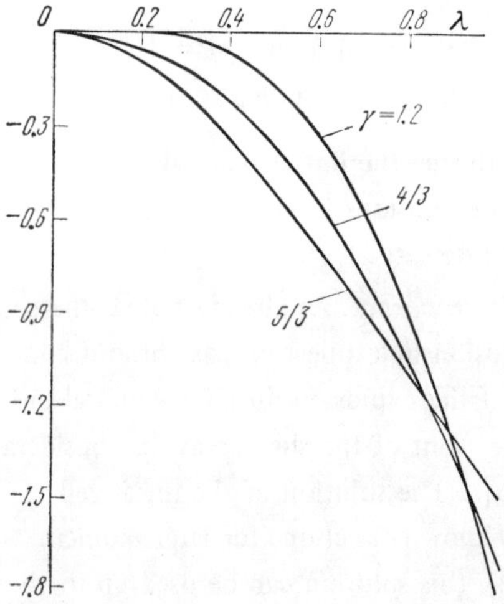

FIGURE 24. Graphs of the functions $h_{11}(\lambda)$ for various γ

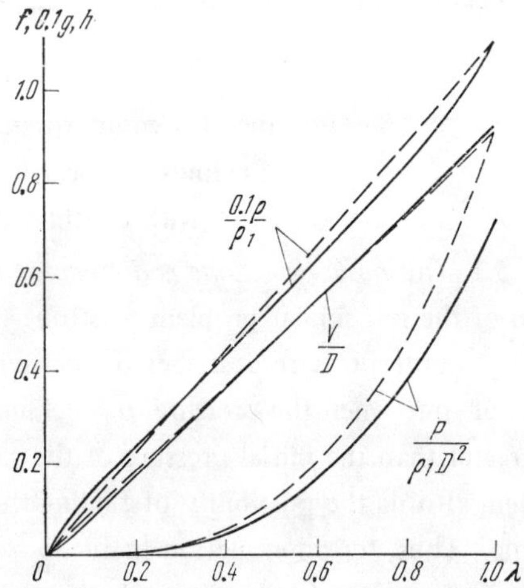

FIGURE 25. Distributions of dimensionless pressures, densities and velocities ($\nu = 0.1$, $\gamma = 1.2$). The dashed curves correspond to the self-similar solution.

We introduce the dimensionless radius $x_2 = r_2/r^*$ and dimensionless time $\tau = t/t^*$, where $r^* = (E_0/AB^2)^{1/(\omega-1)}$ and $t^* = (r^*)^{3-\omega}/B$. Then to determine the relation between x_2 and τ we find

$$\begin{aligned} x_2 &= (\delta\alpha^{-1/2})^{2/(\omega-1)} y^{2/(\omega-1)} \exp \frac{2A_{11}}{\omega-1} y, \\ \tau &= \delta (\delta\alpha^{-1/2})^{2(3-\omega)/(\omega-1)} y^{(5-\omega)/(\omega-1)} \left[1 + \frac{(5-\omega)}{4(\omega-1)} A_{11} y\right]. \end{aligned} \tag{3.66}$$

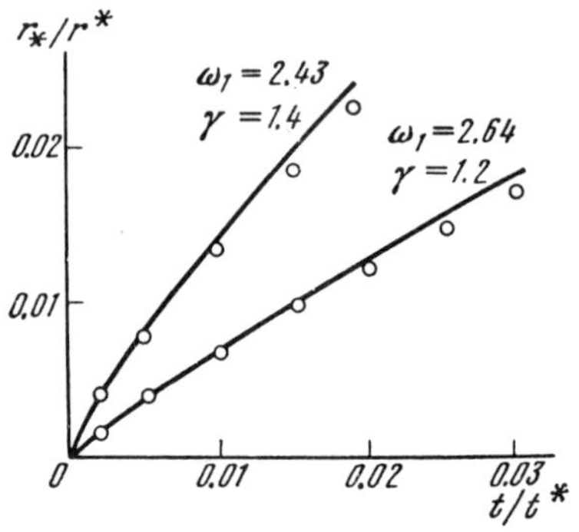

FIGURE 26. The dependence r_2/r^* of the radius of the shock wave on time for $\gamma = 1.33$ and $\gamma = 1.2$. The circles correspond to points of the self-similar solution.

In the self-similar case we have

$$x_2 = \alpha^{-1/(5-\omega)} \tau^{\delta}, \qquad y = \frac{\sqrt{\alpha}}{\delta} x_2^{(\omega-1)/2}. \tag{3.67}$$

The dependence (3.66) for the cases $\omega = \omega_1$, $\gamma = 1.2$ and $\gamma = 4/3$ is represented in Figure 26. Here the circles represent values corresponding to (3.67).

§5. Allowance for counterpressure during the initial stage of plane, cylindrical, or spherical explosions in gases with constant initial density

5.1. *The basic equations and some of their properties*. As already noted, the solution of the self-similar problem of strong explosion in a quiescent gas without considering counterpressure describes the behavior of the explosion during the initial moments of time when the pressure p_2 behind the front of the shock wave is considerably greater than the initial pressure of the gas p_1. The solution of the linearized problem affords the possibility of finding the unknown functions for later moments of time. Thus, for an explosion in air ($\gamma = 1.4$) this solution can be used up to the time when the pressure p_2 exceeds the pressure p_1 by approximately 8 times. In a number of applications of the theory of point explosion it is necessary to have a solution of the problem for media with different adiabatic exponents γ. The results of the linearized solution can be used in calculating the complete problem by approximate analytic or numerical methods (cf. Chapter 4).

The linearized solutions for arbitrary γ are not expressed in terms of elementary functions and can be obtained only by numerical methods. In the linearized formulation the problem of explosion with counterpressure at constant initial density and adiabatic exponent $\gamma = 1.4$ was first considered by N. S. Mel'nikova [1] (cf. also [2]) and independently by A. Sakurai [3]. Calculation of cylindrical and plane explosions for $\gamma = 1.4$ is carried out in [10].

We shall consider numerical methods of solving the linearized problem of an explosion with counterpressure for a wide range of values of γ in cases of plane, cylindrical, and spherical waves. The results of the completed computations have been published in the form of special tables [11] which contain both the self-similar and linearized functions. The computational method was worked out by the author in collaboration with P. I. Čuškin [12].

The defining parameters of the problem are the initial density ρ_1 and pressure p_1 of the gas, the energy E_0, the adiatic exponent γ, the geometrical coordinate r, and the time t. These parameters indicate that for given symmetry the solution will depend on γ and two dimensionless variables which we take to be λ and q. These variables range over the intervals $0 \leqslant \lambda \leqslant 1$ and $0 \leqslant q \leqslant 1$.

We consider the dimensionless velocity f, density g, and pressure h as the unknown functions:

$$f = \frac{v}{D}, \qquad g = \frac{\rho}{\rho_1}, \qquad h = \frac{qp}{\gamma p_1}. \tag{3.68}$$

Moreover, there are also the unknown functions $R_2 = r_2/r^0$ and $\tau = t/t^0$, where r^0 and t^0 are respectively the dynamic length and time, defined by

$$r^0 = \left(\frac{E_0}{p_1}\right)^{1/\nu}, \qquad t^0 = r^0\left(\frac{\rho_1}{p_1}\right)^{1/2}. \tag{3.69}$$

The system of differential equations describing one-dimensional transient motions of a perfect gas in the dimensionless variables adopted has the form (3.12)–(3.14) with $i = 2$ and $\omega = 0$. This system serves to determine $f(\lambda, q), g(\lambda, q)$ and $h(\lambda, q)$. There are the following boundary conditions for this system. On the shock wave (for $\lambda = 1$),

$$f(1, q) = \frac{2}{\gamma+1}(1-q), \quad g(1, q) = \frac{\gamma+1}{\gamma-1+2q}, \quad h(1, q) = \frac{2\gamma-(\gamma-1)q}{\gamma(\gamma+1)}. \tag{3.70}$$

At the center of symmetry (for $\lambda = 0$),

$$f(0, q) = 0. \tag{3.71}$$

The initial conditions (for $q = 0$) are expressed as follows:

$$f(\lambda, 0) = f_0(\lambda), \qquad g(\lambda, 0) = g_0(\lambda), \qquad h(\lambda, 0) = h_0(\lambda),$$

where $f_0(\lambda), g_0(\lambda)$ and $h_0(\lambda)$ are the functions corresponding to the self-similar solution.

The fourth boundary condition (3.71) makes it possible to determine the radius of the shock wave $R_2(q)$. This function can also be found from the integral law of energy conservation (3.25). After this the function $\tau(q)$ is computed by quadrature from the equation

$$\frac{d\tau}{dq}=\left(\frac{q}{\gamma}\right)^{1/2}\frac{dR_2}{dq}.$$

During the initial stage of an explosion with counterpressure (i.e. for small q) the solution may be sought in the form (3.16)–(3.17) with $i = 2$, $\omega = 0$:

$$f=f_0+qf_1+o(q),\qquad g=g_0+qg_1+o(q),\qquad h=h_0+qh_1+o(q),$$
$$\frac{R_2}{q}\frac{dq}{dR_2}=\frac{\nu}{1+A_1q+o(q)},\tag{3.72}$$

where A_1 is a constant which remains to be determined.

If the functions (3.72) are substituted into the system of equations (3.12)–(3.14), terms of order q^2 and higher are neglected, and terms with like powers of q are set equal to zero, we obtain systems of ordinary differential equations and boundary conditions for the self-similar and linearized problems.

In [13] terms of order q^2 were considered (cf. also [14], [15]). In this regard we note the following. The solution (3.72) can be represented in the form $f = f_0 + qf_1 + q^2f_2 + \cdots$ (similarly for g and h), i.e. the functions may be assumed to be series in q. The domain of convergence of representations of (3.72) as series in q has not yet been established. In any case it may definitely be said that the solution in the form of such a series will diverge for $q \geqslant (\gamma - 1)/2$. This follows from the fact that the expansion for the function $g(1, q)$ of (3.70) converges only for $q < (\gamma - 1)/2$. Thus for $\gamma < 2$ such representations for the functions can be used only for small q. But for small values of q consideration of terms of order q^2 hardly enables one to substantially improve the linearized solution. Here, it is true, there is the possibility of using the function $1/g$ in place of g in studying solutions by means of series in q. This function depends linearly on q for $\lambda = 1$, and for it the objections above are not of such fundamental importance.

We introduce in place of λ a new independent variable μ such that

$$\lambda=\mu^{-\delta}u_1^{\beta_2}u_2^{-\beta_1},\tag{3.73}$$

where $\lambda = 0$ corresponds to $\mu = \mu_0 = (\gamma + 1)/2\gamma$, and $\lambda = 1$ corresponds to $\mu = 1$.

As was shown in Chapter 2, for $\gamma \neq 2$ the solution of the self-similar problem for a strong explosion can be written in the following form by means of the parametric variable μ:

$$f_0=\frac{2}{\gamma+1}\mu\lambda,\qquad g_0=\frac{\gamma+1}{\gamma-1}u_1^{\beta_3}u_2^{\beta_4}u_3^{-\beta_5},\qquad h_0=\frac{2}{\gamma+1}\mu^{\nu\delta}u_2^{\beta_4-2\beta_1}u_3^{1-\beta_5}.\tag{3.74}$$

The case $\gamma = 2$ is special, and here for g_0 and h_0 we have

$$g_0 = 3\kappa u_1^{\delta\nu/2} u_3^{\delta(\nu-2)/2}, \qquad h_0 = \frac{2}{3}\kappa\mu^{\delta\nu} u_3^{\delta(\nu-2)}. \tag{3.75}$$

In (3.73)–(3.75) we have used the following notation:

$$u_1 = \frac{2\gamma}{\gamma-1}\left(\mu - \frac{\gamma+1}{2\gamma}\right), \qquad u_2 = \frac{\nu\gamma-\nu+2}{3\nu-2-\gamma(\nu-2)}\left[\frac{(\gamma+1)(\nu+2)}{\nu\gamma-\nu+2} - 2\mu\right],$$

$$u_3 = \frac{2}{\gamma-1}\left(\frac{\gamma+1}{2} - \mu\right), \qquad \kappa = \exp\left(6\delta\frac{\mu-1}{3-2\mu}\right), \qquad \delta = \frac{2}{\nu+2},$$

$$\beta_1 = \beta_2 + \frac{\gamma+1}{\nu\gamma-\nu+2} - \delta, \qquad \beta_2 = \frac{\gamma-1}{2\gamma-2+\nu}, \qquad \beta_3 = 1 - 2\beta_2,$$

$$\beta_4 = \frac{\nu+2}{2-\gamma}\beta_1, \qquad \beta_5 = \frac{2}{2-\gamma}.$$

The quantities $R(q)$ and $\tau(q)$ in the self-similar problem are found from the expressions $R_{20}^{\nu} = \delta^2 q/\gamma\alpha$ and $\tau_{(0)} = (q/\gamma)^{1/2}\delta R_{20}$, where the constant α is defined by (2.25).

The system of linear differential equations for determining the functions $f_1(\lambda)$, $g_1(\lambda)$, $h_1(\lambda)$ and the constant A_1 in the linearized problem has the form (3.18)–(3.20) with $i = 2$, $\omega = 0$ and $A_{21} = A_1$ (cf. §1).

The boundary conditions for this system are

$$f_1(1) = -\frac{2}{\gamma+1}, \quad h_1(1) = -\frac{\gamma-1}{\gamma(\gamma+1)}, \quad g(1) = -\frac{2(\gamma+1)}{(\gamma-1)^2}, \quad f_1(0) = 0. \tag{3.76}$$

It can be assumed that the last of these conditions serves to determine A_1.

For the functions $R_2(q)$ and $\tau(q)$ in the linearized problem we obtain

$$R_2^{\nu} = R_{20}^{\nu}\exp(A, q), \qquad \tau = \tau_{(0)}\left[1 + \frac{(\nu\delta+2)A_1}{2\nu\delta(\nu\delta+1)}q\right]. \tag{3.77}$$

These relations give the law of motion of the shock wave, i.e. the function $R_2(\tau)$, in parametric form. Using (3.77) and the relations on the shock wave (3.76), one can determine the dependence of all the characteristics on the front of the shock wave on the radius or time. For example, for the ratio of pressures p_2/p_1 as a function of R_2 we have

$$\frac{p_2}{p_1} = \frac{\gamma-1}{\gamma+1} + \frac{2\gamma}{\gamma+1}\left(A_1 + \frac{\delta^2}{\gamma\alpha}R_2^{-\nu}\right). \tag{3.78}$$

5.2. *Numerical solution of the problem.* Solution of the linearized problem reduces to numerical integration of the system (3.18)–(3.20) for $i = 2$ and $\omega = 0$ under conditions (3.76). We shall consider only cases where γ is in the interval $1 < \gamma < 7$. We remark that for $\gamma = 7$ ($\nu = 3$) this problem has an exact analytic solution (cf. §2).

In order to bring the linearized system to a form convenient for numerical computations, we introduce new unknown functions $F(\lambda)$, $G(\lambda)$ and $H(\lambda)$ related to $f_1(\lambda)$, $g_1(\lambda)$ and $h_1(\lambda)$ as follows:

$$f_1 = -mF, \qquad g_1 = g_0 G, \qquad h_1 = g_0 H, \qquad m = f_0 - \lambda.$$

After transformations the system (3.18)–(3.20) takes the form

$$\begin{gathered} mF' - \frac{h_0}{g_0 m} H' + \left(2f_0' + \frac{\nu - 2}{2}\right) F - \left(f_0' - \frac{\nu}{2}\frac{f_0}{m}\right)(G - H) - \frac{\nu}{2}\frac{f_0}{m} A_1 = 0, \\ m(F' - G') - \nu(F + G) = 0, \\ m(\gamma F' - H') - \nu[(\gamma - 1)F + H + A_1] = 0. \end{gathered} \tag{3.79}$$

Conditions (3.76) become

$$F(1) = -\frac{2}{\gamma - 1}, \quad G(1) = -\frac{2}{\gamma - 1}, \quad H(1) = -\frac{\gamma - 1}{2\gamma}, \quad F(0) = F_0, \tag{3.80}$$

where the constant F_0 depends only on γ and ν (see formulas (3.83) below).

As in previous problems, the system (3.79) has an integral which via (3.80) can be written in the form

$$F - 2H + (2\gamma - 1)G - 2A_1 = -\left(2A_1 + \frac{3\gamma - 1}{\gamma}\frac{\gamma + 1}{\gamma - 1}\right)\mathcal{J},$$

where

$$\mathcal{J}(\mu) = \frac{2}{(\gamma + 1)h_0}\left(\frac{\gamma - 1}{\gamma + 1} g_0\right)^{\gamma}. \tag{3.81}$$

In the system (3.79) we now go over to the new independent variable μ defined by (3.73). The system of equations obtained as a result can be brought to the following form:

$$\begin{gathered} \frac{dF}{d\mu} = \left[(\gamma - 1)F + H - \frac{\gamma x \Omega}{\nu}\Phi_1 + \left(1 + \frac{\gamma}{\gamma + 1} x\mu\right) A_1\right] \frac{\nu \Phi_2}{\Omega (x\Omega - 1)\gamma}, \\ \frac{dG}{d\mu} = \frac{dF}{d\mu} + \frac{\nu \Phi_2}{\Omega}(F + G), \qquad \frac{dH}{d\mu} = \gamma \frac{dF}{d\mu} + \frac{\nu \Phi_2}{\Omega}[(\gamma - 1)F + H + A_1]. \end{gathered} \tag{3.82}$$

The notation here is as follows:

$$x = \frac{\gamma + 1}{\gamma - 1}\left(\mu - \frac{\gamma + 1}{2\gamma}\right)\frac{1}{\mu^2}, \qquad \Omega = 1 - \frac{2\mu}{\gamma + 1},$$

$$\Phi_1 = -\left[\frac{4}{\gamma + 1}\left(\mu + \frac{1}{\Phi_2}\right) + \frac{\nu - 2}{2}\right] F + \left[\frac{2}{\gamma + 1}\left(\mu + \frac{1}{\Phi_2}\right) + \frac{\nu}{\gamma + 1}\frac{\mu}{\Omega}\right](G - H),$$

$$\Phi_2 = \frac{\gamma + 1}{\gamma - 1}\frac{\beta_2}{x\mu^2} - \frac{\delta\beta_1}{\delta(\mu - 1) - \beta_1 + \beta_2} - \frac{\delta}{\mu}.$$

In a neighborhood of the center of the explosion the following asymptotic formulas can be established for the desired solution of (3.82):

$$
\begin{aligned}
&F = F_1 + B\,(a_{01} + a_{11}\lambda^k) + O\,(\lambda^{2k}),\\
&G = G_1 - \frac{2}{2\gamma - 1}\Big[\Big(A_1 + \frac{2\gamma-1}{2\gamma}\frac{\gamma+1}{\gamma-1}\Big)\mathcal{J}_0 + B\Big(\frac{a_{01}-2}{2}\\
&\quad + \frac{a_{11} - 2b_{11}}{2}\lambda^k\Big)\Big] + O(\lambda^{2k}),\\
&H = H_1 + B\,(1 + b_{11}\lambda^k) + O\,(\lambda^{2k}),
\end{aligned} \tag{3.83}
$$

where F_1, G_1 and H_1 do not depend on λ and represent a particular solution of (3.82), B is an arbitrary constant, $k = 1/\beta_2$, and $\mathcal{J}_0$ is the leading term in the expansion of $\mathcal{J}(\lambda)$ in a neighborhood of $\lambda = 0$. For $1 < \gamma < 7$ (with the exception of the case $\gamma = 2$) the constants in the asymptotic expressions (3.83) have the form

$$
\begin{aligned}
&a_{01} = -\frac{1}{\gamma-1}, \qquad a_{11} = \frac{(\gamma - 1 - n)\,b_{11}}{(\gamma-1)\,(\gamma+n)}, \qquad b_{11} = \Big(m_0 + \frac{l}{\gamma-1}\Big)\frac{D_0\,(\gamma-1)}{\gamma^2 h\,(\mu_0)},\\
&n = \gamma\nu\beta_2, \qquad m = \beta_2\frac{3-2\gamma}{2\gamma-1}\Big(1 + \frac{\nu}{2}\frac{\gamma}{\gamma-1}\Big),\\
&l = \beta_2\Big[\Big(2 + \frac{\nu-2}{2}\gamma + \frac{1}{2\gamma-1}\Big(1 + \frac{\nu}{2}\frac{\gamma}{\gamma-1}\Big)\Big],\\
&D_0 = \frac{\gamma+1}{\gamma-1}\,[u_2\,(\mu_0)]^{\beta_1/\beta_2(2-\gamma)}\,[u_3\,(\mu_0)]^{-\beta_3}\,\mu_0^{\delta\beta_3/\beta_2},\\
&F_1 = -G_1 = \Big[2 + \frac{\nu-2}{2}\gamma - (\gamma-2)\Big(1 + \frac{\nu}{2}\frac{\gamma}{\gamma-1}\Big)\Big]^{-1} A_1,\\
&H_1 = -A_1 - (\gamma - 1)\,F_1.
\end{aligned}
$$

For $\gamma = 2$ only the formula for D_0 is changed, i.e.

$$
D_0 = 3\kappa\,(\mu_0)\,[u_2\,(\mu)]^{\frac{\nu}{\nu+2}\frac{\beta_1}{\beta_2}}\,[u_3\,(\mu_0)]^{\frac{\nu-2}{\nu+2}}\mu_0^{\frac{\delta\nu}{(\nu+2)\beta_2}}.
$$

The following method was employed for the numerical integration of the system of differential equations (3.82). Some initial value was assigned to the constant A_1 (in practice this value was taken close to 2). The differential equations (3.82) were then integrated numerically by the Runge-Kutta method with variable step size from $\mu = 1$ to a certain $\mu = \mu_1$ ($\mu_0 < \mu_1 < 1$). This value μ_1 was chosen as the limit of applicability of the asymptotic formulas (3.83). The constancy (to the required accuracy) as μ_1 was decreased of the constants entering in the asymptotic formulas (3.83) served as the condition for the choice of the quantity μ_1. For a value $\mu = \mu_1$ corresponding to $\lambda = \lambda_1$ the constant B was computed from the asymptotic formula for F. If this quantity is denoted by B_1, then from (3.83) we have

$$
B_1 = \frac{F - F_1}{a_{01} - a_{11}\lambda_1^k}.
$$

At the same time, for $\mu = \mu_1$ the constant B was computed from the asymptotic formula for H; namely,

$$B_2 = \frac{H - H_1}{1 + b_{11}\lambda_1^k}.$$

The difference $B_2 - B_1$ was then determined and correction was made of the initially chosen value of A_1 in order to satisfy the condition

$$\left| \frac{B_2 - B_1}{B_1} \right| < \Delta,$$

where Δ is a small prescribed number. The process for choosing A_1 is easily made automatic in machine computations and proceeds very rapidly. In the computation with the value of A_1 finally chosen the functions F, G and H as well as f_1, g_1, h_1 and all other characteristics of the flow were determined. In addition, computational accuracy was controlled in all calculations by means of the integral (3.81).

This numerical method of solving the linearized problem of explosion with counterpressure differs from the methods applied in [1]–[3], [10] and is quite convenient for machine computations.

5.3. *Results of the computations.* On the basis of a method set forth by K. V. Šarovatova a program was written for the electronic computer BÈSM-2 permitting the computation of the solution of the linearized problem on explosion with counterpressure in gases for various values of the adiabatic exponent γ in the plane, cylindrical, and spherical cases. A long sequence of computations were carried out with this program for $\nu = 1, 2, 3$ and the values $\gamma = 1.1, 1.2, 1.3, 1.4, 5/3, 2, 3$. In these computations, in addition to the functions f_1, g_1, h_1 and constants A_1, the self-similar functions f_0, g_0 and h_0 were determined as were also the following functions:

$$e_0 = \frac{\gamma}{2} f_0^2 g_0 + \frac{\gamma h_0}{\gamma - 1}, \qquad e_1 = \gamma f_0 g_0 f_1 + \frac{\gamma}{2} f_0^2 g_1 + \frac{\gamma h_1}{\gamma - 1},$$

$$\theta_0 = \frac{\gamma h_0}{g_0}, \qquad \theta_1 = \frac{\gamma h_1 - \theta_0 g_1}{g_0}, \qquad \sigma_0 = \frac{\gamma h_0}{g_0^\gamma}, \qquad \sigma_1 = \frac{\gamma (h_1 - \theta_0 g_1)}{g_0^\gamma}.$$

The latter functions serve for the computation of the total energy, the temperature, and the entropy of the gas in the linearized approximation. Moreover, the computed data make it possible to determine the pressure impulse and other integral characteristics and also to find the relation between the Euler and Lagrange coordinates (by means of σ_0 and σ_1). All the functions indicated were computed to five significant digits. The tables obtained in these computations of all the functions enumerated and of the constants entering in the asymptotic formulas for determing these functions near the center of the explosion (in a neighborhood of $\lambda = 0$) have been published in [11].

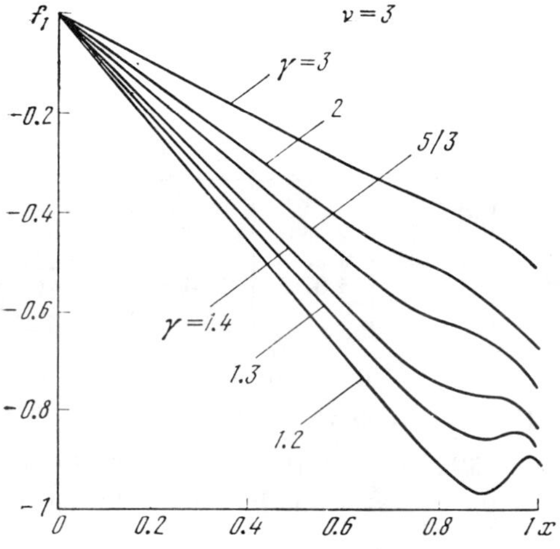

FIGURE 27. Curves of $f_1(\lambda)$ for various γ in the case of spherical symmetry

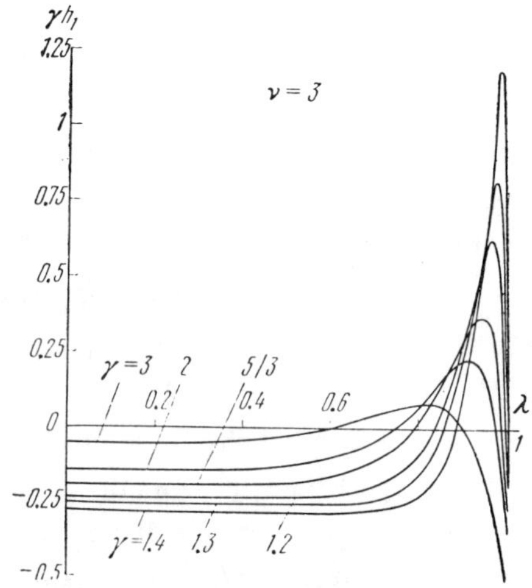

FIGURE 28. Graphs of the curves $\gamma h_1(\lambda)$ characterizing the linear corrections to the pressure distribution for various γ.

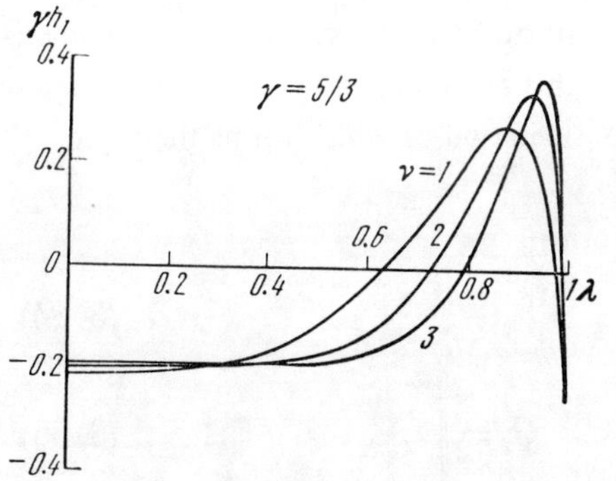

FIGURE 29. Comparison of the distributions of $h_1(\lambda)$ for various symmetries

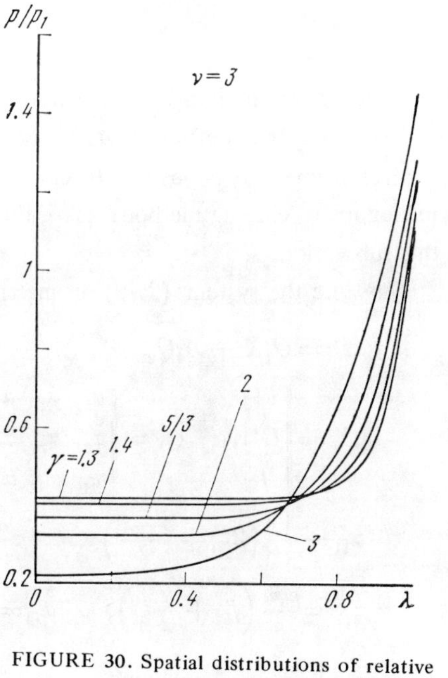

FIGURE 30. Spatial distributions of relative pressures (the case $\nu = 3$, $q = 0.1$)

TABLE 4

Values of the constant A_1

ν	γ = 1.1	1.2	1.3	1.4	5/3	2	3
1	2.3257	2.2437	2.1862	2.1433	2.0683	2.0143	1.9407
2	2.0866	2.0424	2.0092	1.9836	1.9374	1.9043	1.8632
3	2.0010	1.9666	1.9396	1.9182	1.8785	1.8496	1.8141

Here we present for illustration several graphs containing individual results. Figure 27 shows the graphs of the functions $f_1(\lambda)$ for various values of γ in the spherical case, and Figure 28 shows the corresponding graphs for $\gamma h_1(\lambda)$. The change in the functions $\gamma h_1(\lambda)$ for $\gamma = 5/3$ in dependence on ν is shown in Figure 29. Curves of the relative pressure p/p_1 for $q = 0.1$, various γ, and $\nu = 3$ are constructed in Figure 30. Table 4 gives the values of the constant A_1 found for a number of adiabatic exponents γ and $\nu = 1, 2, 3$.

The results given here and in [11] indicate the essential dependence of the solution on the quantity γ. They also imply that $f_1(\lambda), g_1(\lambda)$ and $h_1(\lambda)$ are bounded functions of the linear perturbation, and hence linearization is everywhere meaningful for $0 \leqslant \lambda \leqslant 1$.

5.4. *A version of the chasing method for solving the linearized system.* The method considered for the numerical solution of the linearized system has the drawback that in choosing the coefficient A_1 it is necessary to solve the system (3.82) numerically several times. It is possible to circumvent this shortcoming by using a special method analogous to chasing methods [16], the consideration of which forms the topic of this subsection.

We write the system (3.79) in matrix form:

$$X' = Q_1 X + A_1 Q_2, \tag{3.84}$$

$$X = \begin{Vmatrix} F \\ G \\ H \end{Vmatrix}, \quad Q_1 = \begin{Vmatrix} \alpha_{11} & \alpha_{12} & \alpha_{13} \\ \alpha_{21} & \alpha_{22} & \alpha_{23} \\ \alpha_{31} & \alpha_{32} & \alpha_{33} \end{Vmatrix}, \quad Q_2 = \begin{Vmatrix} a_1 \\ a_2 \\ a_3 \end{Vmatrix}, \tag{3.85}$$

$$\alpha_{11} = \frac{1}{\xi}\left[\left(2f_0' + \frac{\nu-2}{2}\right) g_0 m + \frac{\nu h_0(\gamma-1)}{m}\right], \quad \xi = g_0 m^2 - \gamma h_0,$$

$$\alpha_{12} = \frac{g_0 m}{\xi}\left(f_0' + \frac{\nu}{2}\frac{f_0}{m}\right), \quad \alpha_{13} = \frac{1}{\xi}\left[\frac{\nu h_0}{m} + g_0 m\left(f_0' - \frac{\nu}{2}\frac{f_0}{\Omega}\right)\right],$$

$$\alpha_{21} = \alpha_{11} - \frac{\nu}{m}, \quad \alpha_{12} = \alpha_{12} - \frac{\nu}{m}, \quad \alpha_{23} = \alpha_{13},$$

$$\alpha_{31} = \gamma\alpha_{11} - \frac{\nu}{m}(\gamma - 1), \quad \alpha_{32} = \gamma\alpha_{12}, \quad \alpha_{33} = \gamma\alpha_{13} - \frac{\nu}{m},$$

$$a_1 = m g_0\left(\frac{\nu h_0}{g_0 m^2} - \frac{\nu}{2m} f_0\right), \quad a_2 = a_1, \quad a_3 = \gamma a_1 - \frac{\nu}{m} A_1, \quad m = f_0 - \lambda.$$

Suppose now that the unknown functions satisfy an integral relation

$$\int_0^1 MX\lambda^{\nu-1}d\lambda = \mathcal{J}_1 + A_1\mathcal{J}_2, \tag{3.86}$$

where $\mathcal{J}_1$ and $\mathcal{J}_2$ are (known) numbers and M is a row matrix

$$M = \| M_1,\ M_2,\ M_3 \|. \tag{3.87}$$

We introduce a new column vector Y with unknown components F, G and H, and multiply (3.84) on the left by the row matrix Y^* obtained by transposing Y. We integrate the relation so obtained with respect to λ:

$$\int_0^1 Y^*X'd\lambda = \int_0^1 Y^*Q_1Xd\lambda + A_1\int_0^1 Y^*Q_2d\lambda.$$

Integrating by parts, we obtain

$$\int_0^1 (Y^{*'} + Y^*Q_1)Xd\lambda = [Y^*X] - A_1\int_0^1 Y^*Q_2d\lambda. \tag{3.88}$$

We choose Y such that

$$Y^{*'} + Y^*Q_1 = M\lambda^{\nu-1}. \tag{3.89}$$

Using (3.86) and (3.89), relation (3.88) can be written

$$\mathcal{J}_1 + A_1\mathcal{J}_2 = [Y^*X]\,|_0^1 - A_1\int_0^1 Y^*Q_2d\lambda. \tag{3.90}$$

We now choose Y such that for $\lambda = 0$ the condition

$$[Y^*X]\,|_{\lambda=0} = 0$$

is satisfied, or

$$(\tilde{F}F + \tilde{G}G + \tilde{H}H)\,|_{\lambda=0} = 0. \tag{3.91}$$

If F, G and H are finite for $\lambda = 0$, it may be assumed that

$$\tilde{F}(0) = \tilde{G}(0) = \tilde{H}(0) = 0 \tag{3.92}$$

and condition (3.91) is satisfied.

Equation (3.89) does not contain the constant A_1 and can be integrated from $\lambda = 0$ to $\lambda = 1$ (for example, by numerical integration). On the basis of the values of Y thus found we determine $[Y^*X]|_{\lambda=1}$ and compute the integrals entering in (3.90); we then find A_1 from the formula

$$A_1 = -\frac{\mathcal{J}_1 - [Y^*X]\,|_{\lambda=1}}{\mathcal{J}_2 + \int_0^1 Y^*Q_2d\lambda}.$$

Once the value of A_1 has been found, on integrating (3.84) with the initial data at $\lambda = 1$ we obtain the solution of the problem.

We take the linearized law of mass conservation as the simplest integral relation of the form (3.86). If there is no mass source at the center we have

$$\int_0^1 g\lambda^{\nu-1}d\lambda = 1. \tag{3.93}$$

From this for the linearized solution we find that $\int_0^1 g_1\lambda^{\nu-1}d\lambda = 0$ or $\int_0^1 g_0 G\lambda^{\nu-1}d\lambda = 0$. For the elements of the matrix M and the numbers $\mathcal{J}_1$ and $\mathcal{J}_2$ we have $M_1 = 0$, $M_2 = g_0$, $M_3 = 0$ and $\mathcal{J}_1 = \mathcal{J}_2 = 0$. Other integral relations of the type (3.86) are the linearized relation for the energy integral

$$\gamma\int_0^1\left(-f_0 g_0 mF + \frac{h_0}{\gamma-1}H + \frac{f_0^2 g_0}{2}G\right)\lambda^{\nu-1}d\lambda = -A_1\mathcal{J}_0 + \frac{1}{\nu(\gamma-1)},$$

$$\mathcal{J}_0 = \gamma\int_0^1\left(\frac{g_0 f_0^2}{2} + \frac{h_0}{\gamma-1}\right)\lambda^{\nu-1}d\lambda, \tag{3.94}$$

and the relation obtained by multiplication of the adiabatic integral (3.81) by $\lambda^{\nu-1}$ and integrating from 0 to 1. If (3.93) or (3.94) is used, the adiabatic integral (3.81) can be used for checking the accuracy of the compuatations.

Thus, knowing certain integral and local (near $\lambda = 0$) properties of the solution, one can find A_1 by integrating an auxiliary equation from the center to the shock wave; and then by integrating the basic system of equations from the shock wave to the center it is possible to determine all the unknown functions. In this sense the method considered can be interpreted as a special variant of the chasing method which is used to solve boundary value problems [16]. A method close to that developed above was considered in [14] for the case of a system of two equations of the form (3.84) and in the special case using the integral law of energy conservation.

We point out that the method we have developed can be retailored for a situation in which other independent variables (for example, μ) are used. It can be applied to solve other linearized problems also. Among these we mention the problem of strong explosion in a detonating medium (cf. Chapter 6), the problem of the motion of a piston taking into account the counterpressure (in the integral relation of the method the integration is to be taken not from 0 to 1 but between the limits corresponding to the coordinates of the piston and the shock wave), and the problem of the effect of a magnetic field on the gas flow during an explosion.

§6. On the use of the principle of superposition of linear corrections

As already mentioned, in formulating the basic point explosion problem (PEP) in a perfect gas we neglect such effects as possible variations in the initial velocity, pressure, and equation of state, variability of the energy E_0, etc. Moreover, the effects of these various parameters on the motion are not equivalent, as has been shown by examples. The necessity may arise of considering many small parameters at once. In this case it makes sense to use the principle of superposition of the solutions of the linear equations for the various corrections to the strong explosion.

We shall consider this question in more detail for one-dimensional motions of a gas. Suppose we seek the solution in the form

$$\begin{aligned} v &= v_0 + \varepsilon_1 v_1 + \varepsilon_2 v_2 + \ldots + \varepsilon_n v_n, \\ \rho &= \rho_0 + \varepsilon_1 \rho_1 + \varepsilon_2 \rho_2 + \ldots + \varepsilon_n \rho_n, \\ p &= p_0 + \varepsilon_1 p_1 + \varepsilon_2 p_2 + \ldots + \varepsilon_n p_n, \end{aligned} \tag{3.95}$$

where $\epsilon_1, \cdots, \epsilon_n$ are independent small parameters. We substitute the solution (3.95) into the system (3.5) and assume that v_0, ρ_0, p_0 is a solution of this system.

As a result we obtain, for example, from the momentum equation

$$\frac{d_0}{dt}\Delta v + (\Delta v)\frac{\partial v_0}{\partial r} - \frac{\Delta\rho}{\rho_0(\rho_0+\Delta\rho)}\frac{\partial p_0}{\partial r} + \frac{1}{\rho_0+\Delta\rho}\frac{\partial\Delta p}{\partial r} + \Delta v\frac{\partial\Delta v}{\partial r} = 0, \quad \Delta p = \sum \varepsilon_i p_i,$$

$$\Delta v = \sum \varepsilon_i v_i, \quad \Delta\rho = \sum \varepsilon_i \rho_i, \quad \frac{d_0}{dt} = \frac{\partial}{\partial t} + v_0\frac{\partial}{\partial r}.$$

Since the parameters ϵ_i are independent, on setting them all equal to zero except for one ϵ_k, we arrive at a system of equations of the type (1.31) for the perturbations p_k, v_k, ρ_k. This is true also for the boundary and initial conditions. We thus obtain a linearized system and necessarily arrive at the conclusion that the linear corrections must satisfy systems of linear equations corresponding to a single perturbation with its own parameter ϵ_k and its own boundary and initial conditions. On the other hand, if all the linear systems corresponding to the individual parameters ϵ_k are satisfied by some solutions v_k, ρ_k, p_k, then up to terms $\epsilon_i\epsilon_j$ $(i, j = 1, \cdots, n)$ we also satisfy the original system. Naturally, all corrections must be bounded to justify the application of linearization. On the basis of what has been said, it is possible to formulate a rule for simultaneously considering a number of corrections. We write the solution of the problem in the form

$$\begin{aligned} v &= v_0(r, t) + \sum \varepsilon_i v_i(r, t) + o(l), \\ p &= p_0(r, t) + \sum \varepsilon_i p_i(r, t) + o(l), \\ \rho &= \rho_0(r, t) + \sum \varepsilon_i \rho_i(r, t) + o(l), \quad l = \sqrt{\varepsilon_1^2 + \varepsilon_2^2 + \ldots + \varepsilon_n^2}. \end{aligned} \tag{3.96}$$

We solve the individual linearized problems for the parameters ϵ_i. Substituting these particular solutions into (3.96) we arrive by summation at the complete solution, taking linear corrections to the standard PEP into account. As we have shown in the preceding sections, it is technically simplest to carry out the linearization with respect to some variable parameter $q(t)$ depending on time. Since this almost always means that we are seeking a particular correction in the form $\epsilon_k\varphi_k(r, t) = \epsilon_k\varphi_{1k}(t)\varphi_{2k}(\lambda)$, we obtain from this a means of summing the corrections based on using the constant parameters ϵ_k.

As an example we consider the parameters B and C in the PEP for variable counterpressure p_1 and initial velocity v_1 (§§1, 3 and 4).

An account of the independent corrections here can be given by the solution

$$f=f_0(\lambda)+yf_{11}(\lambda)+qf_{21}(\lambda)+o(y,\ q),$$
$$g=g_0(\lambda)+yg_{11}(\lambda)+qg_{21}(\lambda)+o(y,\ q),$$
$$h=h_0(\lambda)+yh_{11}(\lambda)+qh_{21}(\lambda)+o(y,\ q),$$
$$x=(\delta\alpha^{-1/2})^{2/(\omega-1)}y^{2/(\omega-1)}\exp\left(\frac{2A_{11}}{\omega-1}y+\frac{A_{21}}{3-\omega}q\right).$$

Here for given A, B, and C, the relation between the parameters q and y must be considered in calculating the solution for the same values of time:

$$q=\gamma\,(\delta\alpha^{-1/2})^{-2(\omega-2)}\alpha_3^{\omega-3}y^{-\frac{4(\omega-2)}{\omega-1}+2},$$

where $\alpha_3 = r^0/r^*$ is the ratio of characteristic lengths.

Other examples of superposition of solutions will be considered in subsequent sections. In conclusion, we remark that the principle of superposition of linear solutions is classical and in a general sense is not new. Its use for the linearized equations of a PEP was noted in [11], [12]. The possibilities for using this principle increase the importance of the solution of individual problems obtained by the method of linearization about the exact self-similar solution.

The problem of constructing such solutions was formulated by L. I. Sedov more than fifteen years ago and has now been developed in a number of papers, some of which have been cited in this chapter.

BIBLIOGRAPHY

1. N. S. Burnova (Mel'nikova), *Investigation of the point explosion problem*, Candidate's Dissertation, Moscow State Univ., Moscow, 1953. (Russian)

2. L. I. Sedov, *Similarity and dimensional methods in mechanics*, 4th ed., GITTL, Moscow, 1957; English transl., Academic Press, New York, 1959. MR **19**, 898; **21** #6840.

3. A. Sakurai, *On propagation and structure of the blast wave*. I, II, J. Phys. Soc. Japan **8** (1953), 662–669; **9** (1954), 256–266. MR **15**, 909.

4. V. P. Korobeĭnikov and E. V. Rjazanov, *A theory of linearized explosion problems including back pressure*, Prikl. Mat. Meh. **23** (1959), 749–759 = J. Appl. Math. Mech. **23** (1959), 1066-1080. MR **22** #5296.

5. V. P. Korobeĭnikov, *Application of the theory of explosion to questions of propagation of shock waves during solar flares*, III All-Union Congr. Theoret. Appl. Mech., Abstracts of Reports, "Nauka", Moscow, 1968, p. 171. (Russian)

6. ———, *On the gas flow due to solar flares*, Solar Phys. **7** (1969), 463–469.

7. M. L. Lidov, *On the theory of solutions linearized about one-dimensional self-similar notions of a gas*, Dokl. Akad. Nauk SSSR **102** (1955), 1089–1092. (Russian) MR **17**, 207.

8. V. P. Korobeĭnikov, N. S. Mel'nikova and E. V. Rjazanov, *Theory of point explosions*, Fizmatgiz, Moscow, 1961. (Russian) MR **29** #5494.

9. H. Mirels and J. F. Mullen, *Aerodynamic blast simulation in hypersonic tunnels*, AIAA J. **3** (1965), 2103–2108.

10. D. N. Brušlinskiĭ and G. S. Solomahova, *Investigation of the problem of strong explosion including back pressure*, Teoret. Gidromeh. No. 19, Oborongiz, Moscow, 1956. (Russian)

11. V. P. Korobeĭnikov, P. I. Čuškin and K. V. Šarovatova, *Tables of gas-dynamical functions for the initial stages of a point explosion*, Soobšč. po Vyčisl. Mat., vyp. 2, Izdat. Vyčisl. Centr Akad. Nauk SSSR, Moscow, 1963. (Russian)

12. V. P. Korobeĭnikov and P. I. Čuškin, *Calculation of the initial stage of point explosion in various gases*, PMTF Ž. Prikl. Meh. i Tehn. Fiz. 1963, no. 4, 48–57. (Russian)

13. R. J. Swigart, *Third-order blast wave theory and its application to hypersonic flow past blunt-nosed cylinders*, J. Fluid Mech. 9 (1960), 613–620. MR 22 #7519.

14. A. Sakurai, *Blast wave theory*, Basic Developments in Fluid Dynamics, edited by M. Holt, Vol. 1, Academic Press, New York, 1965, pp. 309–375.

15. G. G. Bach and J. H. Lee, *Higher-order perturbation solutions for blast waves*, AIAA J. 7 (1969), 742–744.

16. I. S. Berezin and N. P. Židkov, *Computing methods*. Vol. II, Fizmatgiz, Moscow, 1959; English transl., Pergamon Press, Oxford; Addidon-Wesley, Reading, Mass., 1965. MR 22 #12686; 30 #4372.

CHAPTER 4

SPHERICAL, CYLINDRICAL AND PLANE EXPLOSIONS FOR CONSTANT INITIAL DENSITY INCLUDING THE EFFECT OF COUNTERPRESSURE

§1. On the asymptotic behavior of the solution near the center of symmetry

In the present chapter we present results of solving the nonlinear point explosion problem taking the counterpressure into account. Before presenting these results, in the first two sections of the chapter we shall consider theoretical questions, the knowledge of which is very useful in studying the point explosion problem (PEP).

We first consider the question of the behavior of the solution of the PEP near the center of the explosion. In the following section asymptotic laws for the damping of shock waves are given.

Qualitative investigation and asymptotic representation of the solution near the center of symmetry can lead to a number of theoretical conclusions; they are also needed in solving the nonlinear problem with counterpressure by numerical methods, since the equations (3.5) have a singularity at the center of symmetry.

The question of the behavior of the solutions of the system (3.5) in a neighborhood of the center of symmetry was first considered by L. I. Sedov in [1]. In Chapter 2 it was shown that in the case of a self-similar problem the desired functions can be represented near the center of symmetry by series (2.26) which in dimensionless variables can be written as follows:

$$\begin{aligned} v_0 &= r\,(\varphi_{00}(t) + \varphi_{10}(t)\,r^{s+2} + \ldots), \\ \rho_0 &= r^s\,(\omega_{00}(t) + \omega_{10}(t)\,r^{s+2} + \ldots), \\ p_0 &= \psi_{-10} + r^{s+2}\,(\psi_{00}(t) + \psi_{10}(t)\,r^{s+2} + \ldots), \end{aligned} \tag{4.1}$$

where $s = \nu/(\gamma - 1)$.

Assuming that the change in the boundary conditions on the front of the shock wave in considering the counterpressure does not alter the form of the representation of the function $\rho(r, t)$ in powers of r indicated in (4.1), we shall suppose that for the non-self-similar problem near the center of symmetry the following asymptotic representation of the gas density is valid:

$$\rho = r^s\,(\omega_0(t) + \omega_1(t)\,r^{s+2} + \omega_2(t)\,r^{2(s+2)} + \ldots). \tag{4.2}$$

In order to obtain asymptotic representations of the other functions, we transform the system (3.5). Using the equation of continuity, the momentum equation can be brought to the form (1.19):

$$\frac{\partial}{\partial t}(\rho v) + \frac{\partial}{\partial r}(\rho v^2) + (\nu - 1)\frac{\rho v^2}{r} + \frac{\partial p}{\partial r} = 0.$$

We take the continuity equation and the last equation of the system (3.5) in the form

$$\frac{\partial}{\partial t}(\rho r^{\nu-1}) + \frac{\partial}{\partial r}(\rho r^{\nu-1} v) = 0,$$

$$\frac{\partial}{\partial t}(p^{1/\gamma} r^{\nu-1}) + \frac{\partial}{\partial r}(p^{1/\gamma} r^{\nu-1} v) = 0.$$

Assuming that $v(0, t) \neq 0$, from the above equations we obtain the following system of integro-differential relations which are equivalent to (3.5):

$$p = p(0, t) - \rho v^2 - (\nu - 1)\int_0^r \frac{\rho v^2}{r}\, dr - \int_0^r \frac{\partial (\rho v)}{\partial t}\, dr, \tag{4.3}$$

$$v = -\frac{1}{r^{\nu-1}\rho}\int_0^r r^{\nu-1}\frac{\partial \rho}{\partial t}\, dr, \tag{4.4}$$

$$v = -\frac{1}{r^{\nu-1} p^{1/\gamma}}\int_0^r \frac{\partial p^{1/\gamma}}{\partial t} r^{\nu-1} dr. \tag{4.5}$$

From (4.3)–(4.5) it follows that if the asymptotic representation for $\rho(r, t)$ has the form (4.2), then for the functions $p(r, t)$ and $v(r, t)$ the expansions indicated in (4.1) are valid. This was first noted in [1], [2]. It is seen by substituting (4.2) into (4.3) and (4.4) and integrating with respect to r.

Thus, recalling the assumption we made regarding the representation of $\rho(r, t)$ near the center, we find that near the center the solution has the asymptotic representation

$$\begin{aligned} v &= r(\varphi_0(t) + \varphi_1(t) r^{s+2} + \dots), \\ \rho &= r^s(\omega_0(t) - \omega_1(t) r^{s+2} + \dots), \\ p &= \psi_{-1}(t) + r^{s+2}(\psi_0(t) + \psi_1(t) r^{s+2} + \dots). \end{aligned} \tag{4.6}$$

We remark that a representation of the form (4.6) was used in [3], [4] for computing the gas flow in a neighborhood of the center of an explosion for the case $\nu = 3$, $\gamma = 1.4$.

According to (4.6), in the expansion of $v(r, t)$ the first term is linear in r. If it is assumed that $p(0, t) \neq 0$ and $(\partial p/\partial t)_{r=0} \neq 0$, then this fact follows from (4.5) independent of the assumptions made on the nature of the expansion of $\rho(r, t)$.

Using (4.6), it is possible to obtain an asymptotic formula giving the relation between the Euler and Lagrange coordinates near the center. According to (1.24), the equation for the amount of motion in Lagrange coordinates has the form

$$\frac{\partial p}{\partial \xi} = -\rho_1 \frac{\xi^{\nu-1}}{r^{\nu-1}} \frac{\partial v}{\partial t}. \tag{4.7}$$

From (4.6) and (4.7) we have

$$(s+2)\, r^{s+\nu-1} \psi_0 \frac{\partial r}{\partial \xi} = -\xi^{\nu-1} \rho_1 \left(\varphi_0^2 + \frac{d\varphi_0}{dt} \right) + \dots,$$

whence we find that

$$r = \xi^{\nu/(s+\nu)} \zeta_1(t) + \zeta_2(t),$$

where $\zeta_1(t)$ and $\zeta_2(t)$ are certain functions of time. Since $r(0, t) = 0$, it follows that $\zeta_2(t) = 0$. Thus the relation between the Euler and Lagrange coordinates is expressed by the formula

$$r = \xi^{1 - \frac{(1)}{\gamma}} \zeta_1(t) + \dots. \tag{4.8}$$

From (4.6)–(4.8) it is possible to make certain qualitative deductions regarding the behavior of the solution. Thus, for example, from these formulas it follows that the temperature $T = p/R\rho$ (R is the gas constant) has order r^{-s} with respect to r. If the function $\omega_0(t)$ is bounded during the entire time of the motion, then the temperature at the center of the explosion is always infinite. It is also possible to obtain an asymptotic formula for calculating the density in a neighborhood of the center for large t. Assuming that the value of the entropy near the center as computed from the self-similar solution differs but little from the true value of the entropy, we have $p_0 \rho_0^{-\gamma} = p^{-\gamma}$. Using (4.1) and (4.6), we can write

$$\rho = \omega_{00}(t) \left(\frac{\psi_{-1}(t)}{\psi_{-10}(t)} \right)^{1/\gamma} r^s.$$

For large values of time the pressure near the center approaches atmospheric pressure; it is therefore possible to set $\psi_{-1} = p_1$. For the values of the density near the center for large t we obtain the asymptotic formula [3], [5]

$$\rho = \omega_{00}(t)\, r^s \left(\frac{p_1}{\psi_{-10}(t)} \right)^{1/\gamma}. \tag{4.9}$$

Since ψ_{00} and ψ_{-10} are known from the self-similar solution, using (4.9) it is possible to calculate approximate values of the density near the center for large t.

Let us make some more remarks on the dependence of the solution near the center on the quantity γ. It is clear from the definition of s that $s(\gamma)$ increases as γ decreases ($\gamma > 1$). Thus for $\nu = 3$ at $\gamma = 1.4$ we have $s = 7.5$, and for $\gamma = 1.2$ we find $s = 15$. Thus for $\gamma = 1.2$ the density near the center is of the order r^{15} with respect to r; the pressure is practically independent of r at a distance from the center which is appreciably greater than for $\gamma = 1.4$. The order with respect to ξ of the Euler coordinate depends only on γ and not on ν.

§2. Laws for the damping of shock waves at large distances from the site of the explosion

Because of the dissipation of energy on the front of the shock wave (and increase in its surface area in the case of spherical and cylindrical symmetry) the shock wave dies out as it propagates away from the center of explosion in a quiescent medium with initial pressure p_1. The laws of damping for a shock wave of a point explosion and the magnitudes of all parameters of the gas flow will be described in detail in subsequent sections of this chapter. The solution of the problem of the damping of a shock wave over the entire time span (from the beginning of the explosion until it almost completely dies out and becomes a wave of small amplitude–an acoustic wave) can be carried out numerically and requires the application of modern computing techniques. Asymptotic formulas which give the limiting laws of the damping of shock waves are very useful for qualitative studies of the behavior of the waves and quantitative estimates of the parameters of the wave front. The laws of degeneration of plane shock waves were established in 1913 by Crussard (in this regard see [6], for example). The laws of the damping of spherical and cylindrical shock waves were first published by L. D. Landau [7]. In establishing these laws it was assumed that the gas motion behind the front of the shock wave, as it dies down, tends to a traveling wave with triangular profile which differs from an acoustic wave only in the precise value of the speed of sound. L. I. Sedov [2] has proposed a method of obtaining the asymptotic laws of shock waves based on using the conditions on the wave front for the basic functions and their derivatives with respect to the coordinate r. We shall present here only a few of the results of these investigations.

In the case of plane waves ($\nu = 1$) and cylindrical waves ($\nu = 2$) the following asymptotic formulas are valid in first approximation for the basic functions and for the law of motion of the shock wave $r_2(t)$:

$$v_2 = \frac{2a_1 C_\nu}{(\gamma+1)\, r_2^{(\nu+1)/4}}, \qquad \rho_2 = \rho_1\left[1 + \frac{2C_\nu}{(\gamma+1)\, r_2^{(\nu+1)/4}}\right], \tag{4.10}$$

$$p_2 = p_1\left[1 + \frac{2\gamma C_\nu}{(\gamma+1)\, r_2^{(\nu+1)/4}}\right], \qquad r_2 = a_1(t - t^*) + \nu C_\nu r_2^{1/2\nu}. \tag{4.11}$$

Here the C_ν are arbitrary constants, t^* is a constant with the dimension of time, and a_1 is the speed of sound.

For spherical shock waves the following asymptotic laws have been obtained:

$$v_2 = \frac{2C_3 a_1}{(\gamma+1)\, r_2 \sqrt{\ln (r_2/r^*)}}, \qquad \rho_2 = \rho_1\left[1 + \frac{2C_3}{(\gamma+1) r_2 \sqrt{\ln (r_2/r^*)}}\right], \tag{4.12}$$

$$p_2 = p_1\left[1 + \frac{2\gamma C_3}{(\gamma+1)\, r_2 \sqrt{\ln (r_2/r^*)}}\right], \qquad r_2 = a_1(t - t^*) - C_3\sqrt{\ln (r_2/r^*)}. \tag{4.13}$$

Under the assumption of smoothness of the solution in some neighborhood of the front of the shock wave, the method of L. I. Sedov was developed by Ju. L. Jakimov [8] and G. M. Šefter [9], who found new asymptotic solutions of the equations of one-dimensional gas dynamics for large values of r. These solutions made it possible to obtain asymptotic formulas for $\nu = 2, 3$ more precise than formulas of the form (4.10), (4.12). In particular, in these papers it is shown that with a suitable definition of the constants C_ν and r^* the more precise formulas for the pressure p_2 on the front of the shock wave have the form

$$\frac{p_2}{p_1} = 1 + \frac{2\gamma}{\gamma+1}\left(\frac{C_{21}}{r_2^{3/4}} + \frac{C_{22}}{r_2^{5/4}}\right) + O\left(r_2^{-3/2}\right) \qquad (\nu = 2), \tag{4.14}$$

$$\frac{p_2}{p_1} = 1 + \frac{2\gamma}{\gamma+1}\left(\frac{C_{31}}{r_2\sqrt{\ln(r_2/r^*)}} + \frac{C_{32}}{r_2(\ln(r_2/r^*))^{2}}\right) + O\left(r_2^{-1}\ln\left(\frac{r_2}{r^*}\right)^{-7/2}\right) \qquad (\nu = 3). \tag{4.15}$$

Here the $C_{\nu 2}$ are new constants, the magnitudes of which depend on the initial wave form. The symbol $O(f(r))$ denotes quantities which tend to zero with increasing r to order $f(r)$. We shall use these asymptotic formulas in the sequel to describe the laws of damping of shock waves arising during a point explosion in a perfect gas. In Chapter 7 we shall consider a generalization of these formulas to the case of the equations of magnetohydrodynamics.

§3. Numerical solution of the problem

3.1. *A brief survey of methods and results.* Various numerical and approximation methods are used to calculate the parameters of the shock waves and transient flows arising during an explosion. In the present section we consider in detail questions of the calculation of a point explosion in a perfect unbounded gas with constant initial density $\rho_1 = \rho_\infty$ in the presence of counterpressure for the spherical, cylindrical, and plane cases. It is assumed that the gas is nonviscous, does not conduct heat, and has a constant adiabatic exponent γ. The formulation of the basic PEP considered has been given in previous chapters. Its complete solution for various ν and γ is of substantial importance both for the theory itself and for various of its applications.

Among the numerical methods used, we mention the method of finite differences and various of its modifications as applied to the calculation of shock waves. Examples of the numerical solution of the PEP with counterpressure by the method of finite differences are given in [3], [4], [10], [11]. Here only the special case of a spherical explosion with $\gamma = 1.4$ was considered. In order to consider the computation of the problem over a time span we introduce the dimensionless time τ defined, as in Chapter 3, by

$$\tau = \frac{t}{t^0}, \qquad t^0 = r^0\left(\frac{\rho_\infty}{p_\infty}\right)^{1/2}, \qquad r^0 = \left(\frac{E^0}{p_\infty}\right)^{1/\nu}, \tag{4.16}$$

where E^0 is the energy of the explosion and p is the initial pressure (in general, quantities with index ∞ will denote parameters of the unperturbed gas, while it will be convenient for us to denote parameters of the gas immediately behind the front of the shock wave by the index n). The authors of the papers indicated prescribed the initial data from the self-similar solution for various values of τ and correspondingly for various ratios of the pressures p_n/p_∞ on passage through the front of a spherical shock wave.

In the work of D. E. Ohocimskiĭ et al. [3] $\tau_{\text{in}} = 0.00039$ $(p_n/p_\infty = 1743)$; in the work of H. Goldstine and J. von Neumann [4] $\tau_{\text{in}} = 0.00425$ $(p_n/p_\infty = 100)$; and H. Brode [10] has $\tau_{\text{in}} = 0.0042$ $(p_n/p_\infty = 1601)$. From these data it follows that the smallest τ_{in} was taken in [3]. Computation was performed to the following values of $\tau = \tau_k$ and correspondingly p_n/p_∞: $\tau_k = 18.8$ $(p_n/p_\infty = 1.008)$ [3]; $\tau_k = 9.526$ $(p_n/p_\infty = 1.017)$ [4]; $\tau_k = 2.905$ $(p_n/p_\infty = 1.06)$ [10].

We note also that a computation for values of $\tau > 18.8$ is given in [11]. The data given indicate that in these papers the computation was carried out to a stage close to the point of degeneration of the shock wave into an acoustic wave. We note that the most complete and most accurate results were obtained in [3], [11]. An analysis of the papers [3]–[10] is given in [2], [5] (cf. also [28]) and will not be reproduced in its entirety here.

In addition to finite-difference methods, the method of characteristics [3], [12] has also been applied to the computation of point explosions (general questions of the method of characteristics are considered in [13]–[15]).

For the computation of transient gas flows it is possible to apply approximation methods [1], [2], [16], [17] based on introducing special interpolation formulas for one or more of the unknown functions. The unknown parameters entering in these interpolation formulas are determined from the integral conservation laws and the equations of gas dynamics.

The approximation methods proposed by L. I. Sedov [1], [2] and based on introducing interpolation formulas and using the integral laws of mass and energy conservation have been applied to point explosions in [23]–[25]. They are characterized by the fact that they are easy to use, but they do not give sufficiently high accuracy. The method we have developed for solving the PEP is close to the approximation methods, but it is based on using the scheme of the numerical method of integral relations proposed by A. A. Dorodnicyn [26] and widely applied to the calculation of stationary flows in the work of D. M. Belocerkovskiĭ, P. I. Čuškin, and other authors (see the surveys [27], [28]). Development of the method of solving problems in the theory of explosion using the scheme of integral relations was begun by the author in 1960. Preliminary results of the first stage of this work were published in the author's paper [19] of 1962 and in a paper [29] written jointly with V. P. Karlikov and E. V. Rjazanov. The second stage of this work, belonging to the years 1962–1967, was carried out jointly with P. I. Čuškin and was characterized by the use of higher-order approximations in the scheme of integral relations. The author has extended this method to problems of explosion taking into account the effect of a magnetic field and for explosion in a detonable gas; this will be discussed in subsequent chapters.

Preliminary results of computing the PEP by the method developed were published in [30] and presented in the reports [31], [32]. A more complete exposition of the method and results is given in [33].

3.2. *The basic equations and conditions of the problem.* We take the equations of transient one-dimensional motion of a gas in the Euler variables in the form

$$\frac{1}{\nu}\frac{\partial}{\partial t}\left(\frac{\rho u}{r^{\nu-2}}\right)+\frac{\partial}{\partial\zeta}(\rho v^2+p)+\frac{\nu-1}{\nu\zeta}\rho v^2=0, \tag{4.17}$$

$$\frac{1}{\nu}\frac{\partial\rho}{\partial t}+\frac{\partial}{\partial\zeta}(\zeta\rho u)=0, \tag{4.18}$$

$$\frac{1}{\nu}\frac{\partial E}{\partial t}+\frac{\partial}{\partial\zeta}[\zeta u(E+p)]=0, \tag{4.19}$$

$$\frac{1}{\nu}\frac{\partial p^{1/\gamma}}{\partial t}+\frac{\partial}{\partial\zeta}(\zeta u p^{1/\gamma})=0, \tag{4.20}$$

where $\zeta = r^\nu$, $u = v/r$, ρ is the density, v is the velocity, and E is the total energy per unit volume of gas, which in the case considered here can be written as follows:

$$E=\frac{\rho v^2}{2}+\frac{p}{\gamma-1}.$$

In the system (4.17)–(4.20) the first equation is the momentum equation, the second is the equation of continuity, the third equation expresses the law of energy conservation, and the last equation gives the condition of adiabatic flow. In this system only three equations are independent. One of the four equations of the system (4.17)–(4.20) can be considered a consequence of the others.

The boundary conditions for the problem are the condition at the center of the explosion

$$v(0,\ t)=0 \tag{4.21}$$

and the condition on the shock wave

$$p_n=p_\infty\frac{2\gamma-(\gamma-1)q}{(\gamma+1)q},\qquad v_n=\frac{2a_\infty}{\gamma+1}\frac{1-q}{\sqrt{q}},\ \rho_n=\rho_\infty\frac{\gamma+1}{\gamma-1+2q},$$
$$E_n=\frac{p_n}{\gamma-1}+\frac{1}{2}\rho_n v_n^2. \tag{4.22}$$

Here a is the speed of sound, $q = a_\infty^2/D^2$, D is the speed of the shock wave, and the index n denotes parameters on the front of the shock wave.

There are also the following integral relations:

$$\rho_\infty r_n^\nu=\nu\int_0^{r_n}\rho r^{\nu-1}dr, \tag{4.23}$$

$$E^0 = \sigma_\nu \int E r^{\nu-1} dr - \frac{\sigma_\nu p_\infty r_n^\nu}{\nu(\gamma-1)}, \tag{4.24}$$

These relations can be obtained either from the differential equations and boundary conditions (cf. [5]) or by applying the integral laws of mass and energy conservation to the mass of gas contained in the shock wave.

In solving the problem it is convenient to introduce a system of dimensionless variables. Solution of the problem in dimensionless variables is adequate for any values of E^0, p_∞ and ρ_∞. In forming dimensionless variables it is possible to relate the unknown functions and independent variables both to corresponding dimensional constants and to certain dimensional functions. We take the following functions as the new dimensionless independent variables and dimensionless functions:

$$q, \quad \xi = \frac{\zeta}{\zeta_n}, \qquad \varphi = u r_n \frac{1}{D}, \qquad g = \frac{\rho}{\rho_\infty}, \qquad e = \frac{Eq}{p_\infty}, \qquad \psi = \left(\frac{pq}{p_\infty}\right)^{1/\gamma}. \tag{4.25}$$

The relation between the dimensionless time $\tau = t/t^0$ and q follows from the kinematic condition $D = dr_n/dt$:

$$\tau' = R_n \frac{\mu}{\nu} \sqrt{\frac{q}{\gamma}} \qquad \left(\mu = \frac{\delta'}{\delta} + \frac{1}{q}\right). \tag{4.26}$$

Here and below the prime denotes the derivative with respect to q. We have also introduced the notation $R = r/r^0$ and $\delta = \sigma_\nu R_n^\nu / \nu q$.

In the new variables (4.25) the system (4.17)–(4.20) takes the form

$$\frac{\partial m}{\partial q} - \frac{m}{2q} + \frac{\mu}{\nu}\left[m + \nu \frac{\partial}{\partial \xi}\left(g f^2 + \frac{\psi^\gamma}{\gamma} - m\xi\right) + \frac{\nu-1}{\xi} g f^2\right] = 0, \tag{4.27}$$

$$\frac{\partial g}{\partial q} + \mu\left[\frac{\partial}{\partial \xi}(g\varphi\xi) - \frac{\partial}{\partial \xi}(g\xi) + g\right] = 0, \tag{4.28}$$

$$\frac{\partial e}{\partial q} - \frac{e}{q} + \mu\left[\frac{\partial}{\partial \xi}(e\varphi\xi) - \frac{\partial}{\partial \xi}(e\xi) + \frac{\partial}{\partial \xi}(\psi^\gamma \varphi\xi) + e\right] = 0, \tag{4.29}$$

$$\frac{\partial \psi}{\partial q} - \frac{\psi}{\gamma q} + \mu\left[\frac{\partial}{\partial \xi}(\psi\varphi\xi) - \frac{\partial}{\partial \xi}(\psi\xi) + \psi\right] = 0, \tag{4.30}$$

where

$$m = g\varphi\xi^{(2-\nu)/\nu}, \quad f = \varphi\xi^{1/\nu}.$$

The connection between the functions g, φ, ψ and e is given by the relation

$$e = \frac{\gamma}{2} g\varphi^2\xi^{2/\nu} + \frac{\psi^\gamma}{\gamma-1}. \tag{4.31}$$

The boundary conditions (4.21) and (4.22) can be brought to the form

$$f(0, q) = 0, \tag{4.32}$$

$$\varphi_n = \frac{2}{\gamma+1}(1-q), \quad g_n = \frac{\gamma+1}{\gamma-1+2q}, \quad \varphi_n^\gamma = \frac{2\gamma-(\gamma-1)q}{\gamma+1},$$
$$e_n = \frac{2\gamma}{\gamma+1}\left[\frac{(1-q)^2}{\gamma-1+2q}+\frac{1}{\gamma-1}\right]-\frac{q}{\gamma+1}. \tag{4.33}$$

Obviously, of the four unknown functions g, e, φ and ψ defined by (4.25) it sufficies to obtain from the system (4.27)–(4.30) only three in order to solve the problem. The fourth unknown function can be found from the algebraic relation (4.31).

To solve the problem in question it is necessary to integrate the system of equations of gas dynamics (4.27)–(4.30) in the domain $0 \leqslant q \leqslant 1$, $0 \leqslant \xi \leqslant 1$ with boundary conditions (4.32) and (4.33). We note that to the value $q = 0$ there corresponds a known solution of the strong explosion problem, and for small q the solution of the linearized problem is valid.

In studying the problem in the Lagrange variables the original system of equations of gas dynamics (cf. §1 of Chapter 1) can be taken in the following form:

$$g\frac{\partial R^\nu}{\partial \eta} = 1, \quad \frac{\partial R}{\partial \tau} = \sqrt{\gamma}F, \quad \frac{\partial F}{\partial \tau} = -\frac{\nu}{\sqrt{\gamma}}R^{\nu-1}\frac{\partial P}{\partial \eta},$$
$$\frac{\partial}{\partial \tau}(\chi(\theta)) = 0 \quad (0 \leqslant \tau < \infty, \quad 0 \leqslant \eta < \infty), \tag{4.34}$$

where $\eta = r_0^\nu / r^{0\nu}$, $F = v/a_\infty$, $P = p/p_\infty$, $\theta = g/P^{1/\gamma}$, $\chi(\theta)$ is the entropy function, and r_0 is the initial coordinate of the particle. In the Lagrange variables, in addition to (4.21) and (4.22) we have

$$R(0, \tau) = 0, \qquad R(\eta_n, \tau) = \eta_n^{1/\nu}. \tag{4.35}$$

From the first equation of (4.34) we find the relation

$$\eta = R_n^\nu\left(1 - \int_\xi^1 g\,d\xi\right), \tag{4.36}$$

which gives the connection between the Lagrange coordinate η and the relative Euler coordinate ξ.

3.3. *Scheme of calculating in the Euler variables.* From the results of Chapter 1, §2 of Chapter 2, and §1 of Chapter 4 it follows that for the unknown functions g, e, ψ and φ we have the asyptotic formulas

$$g = \frac{\gamma+1}{\gamma-1}(C\delta\psi_0^\gamma\xi)^{1/(\gamma-1)} + O(\xi)^{2d\nu-1/\nu}), \quad e = e_0 + O(\xi^d), \quad \psi = \psi_0 + O(\xi^d),$$
$$\varphi = \varphi_0 + O(\xi^d), \tag{4.37}$$

where

$$e_0=\frac{\psi_0^\gamma}{\gamma-1},\qquad \varphi=\frac{1}{\mu}\left(\frac{1}{\gamma q}-\psi_0^{-1}\psi_0'\right),\qquad d=\frac{\nu+2(\gamma-1)}{\nu(\gamma-1)},\qquad C=\frac{\alpha\nu(\nu+2)^2(\gamma+1)^2}{8\gamma\sigma_\nu},$$

where α is a constant defined by the integral law of energy conservation in the self-similar case (cf. Chapter 2).

In the domain of integration $0\leqslant\xi\leqslant 1$, $0\leqslant q\leqslant 1$ we single out a central interval bounded by the lines $\xi=0$ and $\xi=\xi_0(q)$ in which the asymptotic formulas (4.37) are applicable. For the choice of ξ_0 we fix the Lagrange coordinate $\eta=\eta_0=(r_{00}/r^0)^\nu$ of the particle such that in the initial stage of the explosion the corresponding asymptotic formula in (4.37) gives ψ_0 with prescribed accuracy. Then, using the equation of continuity of the system (4.34) and (4.37) we obtain for ξ_0 the expression

$$\xi_0=\frac{1}{C\delta\psi_0}\left[\frac{\gamma\sigma_\nu C\eta_0}{\nu q(\gamma+1)}\right]^{\frac{\gamma-1}{\gamma}}. \tag{4.38}$$

We remark that other methods of choosing the central interval are possible.

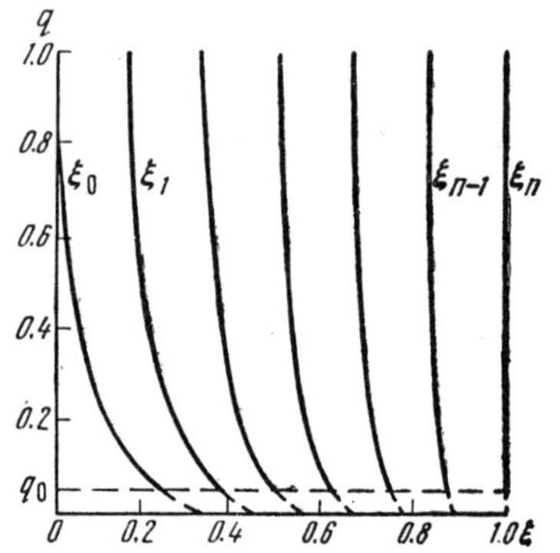

FIGURE 31. Scheme of partitioning the domain of integration into strips

In the domain between the boundary of the central interval $\xi=\xi_0(q)$ and the shock wave $\xi=\xi_n=1$ we define the solution by the method of integral relations. At the nth approximation we partition this domain into n strips by the $n-1$ intermediate lines (Figure 31)

$$\xi_i(q)=\frac{n-i}{n}\xi_0+\frac{i}{n}\qquad (i=1,\ 2,\ \ldots,\ n-1).$$

If each equation of the system (4.27)–(4.30) is integrated on ξ from the value $\xi_l(q)$ $(l=0,1,\cdots,n-1)$ to the shock wave $\xi_n=1$, then from this system, taking the boundary conditions into account, we obtain the $4n$ integral relations

$$\mathcal{J}'_{1l} + m_l\xi'_l - \frac{1}{2q}\mathcal{J}_{1l} + \mu\Big[\frac{1}{\nu}\mathcal{J}_{1l} + m_l\xi_l - g_l f_l^2 + \frac{1}{\gamma}(q - \psi_l^\gamma)\Big] + \frac{\nu-1}{\nu}\mathcal{J}_{0l} = 0, \qquad (4.39)$$

$$\mathcal{J}'_{2l} + g_l\xi'_l - \mu[1 + g_l\xi_l(\varphi_l - 1) - \mathcal{J}_{2l}] = 0, \qquad (4.40)$$

$$\mathcal{J}'_{3l} + e_l\xi'_l - \frac{1}{q}\mathcal{J}_{3l} - \mu\Big[-\mathcal{J}_{3l} + e_l\xi_l(\varphi_l - 1) + \psi_l^\gamma\varphi_l\xi_l + \frac{q}{\gamma-1}\Big] = 0, \qquad (4.41)$$

$$\mathcal{J}'_{4l} + \psi_l\xi'_l - \frac{1}{\gamma q}\mathcal{J}_{4l} - \mu\Big[-\mathcal{J}_{4l} + \psi_l\xi_l(\varphi_l - 1) + \frac{\psi_n}{g_n}\Big] = 0. \qquad (4.42)$$

Here we have written

$$\mathcal{J}_{0l} = \int_{\xi_l}^{1} gf^2\frac{d\xi}{\xi}, \qquad \mathcal{J}_{1l} = \int_{\xi_l}^{1} m\,d\xi, \qquad \mathcal{J}_{2l} = \int_{\xi_l}^{1} g\,d\xi,$$

$$\mathcal{J}_{3l} = \int_{\xi_l}^{1} l\,d\xi, \qquad \mathcal{J}_{4l} = \int_{\xi_l}^{1} \psi\,d\xi.$$

We denote the unknown functions generally by Ω_s and take $\Omega_1 = m$, $\Omega_2 = g$, $\Omega_3 = e$ and $\Omega_4 = \psi$.

In addition, we denote the values of all the unknown functions on each boundary of the strips in question by the corresponding index. Then the system of integral relations (4.39)–(4.42) by means of expressions (4.33) for the functions on the shock wave and the asymptotic formulas (4.37) can be brought to the form

$$\mathcal{J}'_{s0} + k_s\Omega_{0}\xi'_0 + b_s\xi_0\Omega'_{s0} + \kappa_s\Omega'_{40} + \delta^{-1}\delta' X = q^{-1}Y_s, \qquad (4.43)$$

$$\mathcal{J}'_{si} + \Omega_{si}\xi'_i - \delta^{-1}\delta' W_{si} = (W_{si} + \beta_s\mathcal{J}_{si})\,q^{-1}, \qquad (4.44)$$

where

$$\mathcal{J}_{sl} = \int_{\xi_l}^{1} \Omega_s\,d\xi,$$

$$X_1 = \frac{1}{\nu}\mathcal{J}_{10} - \frac{\psi_0{}^\gamma}{\gamma} + \frac{q}{\gamma} + \frac{\nu-1}{\nu}\mathcal{J}_{00} + \frac{k_1 m_0\xi_0}{\nu}[1 + (\nu-1)\varphi_0],$$

$$X_2 = g_0\xi_0 - 1 + \mathcal{J}_{20}, \qquad X_3 = e_0\xi_0 - Y_3 + \mathcal{J}_{30},$$

$$X_4 = \psi_0\xi_0 - \psi_n g_n^{-1} + \mathcal{J}_{40}, \qquad Y_1 = \frac{1}{2}(\mathcal{J}_{10} + k_1 m_0\xi_0) - X_1,$$

$$Y_2 = \beta_4 g_0\xi_0 - X_2, \qquad Y_3 = \frac{q}{\gamma-1}, \qquad Y_n = \beta_4(\psi_0\xi_0 + \mathcal{J}_{40}) - X_4,$$

$$W_{1i} = g_i f_i^2 + \frac{\psi_i^\gamma}{\gamma} - \frac{q}{\gamma} - \frac{\nu-1}{\nu}\mathcal{J}_{0i} - \frac{\mathcal{J}_{1i}}{\nu} - \xi_i m_i,$$

$$W_{2i}=g_i(\varphi_i-1)\xi_i+1-\mathcal{J}_{2i},$$
$$W_{3i}=e_i(\varphi_i-1)\xi_i+Y_3-\mathcal{J}_{31}+\psi_i^\gamma\varphi_i\xi_i,$$
$$W_{4i}=\psi_i(\varphi_i-1)\xi_i+\psi g_n^{-1}-\mathcal{J}_{4i},$$

$$k_1=\frac{\nu(\gamma-1)}{\nu+2(\gamma-1)(\nu-1)},\qquad k_2=k_3=k_4=1,$$
$$b_i=k_1,\qquad b_2=b_3=b_4=0,$$
$$\kappa_1=0,\qquad \kappa_2=\frac{g_0\xi_0}{\psi_0},\qquad \kappa_3=\frac{\gamma\xi_0 e_0}{\psi_0},\qquad \kappa_4=\xi_0,$$
$$\beta_1=\frac{1}{2},\qquad \beta_2=0,\qquad \beta_3=1,\qquad \beta_4=\frac{1}{\gamma}.$$

The relations (4.43) and (4.44) contain the quantities m_l, g_l, φ_l and e_l, between which there are the relations

$$\varphi_l=\frac{m_l}{g_l}\xi_l^{(\nu-2)/\nu}, \tag{4.45}$$

$$e_0=\frac{\psi_0^\gamma}{\gamma-1},\qquad e_i=\frac{\gamma}{2}\varphi_i^2 g_i\xi_i^{2/\nu}+\frac{\psi_i^\gamma}{\gamma-1}. \tag{4.46}$$

Equations (4.43)–(4.44) are basic for the following construction of the method of calculating the problem. It follows from the foregoing that in the system (4.43), (4.44) there are only $3n$ independent integral relations. In order to obtain from (4.43), (4.44) an approximate system of ordinary differential equations we shall approximate the integrands by Lagrange interpolation polynomials with interpolation nodes on the lines $\xi=\xi_k$ $(k=0,1,\cdots,n)$. It is possible to apply either "global" approximations in which the degree of the polynomials is equal to the number of strips n, or "piecewise" approximations (for $n>1$) in which the degree of the polynomials is less than n.

If the degree of the interpolation polynomials is n, we have

$$\Omega_s=\sum_{k=0}^{n}\frac{(-1)^{n-k}z(z-1)(z-2)\dots(z-n)}{k!(n-k)!(z-k)}\Omega_{sk}, \tag{4.47}$$

where $z=(\xi-\xi_0)n/(1-\xi_0)$. For the integrals $\mathcal{J}_{sl}$ we obtain the expression

$$\mathcal{J}_{sl}=(1-\xi_0)\sum_{k=0}^{n}A_{lk}\Omega_{sk}, \tag{4.48}$$

where the coefficients A_{lk} for any n were introduced in [30] and called the generalized Cotes coefficients:

$$A_{kl}=\frac{(-1)^{n-k}}{k!(n-k)!n}\int_l^n\frac{z(z-1)\dots(z-n)}{z-k}dz. \tag{4.49}$$

Numerical values of the A_{lk} for several n are given in the Appendix.

If the approximations are carried out by choosing interpolation polynomials having degree less than n, then the expressions for the approximate representation of the integrals can, as before, be written in the form (4.48). We mention that application of interpolation polynomials of this type (polynomials of second degree, for example) is expedient in computing schemes with approximations of high order $(n > 4)$ when the required accuracy can be achieved by using a large number of strips. Such "piecewise" approximations simplify the structure of the approximating system and may well be more suitable then "global" approximations in regions where there is abrupt variation of the functions represented. Questions of the optimal choice of approximations can be resolved in the process of experimental computations.

We note also that in a number of cases it is possible to use overlapping approximations. This means that in computing the integrals $\mathcal{J}_{si}$ values of the corresponding integrands are chosen not only on the boundaries of the strips with numbers $i, i+1, \cdots, n$ but also on the boundaries $i-1, i-2, \cdots$.

Substituting the quadrature formulas (4.48) into the system of integral relations (4.43), (4.44), we arrive at an approximating system of $4n$ ordinary differential equations with respect to q for $4n$ unknown functions. However, we shall find g_0 not from this system but from the corresponding asymptotic formula (4.37) and thus drop one of the equations of the approximating system.

To determine the $4n-1$ unknown functions $m_0, \varphi_0, \delta, m_i, g_i, e_i, \psi_i$ we then obtain a system of differential equations which can be written as follows:

$$\left[b_\lambda \xi_0 + (1-\xi_0)\left(A_{00} - \sum_{i=1}^{n-1} \omega_i A_{i0}\right)\right] \underline{\Omega}'_{\lambda 0} + K_\lambda \xi'_0$$

$$+ \kappa_\lambda \underline{\Omega}'_{40} + \left(X_\lambda + \sum_{i=1}^{n-1} \omega_i W_{\lambda i}\right) \frac{\delta'}{\delta}$$

$$= \frac{1}{q}\left[Y_\lambda - \sum_{i=1}^{n-1} \omega_i (W_{\lambda i} + \beta_\lambda \mathcal{J}_{\lambda i})\right] - (1-\xi_0)\left(A_{0n} - \sum_{i=1}^{n-1} \omega_i A_{in}\right) \underline{\Omega}'_{\lambda n}, \tag{4.50}$$

$$\underline{\Omega}'_{sj} = \frac{1}{1-\xi_0}\left[\underline{\Omega}_{sj} \xi'_0 + \sum_{i=1}^{n-1} \frac{\Delta_{ij}}{\Delta_n} F_{si}\right]$$

$$(\lambda = 1, 3, 4), \qquad (s = 1, 2, 3, 4), \qquad (j = 1, 2, \ldots, n-1). \tag{4.51}$$

Here we have introduced the notation

$$K_\lambda = (k_\lambda - A_{00}) \underline{\Omega}_{\lambda 0} - A_{0n} \underline{\Omega}_{\lambda n} + \sum_{i=1}^{n-1} \omega_i z_{\lambda i},$$

$$z_{si} = A_{i0} \underline{\Omega}_{s0} + A_{in} \underline{\Omega}_{sn} - \frac{n-i}{n} \underline{\Omega}_{si},$$

$$F_{si} = z_{si} \xi'_0 - (1-\xi_0)(A_{i0} \underline{\Omega}'_{s0} + A_{in} \underline{\Omega}'_{sn}) + \frac{\beta_s}{q} \mathcal{J}_{si} + \mu W_{si},$$

$$\omega_i = \frac{1}{\Delta n} \sum_{j=1}^{n-1} A_{0j} \Delta_{ij}, \quad \Delta_n = |A_{ij}|;$$

the Δ_{ij} are the algebraic complements of the elements A_{ij} of the determinant Δ_n. The derivative ξ_0' contained in (4.50) and (4.51) is eliminated by means of (4.38).

We adjoin to the system (4.50)–(4.51) still another equation which follows from (4.37):

$$\gamma\varphi_0 (1 + q\delta^{-1}\delta') - 1 + \gamma q \psi_0^{-1} \psi_0' = 0. \tag{4.52}$$

This equation may be chosen in place of one of the equations (4.50), (4.51). The system (4.50)–(4.51) then affords the possibility of constructing different variants of the computing schemes. In all these schemes the function $\tau(q)$ is determined from the differential equations (4.26).

General control of the accuracy of the solution is achieved by satisfying the adiabatic condition in the particle with coordinate η_0 and by satisfying the integral conservation laws of mass and energy (4.23), (4.24) which in the present notation can be written as follows:

$$Y_2 = 0, \qquad X_3 - \frac{1}{\delta} = 0.$$

These equations are satisfied with the respective errors ϵ_{1n} and ϵ_{2n} defined by

$$\varepsilon_{1n} = Y_2, \qquad \varepsilon_{2n} = \left(X_3 - \frac{1}{\delta}\right)\left(\frac{1}{\delta} + \frac{q}{\gamma - 1}\right)^{-1}.$$

We mention that certain computing schemes make it possible to introduce other means of accuracy control.

We have analyzed the following variants of the computing schemes.

VARIANT I. The following system of equations is chosen: (4.52), the first and second equations of (4.50), i.e. the equations with indices $\lambda = 1$ and $\lambda = 3$, and the system (4.51). The quantities $\tau, \delta, m_l, \psi_l, g_i, e_i$ are determined by integration. The quantities φ_l are found from (4.45). Formulas (4.46) serve as auxiliary control of the accuracy of the computations.

VARIANT II. The following system of equations is chosen: (4.52), the first and third equations of (4.50), i.e. the equations with indices $\lambda = 1$ and $\lambda = 4$, and the system (4.51). The quantities $\tau, \delta, m_l, \psi_l, g_i, e_i$ are determined by integration. The quantities φ_l are found from (4.45). The quantities e_i are also determined according to (4.46), which provides the possibility for additional control of accuracy.

VARIANT III. The following system of equations is chosen: the second and third equations of (4.50), i.e. the equations with indices $\lambda = 3$ and $\lambda = 4$, and the equations of (4.51) with indices $s = 2, 3, 4$. The quantities $\tau, \delta, \psi_l, g_i, e_i$ are determined by integration. The relative dimensionless velocities φ_i are found from the algebraic equation following from (4.46):

$$\varphi_i = \pm\left[\left(e_i - \frac{\psi_i^\gamma}{\gamma - 1}\right)\frac{2}{\gamma g_i}\right]^{1/2} \xi^{-1/\nu}. \tag{4.53}$$

The quantity φ_0 is computed from (4.52).

We remark that the computing scheme corresponding to Variant III was considered in [30].

In all these variants of the computing scheme the approximating system of ordinary differential equations is integrated by some numerical method on a computer. For this system there is a Cauchy problem in which the initial data for some sufficiently small $q = q_0$ are taken from the linearized solution of the explosion problem presented in Chapter 3. Integration of the approximating system is carried out to a value of q close to one.

In the method under consideration the solution of the problem is determined for various numbers of strips n. Comparison of results serves as an internal control of the method and characterizes the accuracy of the solution computed. Computations show that in practice satisfactory accuracy is obtained with $n = 4$.

For a more precise description of the behavior of parameters of the front of the shock wave as it degenerates into an acoustic wave it is expedient to consider the asymptotic laws of the damping of shock waves (4.10)–(4.13) discussed in the preceding section. Using these damping laws, it is possible to obtain the following formulas for $\delta(q)$ for q close to one:

$$\begin{aligned}
&\frac{\delta'}{\delta} = \frac{2\sqrt{R_n}}{\alpha_{1*}} - \frac{1}{q} && (\nu = 1),\\
&\frac{\delta'}{\delta} = \frac{8}{3}\alpha_{2*}R_n^{3/4} - \frac{1}{q} && (\nu = 2),\\
&\frac{\delta'}{\delta} = \frac{6(\ln R_n + \alpha_{3*}) - 3}{2(\ln R_n + \alpha_{3*})(1-q)} - \frac{1}{q} && (\nu = 3),
\end{aligned} \tag{4.54}$$

where the $\alpha_{\nu *}$ are constants which are determined for sufficiently large q from the condition that the function $\delta(q)$ found from system (4.50)–(4.52) match smoothly with that found from the asymptotic formulas (4.54).

No special analysis of the stability of the computing scheme was made. Selective verification of the Courant stability condition indicated that in the computations this condition was satisfied but without much margin. It is also to be noted that for values of q close to the initial value q_0 in some cases "oscillation" of the solution of the approximating system about the known linearized solution of the problem was observed. This phenomenon disappeared in considering approximating systems with a higher number n. In all the cases computed, moreover, these oscillations died out with increasing q. It was found that by suitable choice of the quantities q_0 and

$\xi_0(q_0)$ and by choice of the type of approximation of the integrals it was possible to considerably reduce this initial oscillation of the solution, which is probably related to insufficiently precise correlation of the initial data.

We consider the method of computing the flow field and the basic parameters of the blast wave. The unknown functions on the lines $\xi_i(q)$ are found by numerical solution of the differential equations. After this the values of all the unknown functions for $\xi > \xi_0$ with fixed τ or q can be computed by the interpolation polynomials (4.47) or by using interpolation of lower order for large n.

To compute the variation of the flow parameters at fixed points of space the following method can be used. At some moment of time we fix the quantity $R_n = R^*$ and accordingly the quantity $R = r/r^0$; namely, we take $R = R^*$. Noting that $\xi = (R/R_n)^\nu$, for fixed $R = R^*$ the values $\xi^* = (R^*/R_n)^\nu$ are determined at any other moment of time. On the basis of these values of ξ^* the quantities $z = n(\xi^* - \xi_0)/(1 - \xi_0)$ are computed, and the variations of the functions in time at fixed points of space are found from the values of the unknown functions on the boundaries of the strips by (4.47) or for large n by interpolation of lower order.

In particular, the determination of the magnitudes of the dimensionless pressure $P = (p/p_\infty) = \psi^\gamma/q$ is of major interest. From the values $P(R^*, \tau) = P^*(\tau)$ it is possible to compute the total impulses of overpressures

$$J_p = \int_{\tau_*}^{\tau_k} (P^* - 1)\, d\tau,$$

and also the positive impulses

$$J_p^+ = \int_{\tau_*}^{\tau_k} (P^* - 1)\, d\tau, \qquad P^* - 1 > 0,$$

where τ_* is the time of arrival of the shock wave at the given point and τ_k is a finite computed time.

The dimensionless impulse J_p and the dimensional impulse I_p are related by the formulas

$$J_p = \frac{I_p}{t^0 p_\infty}, \qquad I_p = \int_{t_*}^{t_k} (p^* - p_\infty)\, dt.$$

It is moreover possible to compute the dimensionless impulse of pressure head J_v by

$$J_v = \int_{\tau_*}^{\tau_k} \frac{g^* f^{*2}}{2q}\, d\tau,$$

and

$$J_v = \frac{I_v}{t^0 a_\infty^2 \rho_\infty}, \qquad I_v = \int_{t_*}^{t_k} \frac{\rho v^{*2}}{2}\, dt.$$

The quantity I_v is sometimes called the impulse of mass transport.

The action time of the positive phase of pressure τ^* is found from the relation $\tau^* = \bar{\tau} - \tau_*$, where $\bar{\tau}$ is the time of passage of the overpressure through zero.

By means of (4.46) it is possible to determine the relation between the Lagrange coordinate η and the variables ξ and τ. The formulas presented in this section make it possible to determine the basic unknown quantities.

In constructing the method of computing in the Euler variables we made use of the system of dimensionless variables (4.25). Naturally, an analogous method can be constructed for other types of dimensionless variables as well. For example, one may introduce the independent variables R, τ and unknown functions v/a_∞, p/p_∞, ρ/ρ_∞, E/E_∞. In this case the approximation is carried out on R, while the approximating system will contain differential equations where τ (or q) can be taken as the independent variable. A scheme of computing in the Lagrange variables is presented in [33].

We remark that when the problem is considered in either the Euler or Lagrange variables it is possible to use interpolation with respect to the time rather than with respect to the spatial coordinate.

3.4. *Results of the computations.* In order to solve the problem according to the method proposed by E. Bišimov and K. V. Šarovatova programs were written for the numerical integration of the approximating system of differential equations for the computers **BÈSM**-2 and **BÈSM-3M** of the Computing Center of the Academy of Sciences, USSR. These programs made it possible to compute the basic functions φ_k, g_k, e_k, ψ_k for $n = 1, 2, 3, 4, 8$ according to the three variants of the computing schemes described above. The computations were made with through approximation of the integrands by Legendre polynomials, and for $n = 8$ piecewise-linear and parabolic approximations were also considered.

In the initial stage of the work computations were made according to the scheme of Variant III. However, in this scheme a loss of accuracy was found in computing φ_i in the region of the zero of the expression appearing to the one half power in (4.53). Variant I was later used as basis. Separate computations were also made according to Variant II. The greater n is, the better the results of computations with respect to different variants agree with one another.

The initial data were prescribed as follows. The quantities q_0 and $\xi_0(q_0)$ were chosen such that in the domain of validity of the linearized solution, i.e. for $q_0 < (\gamma - 1)/2$ (cf. §5 of Chapter 3) the following conditions were satisfied:

$$\varepsilon_{1n}(q_0) < \varepsilon_{1n}^0, \qquad \varepsilon_{2n}(q_0) < \varepsilon_{2n}^0,$$
$$|\psi(\xi_0, q_0) - \psi(0, q_0)| < \varepsilon_{3n}^0 \psi(0, q_0),$$

where ϵ_{1n}^0, ϵ_{2n}^0 and ϵ_{3n}^0 are given numbers which can be chosen depending on the accuracy required and the number of the approximation n. Moreover, the condition that the derivative $\varphi_0'(q_0)$ computed by the approximating system be close to its exact value at $q = 0$ known from the solution of the linearized problem was also checked.

If it is assumed that the initial value of the central interval $\xi_0(q_0) = \xi_{00}$ in the self-similar problem and in the linearized problem coincide, then from (3.72) we have for computing the unknown functions at $q = q_0$

$$g_k = g_{k0} + q_0 g_{k1} \quad (k \neq 0), \qquad \psi_k = \psi_{k0} + q_0 \psi_{k1},$$

$$e_k = e_{k0} + q_0 e_{k1} \quad (k \neq 0), \quad \varphi_k = \frac{f_k}{\xi_k^{1/\nu}}, \qquad m_k = \frac{\varphi_k g_k}{\xi_k^{(\nu-2)/\nu}}.$$

The quantities Ω_{sk0} and Ω_{sk1} are taken at the points $\xi_{k0} = (n-k)/n\xi_{00} + k/n$ from the tables [34] using interpolation. The quantities g_0 and e_0 are computed from the asymptotic formulas (4.37). The initial data for τ, R_n and δ are found by means of (3.78).

Computations were made mainly for the cases $n = 4$ and $n = 8$. In computations with $n = 4$ for $\gamma = 1.4$ and all ν the values $\epsilon_{14}^0 = 0.015$, $\epsilon_{24}^0 = 0.01$ and $\epsilon_{34}^0 = 0.01$ were taken. For $\nu = 1$ the computation was also carried out with $\epsilon_{34}^0 = 0.04$. This computation indicated that acceptance of less precise initial data does not appreciably affect the magnitudes of the basic functions for $q > 0.4$.

TABLE 5

Initial errors

Errors	γ							
	$\nu=1$		$\nu=2$			$\nu=3$		
	1.4	5/3	1.3	1.4	5/3	1.3	1.4	5/3
ε_{1n}	0.043	0.007	0.014	0.009	0.003	0.016	0.008	0.009
ε_{2n}	0.021	−0.002	−0.001	−0.002	0.006	0.004	−0.002	0.003
ε_{3n}	0.078	0.028	0.023	0.025	0.013	0.032	0.026	0.076

Systematic computations of the problem for $n = 8$ were carried out with the value $q_0 = 0.05$. The magnitudes of the initial errors ϵ_{1n}, ϵ_{2n} and ϵ_{3n} for the cases computed $(n = 8)$ are indicated in Table 5.

The numerical integration of the approximating system of ordinary differential equations was carried out, as a rule, by the Runge-Kutta method with variable step size and by the modified Euler method.

Mainly the cases of explosion for the values $\gamma = 1.3, 1.4, 5/3, 2$ and for the different ν were computed.

The effect of the number of strips n on the variation of the pressure at the center of the explosion p_0/p_n is represented in Figure 32 for $\nu = 3$ and $\gamma = 1.4$.

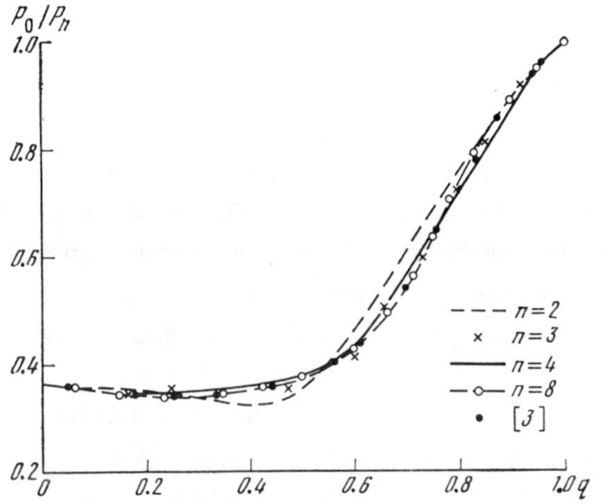

FIGURE 32. Variation of the relative pressure at the center of the explosion for various n ($\nu = 3$, $\gamma = 1.4$)

These data were obtained with $q_0 = 0.5$ and $\xi_0(q_0) = 0.206$ for $n = 2, 3, 4$, and $\xi_0(q_0) = 0.350$ for $n = 8$. The graph does not show the curve for $n = 1$ due to its low accuracy; data computed by the method of finite differences [3] are indicated by dots. We point out that for the law of motion of the shock wave the convergence of the solution proceeds much more rapidly. Here already the first approximation gives practically the required accuracy, differing little from results for large n.

Figure 33 shows the variation of pressure p_0/p_n obtained for the case $\nu = 3, \gamma = 1.4, n = 8$ with various types of interpolation polynomials. The continuous line indicates results for linear interpolation across each strip, while the

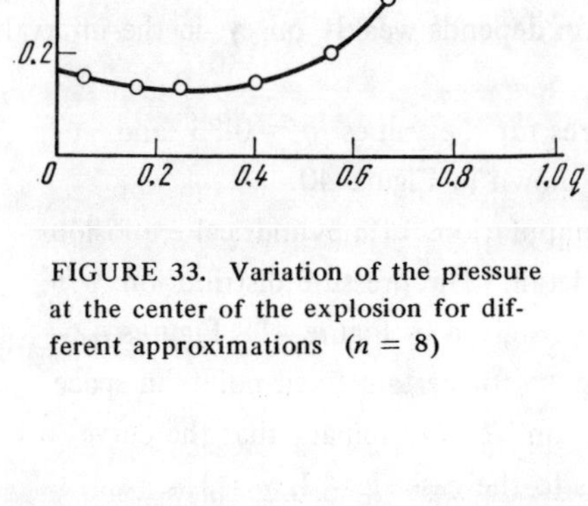

FIGURE 33. Variation of the pressure at the center of the explosion for different approximations ($n = 8$)

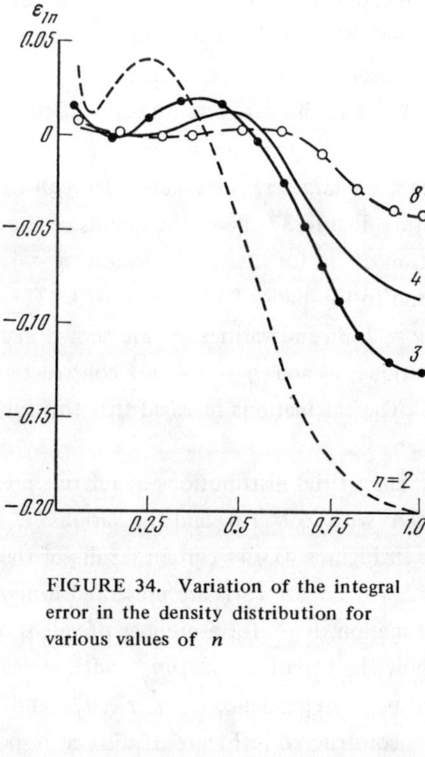

FIGURE 34. Variation of the integral error in the density distribution for various values of n

circles indicate results for piecewise interpolation by two matching polynomials of fourth order. Further results for $n = 8$ refer to the case of interpolation of the integrands by two polynomials of fourth order.

The graph of the variation of the integral error ϵ_{1n} depending on q is shown in Figure 34 for various n with $\nu = 3$ and $\gamma = 1.4$. Here a decrease in the error with increasing degree of the polynomial n is evident. The best accuracy, defined in terms of satisfying the integral law of mass conservation, occurs for intermediate values of q, while for large q the error ϵ_{1n} increases. The latter is caused by the complicated, nonmonotonic variation of the functions for large q and the diminishing of the central interval ξ_0, i.e. the increase in the region of approximation of the solution by polynomials in ξ. In all cases computed for $\nu = 1$ and $\nu = 2$ the behavior of the errors ϵ_{1n} had approximately the same character as for $\nu = 3$. Graphs of the dependence of ϵ_{1n} on q for $n = 8$ and $q_0 = 0.05$ with various γ and $\nu = 1$ (solid line), $\nu = 2$ (dots), and $\nu = 3$ (circles) are shown in Figure 35. It is evident that the integral error in the greater part of the interval of variation of q is small and changes within the limits 0.002 to 0.02. The absolute value of the error ϵ_{2n} in all cases computed fell from the greatest value at $q = q_0$ to values close to zero for q close to one.

The supplementary control of accuracy by means of the "extra" equation for ψ_l indicated that for $\gamma = 1.4$ and the various ν the quantities ψ_l are determined from the differential equations and from the algebraic relations following from (4.46) with a discrepancy of less than 1%.

We shall discuss certain results of the computations for $n = 4$. Figure 36 shows the function $R_n(\tau)$ for $\nu = 1, 2, 3$ and $\gamma = 1.4$. Since the curves $R_n(\tau)$ almost coalesce for large τ, they have been constructed with a shift in the axis of the abscissa. Figure 37 shows the graphs of the dimensionless pressures p_0/p_n as a function of τ for $\nu = 1, 2, 3$ and $\gamma = 1.4$. Here and in Figure 36 the dots correspond to the data of [3] for $\nu = 3$. The variations of p_0/p_n with increasing τ for $\nu = 2, 3$ and various γ are shown in Figure 38 a, b. The functions $q(\tau)$ for the various ν and $\gamma = 1.4$ are constructed in Figure 39 (the dots represent data of [3]). The calculations revealed that this function depends weakly on γ in the interval $1.4 \leqslant \gamma \leqslant 2$.

The spatial distributions of relative pressures for the values $q = 0.33$ and $q = 0.8$ with $\gamma = 1.4$ and the various ν are shown in Figure 40.

In Figures 41–44 certain results of the computations of a cylindrical explosion in air ($\nu = 2, \gamma = 1.4$) are presented in more detail. The pressure distribution p/p_∞ as a function of ξ for a number of values of τ is given in Figure 41. Figures 42–44 show the variations of p/p_∞ with increasing τ for certain fixed points in space and also the dependence of $\bar{\tau}, \tau^+, J_p^+$ and J_p on R. We remark that the curve $\xi_0(q)$ constructed in Figure 31 also corresponds to the case $\nu = 2, \gamma = 1.4$.

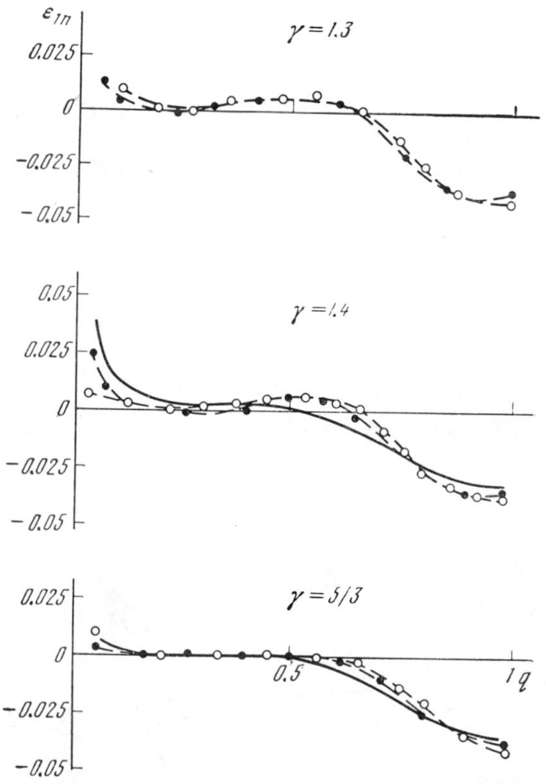

FIGURE 35. Variation of the integral error ϵ_{1n} for $n = 8$ and various γ

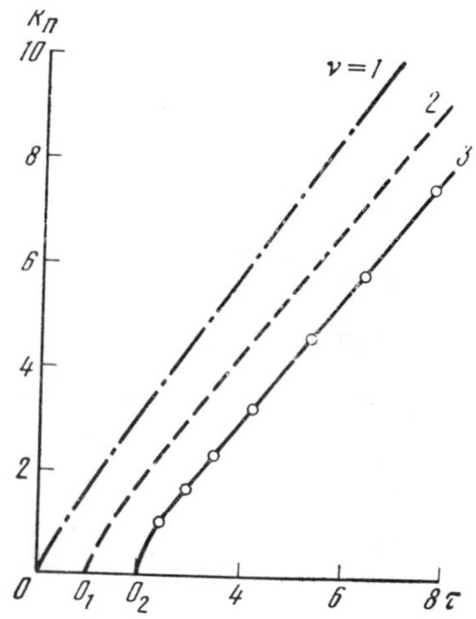

FIGURE 36. Law of motion of the shock wave for different ν with $\gamma = 1.4$

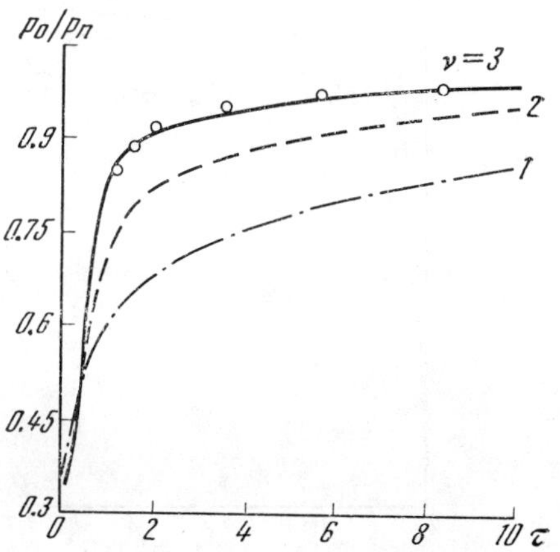

FIGURE 37. Variation of the relative pressure at the center of the explosion for different ν and $\gamma = 1.4$

The computational results presented for the cylindrical explosion and a comparative analysis of a number of other parameters of the gas flow indicate that in the dimensionless variables we have chosen the picture of the flow for $\nu = 2$ is basically close to that in the corresponding case of spherically symmetric motion which has been investigated in detail in [3], [5]. We studied the case of plane explosion in less detail. From the analysis of certain data for this flow it follows that here the general quantitative characteristics of the motion vary in a more substantial way with change in ν.

For illustration we present in still more detail the solution for a spherical point explosion in a monatomic gas. In computing this flow we took $\epsilon^0_{14} = 0.005$, $\epsilon^0_{24} = 0.005$ and $\epsilon^0_{34} = 0.01$. The absolute value of the maximum integral error according

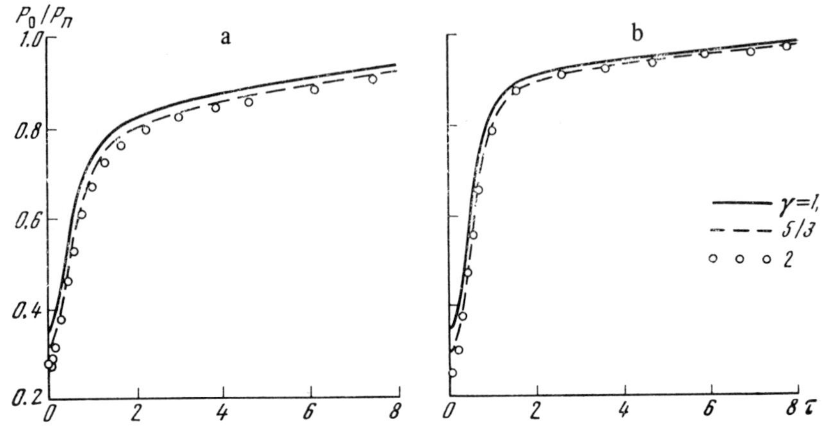

FIGURE 38. Variation of the relative pressure at the center of the explosion for different γ: a) cylindrical case ($\nu = 2$); b) spherical case ($\nu = 3$)

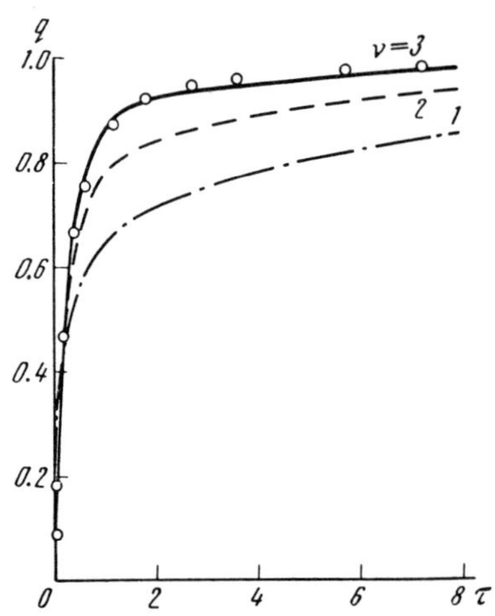

FIGURE 39. The function $q(\tau)$ for different ν with $\gamma = 1.4$

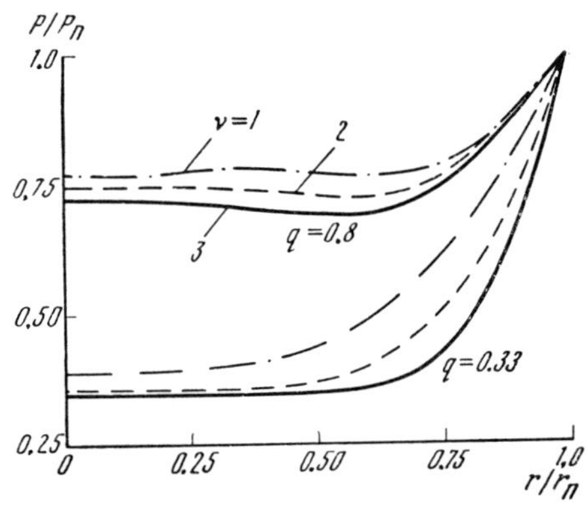

FIGURE 40. Spatial distribution of relative pressures for fixed q and different ν ($\gamma = 1.4$ and $n = 4$)

to the law of mass conservation ϵ_{14} here did not exceed 0.086. The dependence on τ of the relative quantities p_l/p_n, v_l/D, ρ_l/ρ_n and T_l/T_n, i.e. the values of these quantities on the boundaries of the strip ξ_l $(l = 0, 1, 2, 3)$ for $\nu = 3$ and $\gamma = 5/3$ are given in Figures 45 and 46. Here the temperature is denoted by T_0, T_i, and $(T_l/T_n) = (p_l\rho_n/p_n\rho_l)$. The relations between the coordinate of the front of the shock wave R_n, the dimensionless time τ, and the quantity q are represented in Figure 47. Here also is constructed a graph of the variation of the quantity of the central

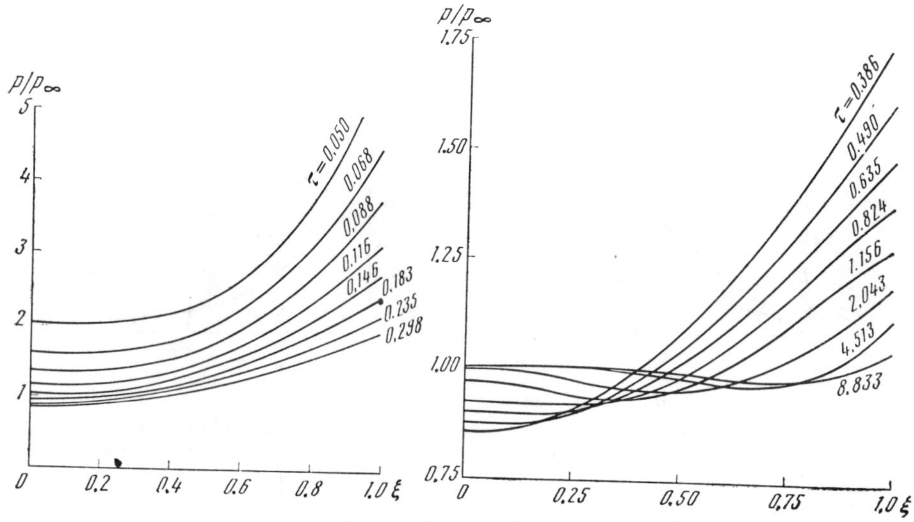

FIGURE 41. Field of pressures for the cylindrical explosion for τ from 0.05 to 8.833 ($\gamma = 1.4$)

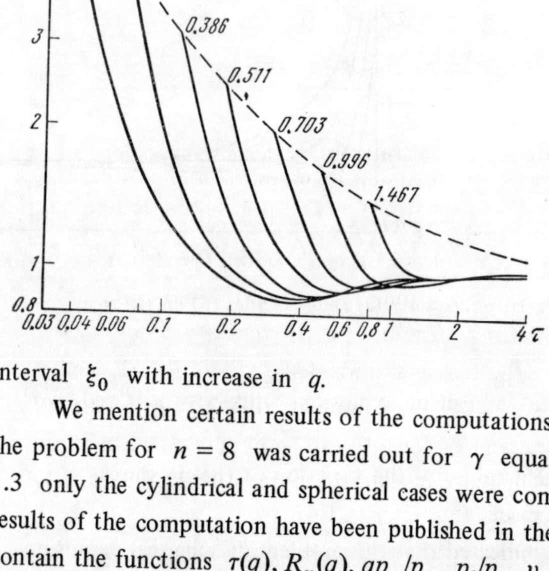

FIGURE 42. Pressure at fixed points of space for the cylindrical explosion ($\gamma = 1.4$)

interval ξ_0 with increase in q.

We mention certain results of the computations for $n = 8$. Computation of the problem for $n = 8$ was carried out for γ equal to 1.3, 1.4 and 5/3 (for $\gamma = 1.3$ only the cylindrical and spherical cases were computed in detail). The main results of the computation have been published in the form of special tables which contain the functions $\tau(q)$, $R_n(q)$, qp_n/p_∞, p_l/p_n, v_n/D, v_l/D, ρ_n/ρ_∞, ρ_l/ρ_n, qT_n/T_∞, T_l/T_n, qE_n/E_∞ and E_l/E_n for values of q in the range $0.1 \leqslant q \leqslant 0.9$ [35]. In Figures 48–50 comparisons are given of the distribution of relative pressures, velocities,

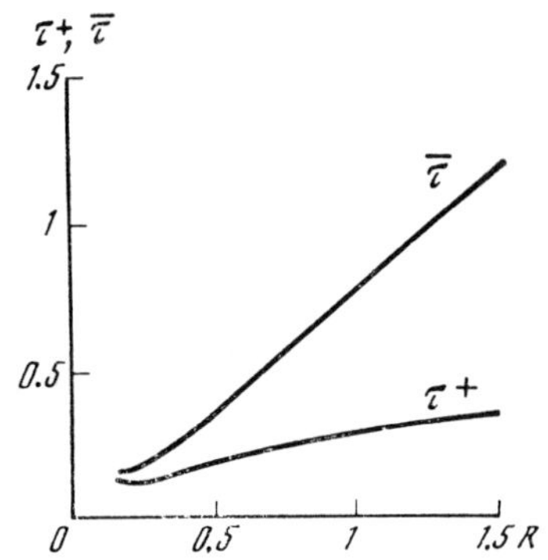

FIGURE 43. Duration of the positive phase τ^+ and time of transition of the excess pressure to the negative phase $\bar{\tau}$ ($\nu = 2; \gamma = 1.4$)

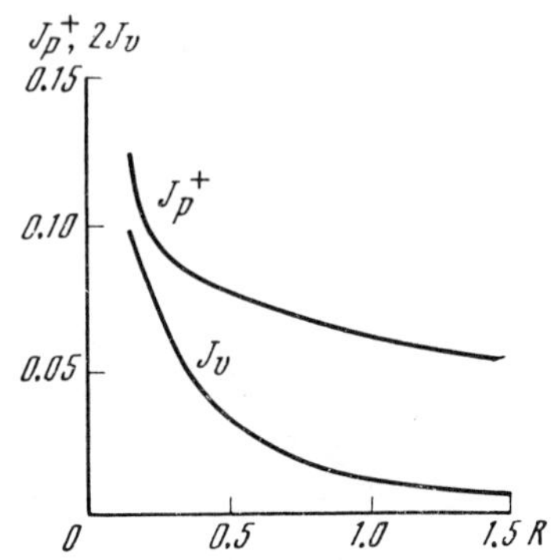

FIGURE 44. Positive impulse of pressures J_p^+ and impulse of velocity head ($\nu = 2; \gamma = 1.4$)

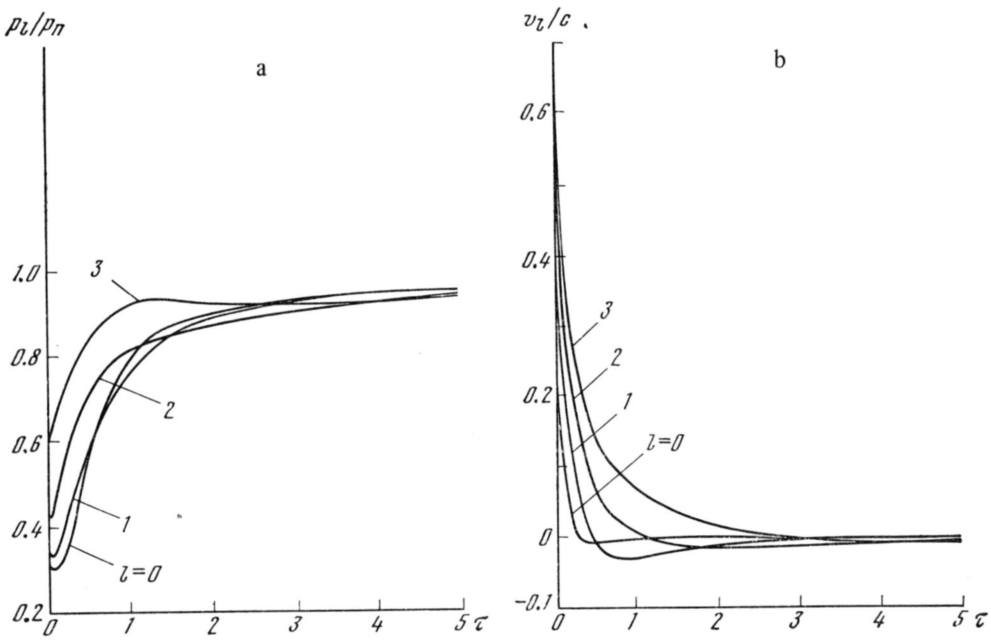

FIGURE 45. Variations of the relative pressures (a) and velocities (b) on the strip boundaries ($\nu = 3$, $\gamma = 5/3$)

and densities for spherical, cylindrical, and plane explosions with $\gamma = 1.4$ and two values of q ($q = 0.3$ and 0.75).

In the Appendix we also present tables of the variation of the pressures p/p_∞ and $\rho v^2/p_\infty D^2$ at fixed points of space ($\nu = 2; \gamma = 1.4$).

The computations performed indicated that the problem of spherical explosion

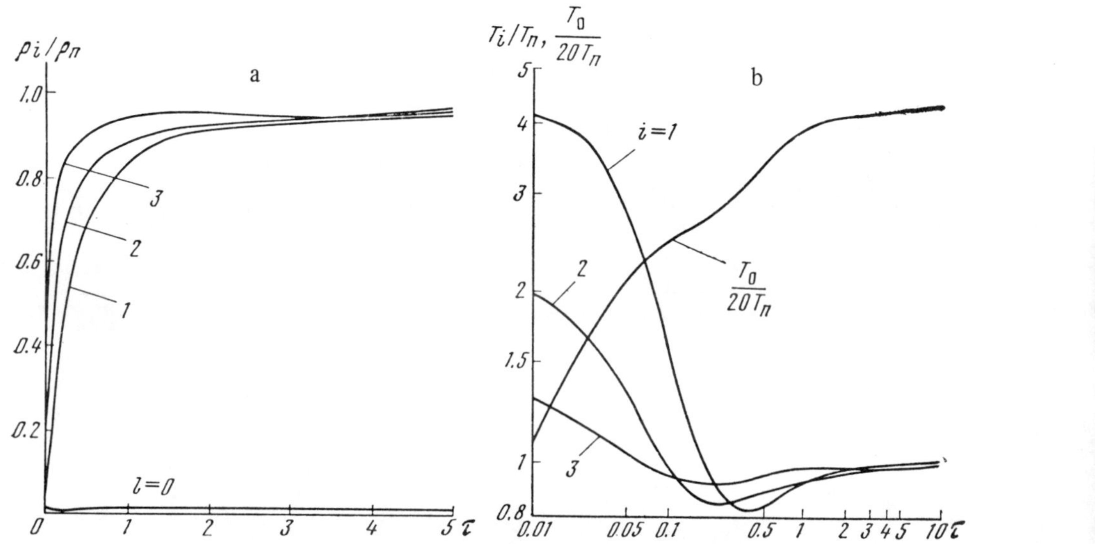

FIGURE 46. Variations of relative densities (a) and temperatures (b) on the strip boundaries ($\nu = 3$; $\gamma = 5/3$)

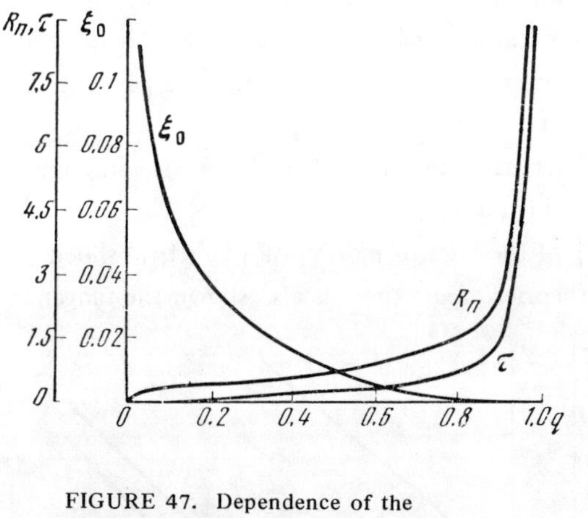

FIGURE 47. Dependence of the coordinate of the shock wave R_n, time τ, and the quantity of the central interval ξ_0 on q ($\nu = 3$; $\gamma = 5/3$)

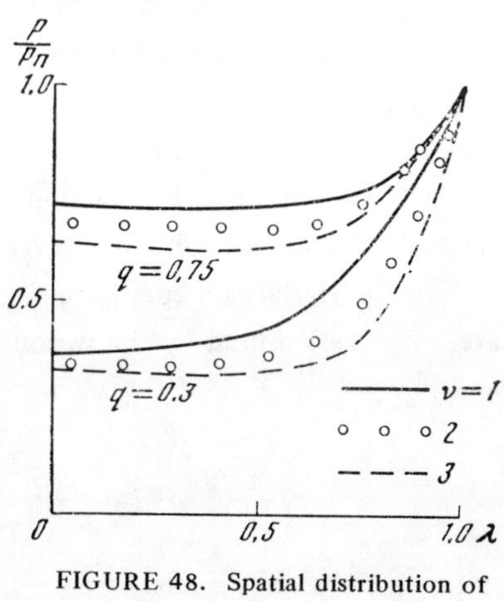

FIGURE 48. Spatial distribution of relative pressure for fixed q and different ν ($\gamma = 1.4$; $n = 8$)

is most simply computed by the method proposed. The worst results with regard both to accuracy and loss of machine time were obtained for the planar case. Here more abrupt variations in the functions occur; this reduces the accuracy of the computation. In addition, the accuracy of the asymptotic formulas (4.37) drops with decreasing ν. To raise the accuracy it is possible to take asymptotic formulas for computing the pressure in the zone of the central interval in the form

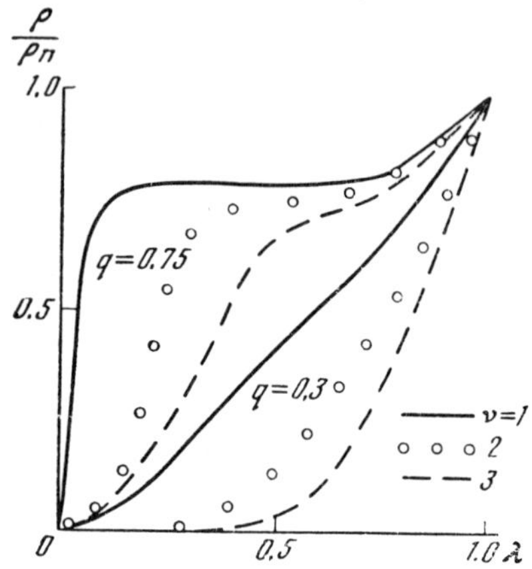

FIGURE 49. Spatial distribution of the relative density $(\gamma = 1.4)$

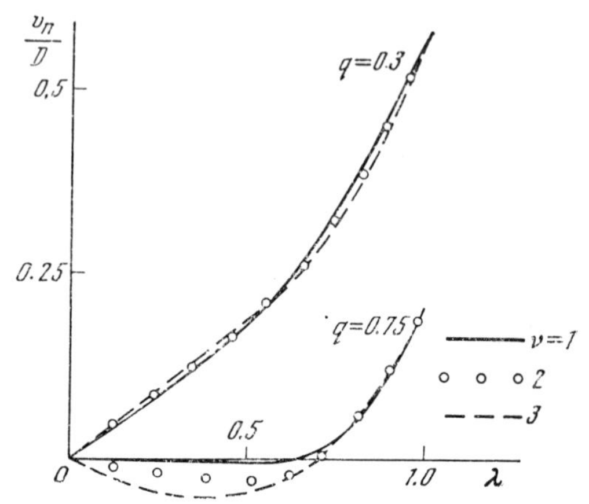

FIGURE 50. Spatial distribution of the relative velocity for the values $q = 0.3$ and $q = 0.75$ $(\gamma = 1.4)$

$$p = p_{0*}(t) + p_{1*}(t) r^d + O(r^d), \quad d = \nu/(\gamma - 1) + 2.$$

§4. Parameters of the shock wave front and comparison with other computations and experimental data

We again denote the parameters of the front by the index 2 and the parameters of the unperturbed medium by the index 1.

As already noted, the functions $R_2(\tau)$ for $\nu = 1, 2, 3$ and $\gamma = 1.4$ obtained as a result of computations were given in Figure 36.

Figure 51 shows a comparison [5] of the laws of motion of a spherical shock wave $(\gamma = 1.4)$ found by the various theories. Here there is also shown the tangent

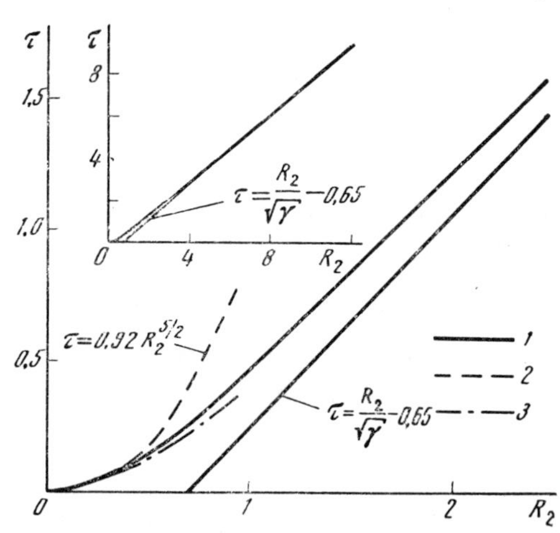

FIGURE 51. Law of motion of the shock wave. Comparison according to different theories: 1) computed curve; 2) self-similar curve; 3) linearized theory

line (with slope equal to the speed of sound) passing through the point $\tau = 0.527$, $R_2 = 12.04$ lying on the curve $R_2(\tau)$.

The function $q(\tau)$ for the different ν with $\gamma = 1.4$ was shown in Figure 40. The variation of excess pressures with increasing R_2 is shown in Figures 52 and 53 (solid lines) for $\nu = 2$ and 3 respectively.

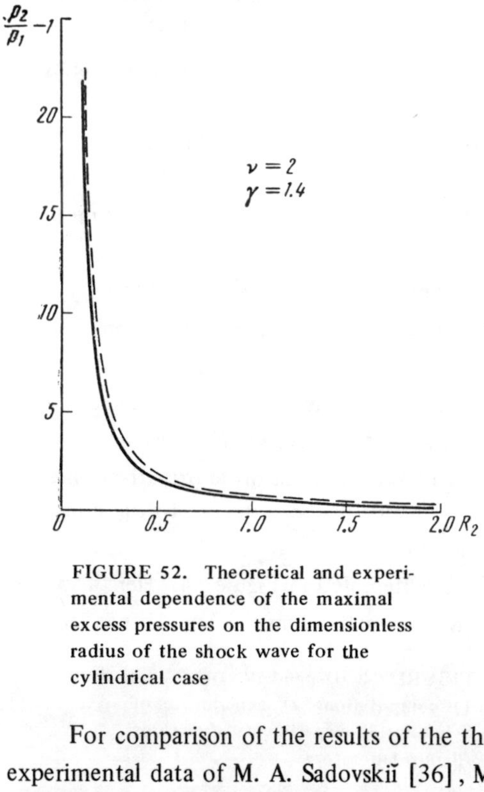

FIGURE 52. Theoretical and experimental dependence of the maximal excess pressures on the dimensionless radius of the shock wave for the cylindrical case

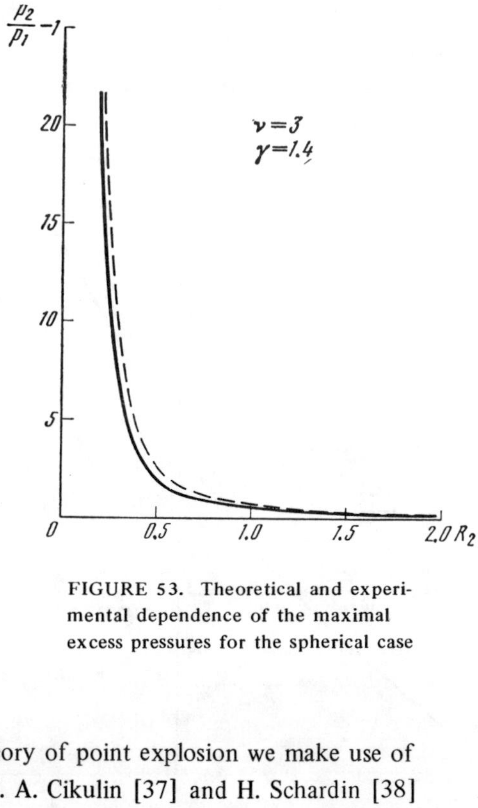

FIGURE 53. Theoretical and experimental dependence of the maximal excess pressures for the spherical case

For comparison of the results of the theory of point explosion we make use of experimental data of M. A. Sadovskiĭ [36], M. A. Cikulin [37] and H. Schardin [38] on explosions of chemical explosives, and experimental-theoretic data of Los Alamos laboratory on atomic explosions published in [39]. Comparison with other experimental data is given in [5]. The experimental data were first processed and converted to our dimensionless variables. From the experiments of Sadovskiĭ and Cikulin we used the dependence of $p_2/p_1 - 1$ on R_2.

For the case of a cylindrical explosion the formula corresponding to the experimental data is presented in [37]:

$$\frac{p_2}{p_1} - 1 = \frac{0.24}{R_2^2} + \frac{0.48}{R_2^{3/4}}. \qquad (4.55)$$

If Sadovskiĭ's formula for spherical explosion of a charge of trinitrotoluene (TNT) [40] is converted to our variables, it takes the form

$$\frac{p_2}{p_1} - 1 = \frac{0.21}{R_n} + \frac{0.21}{R_n^2} + \frac{0.15}{R_n^3}. \tag{4.56}$$

We remark that in [37] Sadovskiĭ's formula (4.56) is given with somewhat larger coefficients, differing approximately by a factor of 1.1 from the coefficients of (4.56). This discrepancy is apparently explained by coefficients in the original (dimensional) formula which are different from those we took for TNT from [40] (cf. also [41]).

Graphs of $p_2/p_1 - 1$ corresponding to (4.55) and (4.56) are shown in Figures 52 and 53 (dashed lines).

Experimental data on the law of motion of spherical waves are taken from [38]. From Schardin's experiments [38] we used the results of determining the law of motion of the shock wave $r_2(t)$ during explosion of lead nitride. Here we took the caloricity of lead nitride equal to 400 kcal/kg (360–380 kcal/kg is thought to be more accurate). Starting with the weight of the charge, the energy of the explosion E_0 was computed and then conversion was made from the dimensional variables r_2 and t to the dimensionless variables R_2 and τ. These data are represented in Figure 54. Here also are given the data from the Los Alamos laboratory. In reducing these data to dimensionless form it was assumed that $E_0 = 8.5 \times 10^{12}$ kG · m and $p_1 = 1$ atm. The experimental data on $r_2(t)$ for the cylindrical wave [37] agree well with the computed data (for the same τ the experimental values of R_2 exceed the theoretical values by 3–4%).

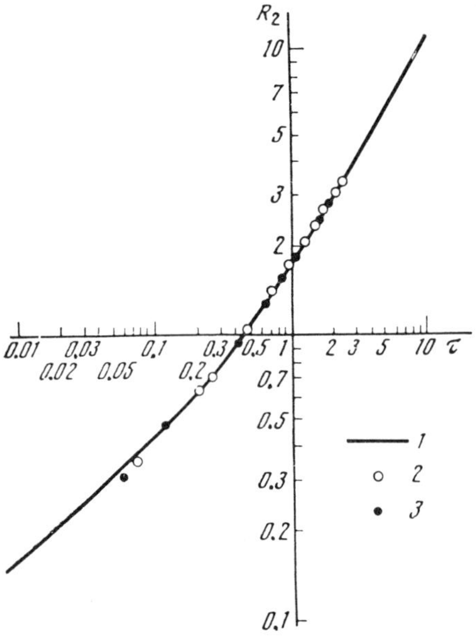

FIGURE 54. Dependence of R_2 on τ: 1) computations; 2) experiments of Schardin; 3) experiments of the Los Alamos laboratory

The above comparison of the results of the theory of point explosions with experiment (see also [5]) allows us to draw the following conclusions.

1. Comparison with experimental data on explosions of chemical explosives shows that the experimental excess pressures are higher than the theoretical values and differ from them strongly near the center of the explosion. At large distances from the site of the explosion the experimental data agree better with the theoretical data. The values of the excess pressures found from the approximate formulas (4.55) and (4.56) agree with the theoretical results if the values of R_n are increased by a

factor of 1.1–1.5, which is equivalent to reducing the explosive charge.

2. The experimental relations for $R_2(\tau)$ agree well with the theoretical ones.

3. The data on atomic explosions give good agreement with the theoretical results for the case $\gamma = 1.4$ if the energy of the explosion E_0 is taken equal to 8.5×10^{12} kG · m (this amount of energy is close to the energy of explosion of a so-called nominal atomic bomb).

In the work of A. S. Fonarev and S. Ju. Černjavskiĭ [42] comparison was made of the data from the theory of point explosions with the computation of an explosion of charges of TNT in air, taking into account the motion of the products of the explosion. The results of this work show that approximately with $\Delta p/p_1 < 20$ the theoretical results agree well with the computation of an explosion of TNT. When the specific energy of the explosive is increased the agreement with the theoretical results improves; this also follows from simple physical considerations (the concentration of energy in a given volume is increased). On the basis of the computations it was concluded in [42] that at large distances the parameters of the shock wave are determined mainly by the total energy of the explosion and the initial air pressure.

These data afford the conclusion that the theoretical results on the parameters of the wave front may successfully be applied after certain distances to determine the characteristics of blast waves from plane, cylindrical, and spherical charges independent of their particular nature.

§5. Approximate formulas for determining the parameters of the shock wave front

5.1. *Approximate asymptotic formulas for large distances.* The analytic dependence of the parameters of the shock wave front is of major importance both in theoretical investigations and in practical applications. Since the exact value of these relations cannot be obtained, the problem arises of finding effective approximation formulas. This question is considered in the present section.

According to the results of §2 the asymptotic formulas (4.10)–(4.13), or the more exact formulas (4.14), (4.15), can be used to determine the parameters of the front of shock waves at large distances.

The constants C_ν and r^* entering in these formulas are to be found from data on the computation of the problem for times when $(\Delta p/p_1) \sim 0.1–0.3$ $(\Delta p = p_2 - p_1)$.

As an example, we consider the case $\nu = 3, \gamma = 1.4$ in detail. For p_2/p_1 in the case $\nu = 3$ we have

$$\frac{p_2}{p_1} = 1 + \frac{4\gamma}{5(\gamma+1)} \sqrt{\frac{E_0}{\alpha\gamma p_1}} \frac{k_3}{r_2 \sqrt{\ln (r_2/r^*)}}.$$

In dimensionless variables this formula appears as follows:

$$\frac{p_2}{p_1}=1+\frac{4\gamma}{5(\gamma+1)}\frac{1}{\sqrt{\alpha\gamma}}\frac{\bar{k}_3}{\sqrt{\ln(R_2/R^*)}}. \tag{4.57}$$

The constants $\bar{k}_3$ and R^* are found from the conditions that p_2/p_1 coincide with computed values taken from [4] at certain selected points for $R_2 > 2$. In the formulas for ρ_2/ρ_1 and v_2/a_1 these constants are assumed known. The following asymptotic formulas were obtained as a result of calculations for $\gamma = 1.4$:

$$\frac{p_2}{p_1}=1+\frac{0.227}{R_2\sqrt{\lg R_2+0.158}}, \tag{4.58}$$

$$\frac{\rho_2}{\rho_1}=1+\frac{0.163}{R_2\sqrt{\lg R_2+0.158}}, \tag{4.59}$$

$$\frac{v}{v_2}=\frac{0.163}{R_2\sqrt{\lg R_2+0.158}}. \tag{4.60}$$

These formulas were first obtained by us and were published in [2], [5]. They provide sufficiently precise values of the parameters of the front for values of $R_2 >$ 1.5 (for details see [5]).

Clarification of the precision of formulas of the type (4.58)–(4.60) and further development of the method we proposed are given in the work of Miles [43]. There he proposes the following formula for the positive impulse of pressures:

$$\frac{a_1}{p_1 r^0} I_p^+=\frac{1}{4}\gamma(\gamma+1)A^2 R^{-1},$$

where the constant A is proportional to $\bar{k}_3$ and gives the best results for $A = 0.23$.

Using the results of the computations for plane and cylindrical explosions presented in the preceding section, we can determine the corresponding constant in the asymptotic formulas (4.10)–(4.13) and obtain an analogue of the approximate formulas (4.58)–(4.60).

5.2. *Dependence of the velocity of gas particles on the front of a shock wave on the coordinate.* We shall consider in detail the dependence of the velocity of particles on the front of a shock wave on the coordinate of the shock wave from a point explosion.

In the self-similar case we have, according to (2.12),

$$D_0=\frac{2}{\nu+2}\sqrt{\frac{E_0}{\alpha\rho_1}}\, r_{20}^{-\nu/2}. \tag{4.61}$$

From the condition on the shock wave (1.62) we obtain for the function v_2

$$v_{20}=\frac{2}{\gamma+1}D_0, \tag{4.62}$$

$$v_2=\frac{2}{\gamma+1}(1-q)D. \tag{4.63}$$

If $r_2 \longrightarrow \infty$, then $q \longrightarrow 1$ and hence $v_2 \longrightarrow 0$ as $r_2 \longrightarrow \infty$. From (4.61) and (4.62) we see that also $v_{20} \longrightarrow 0$ as $r_{20} \longrightarrow \infty$. Thus the limits of v_2 as $r_2 \longrightarrow \infty$ are the same both for a strong explosion and for an explosion with counterpressure. This limiting property is not satisfied for other characteristics of the motion. We further compare the dependence of v_2 on r_2 for a strong explosion and for the case of degeneration of the shock wave into an acoustic wave. For a strong explosion we find from (4.61) and (4.62) that

$$v_{20} = \frac{4}{(\gamma+1)(\nu+2)} \sqrt{\frac{E_0}{\alpha \rho_1}}. \tag{4.64}$$

For the degeneration of a shock wave into an acoustic wave we have the asymptotic formulas (cf. (4.10) and (4.12))

$$\begin{aligned} v_{2\text{as}} &= \frac{2a_1 C_\nu}{(\gamma+1)\, r_2^{(\nu+1)/4}} \qquad (\nu = 1,2), \\ v_{2\text{as}} &= \frac{2a_1}{(\gamma+1)} \frac{C_3}{r_2 \sqrt{\ln (r_2/r^*)}}. \end{aligned} \tag{4.65}$$

Using the expression for the acoustic velocity $a_1 = (\gamma p_1/\rho_1)^{1/2}$ in (4.64) and introducing in (4.65) in place of the C_ν the constants k_ν, we obtain

$$v_{20} = \frac{4a_1}{(\nu+2)(\gamma+1)} \sqrt{\frac{E_0}{\alpha \gamma p_1}}\, r_2^{-\nu/2}, \qquad v_{2\text{as}} = \frac{4a_1}{(\nu+2)(\gamma+1)} \sqrt{\frac{E_0}{\alpha\gamma p_1}} \frac{k_\nu}{r_2^{(\nu+1)/4}} \qquad (\nu = 1,2),$$

$$v_{20} = \frac{4a_1}{5(\gamma+1)} \sqrt{\frac{E_0}{\alpha\gamma p_1}}\, r_2^{-3/2}, \qquad v_{2\text{as}} = \frac{4a_1}{5(\gamma+1)} \sqrt{\frac{E_0}{\alpha\gamma p_1}} \frac{k_3}{r_2^{3/2}} \left(\frac{r_2}{\ln \frac{r_2}{r^*}} \right)^{1/2} \qquad (\nu = 3),$$

where the k_ν are related to the C_ν by

$$C_\nu = \frac{2}{\nu+2} \sqrt{\frac{E_0}{\alpha\gamma p_1}}\, k_\nu.$$

From these formulas it is evident that the functional dependence of v_2 on r_2 changes mildly on passage from small r_2 (strong explosion) to large r_2, and in the case of plane waves it does not change at all.

Since the qualitative character of the variation of v_2 with increasing r_2 remains the same as for v_{20} (from physical considerations it is clear that v_2 decreases as r_2 increases) and the asymptotic laws differ little from the self-similar laws, it may be assumed that approximately

$$v_2 = \frac{4}{(\gamma+1)(\nu+2)} \sqrt{\frac{E_0}{\alpha\gamma p_1}} \frac{a_1}{r_2^{\nu/2}} \tag{4.66}$$

also for sufficiently large values of r_2. We point out that we do not consider here possible interactions of the basic shock wave with secondary (weak) waves which may arise in the gas flow. A more precise dependence of $v_2(r_2)$ with the correct asymptotic behavior at infinity is given by the formulas [44]

$$v_2 = \begin{cases} \dfrac{4a_1}{(\gamma+1)(\nu+2)}\sqrt{\dfrac{E_0}{\alpha\gamma p_1}}\, r_2^{-\nu/2}, & 0 \leqslant r_2 \leqslant r_2^*, \qquad (4.67)\\ v_{2\,\mathrm{as}}, & r_2^* \leqslant r_2 < \infty. \qquad (4.68)\end{cases}$$

The choice of r_2^* and the constants k_ν and r^* will be made in the sequel. These properties for the dependence $v_2(r_2)$ are corroborated by computation and illustrated in Figure 55 for $\nu = 3$ and $\gamma = 1.4$.

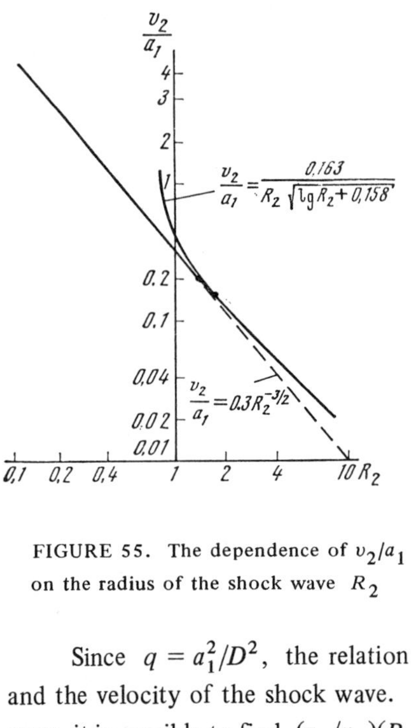

FIGURE 55. The dependence of v_2/a_1 on the radius of the shock wave R_2

The nature of the dependence $v_2(r_2)$ expressed by (4.66) and (4.67) is also corroborated by data taken from American work [39] (for more details on this see [5], Chapter 5).

5.3. *The law of motion of a sufficiently strong shock wave and its parameters.* Using (4.66), it is possible to give formulas for computing all parameters of the front of a shock wave. From the condition on the shock wave for v_2 and (4.66) we obtain

$$\frac{2}{\nu+2}\sqrt{\frac{E_0}{\alpha\gamma p_1}}\, r^{-\nu/2} = \frac{1-q}{\sqrt{q}}. \qquad (4.69)$$

Introducing the dimensionless radius R_2, we find

$$R_2 = \left(\frac{2}{\nu+2}\right)^{2/\nu}\left(\frac{1}{\alpha\gamma}\right)^{1/\nu}\frac{q^{1/\nu}}{(1-q)^{1/\nu}}. \qquad (4.70)$$

Since $q = a_1^2/D^2$, the relation (4.70) gives the connection between the radius and the velocity of the shock wave. Using (4.70) and the conditions on the shock wave, it is possible to find $(p_2/p_1)(R_2)$ and $(\rho_2/\rho_1)(R_2)$. It is also possible to find the law of motion of the shock wave $R_2(\tau)$. From (4.70) and the relation $(dr_2/dt)^2 = a_1^2 q^{-1}$ we have

$$\frac{dR_2}{d\tau} = \frac{1}{(\nu+2)\sqrt{\alpha}} R_2^{-\nu/2}\left[1+\sqrt{\theta_\nu}\right], \qquad \theta_\nu = 1 + \gamma\alpha(\nu+2)^2 R_2^\nu. \qquad (4.71)$$

For the self-similar motion the function $R_{20}(\tau)$ is known and has the form

$$R_{20}(\tau) = \alpha^{-1/(\nu+2)}\tau^{2/(\nu+2)}. \qquad (4.72)$$

For the cases $\nu = 1$ and $\nu = 2$ equation (4.71) can be integrated in terms of elementary functions. We shall consider these cases. From (4.71) we find

$$d\tau = \frac{(\nu+2)\sqrt{\alpha}\, R_2^{\nu/2} dR_2}{1+\sqrt{1+\gamma\alpha(2+\nu)^2 R_2^\nu}}. \qquad (4.73)$$

From this $\tau(R_2)$ is found by quadrature. After integrating we obtain

$$\tau = \frac{1}{4\gamma\sqrt{\alpha}}[\sqrt{\theta_2} - \ln(1+\sqrt{\theta_2})] + C_2 \quad (\nu=2),$$

$$\tau = \frac{1}{3\gamma\sqrt{\alpha}}\left\{-2\sqrt{R_2} + \sqrt{R_2\theta_1} + \frac{1}{3\sqrt{\gamma\alpha}}\ln[\sqrt{\theta_1} + \sqrt{9\gamma\alpha R_2}]\right\} + C_1 \quad (\nu=1),$$

where C_2 and C_1 are constants of integration. The constants C_2 and C_1 can be chosen from the condition $R_2(\tau_0) = R_{20}$, where R_{20} is related to τ_0 by the self-similar dependence (4.72). A small value may hereby be taken for τ_0, corresponding to that stage of the explosion where the self-similar solution applies.

The final formulas giving the relation between R_2 and τ can be written as follows:

$$\tau = \frac{1}{4\gamma\sqrt{\alpha}}\left[\sqrt{1+16\gamma\alpha R_2^2} - \sqrt{1+16\gamma\alpha R_{20}^2} - \ln\frac{1+\sqrt{1+16\gamma\alpha R_2^2}}{1+\sqrt{1+16\gamma\alpha R_{20}^2}}\right] + \tau_0 \quad (4.74)$$

$$(\nu=2),$$

$$\tau = \frac{1}{3\gamma\sqrt{\alpha}}\left[-2\sqrt{R_2} + 2\sqrt{R_{20}} + \sqrt{R_2(1+9\gamma\alpha R_2)}\right.$$
$$\left.- \sqrt{R_{20}(1+9\gamma\alpha R_{20})} + \sqrt{\frac{1}{9\gamma\alpha}}\ln\frac{\sqrt{1+9\gamma\alpha R_2} + \sqrt{9\gamma\alpha R_2}}{\sqrt{1+9\gamma\alpha R_{20}} + \sqrt{9\gamma\alpha R_{20}}}\right] + \tau_0 \quad (\nu=1).$$

In the case of spherical symmetry we obtain from (4.71)

$$\frac{dR_2}{d\tau} = \frac{R_2^{-3/2}}{5\sqrt{\alpha}}[1+\sqrt{1+25\gamma\alpha R_2^3}]. \quad (4.75)$$

This equation can be integrated numerically by prescribing initial data from the self-similar solution. Tables of elliptic integrals can also be used.

Formulas (4.70), (4.74), (4.75) and (4.22) (or (1.62)) make it possible to determine all the parameters of the shock wave with a sufficient degree of accuracy when it is still sufficiently strong ($r_2 \leqslant r_2^*$). The question of constructing more precise formulas valid for computing the flow characteristics at all distances from the center as in the dependence (4.67) for $v_2(r_2)$ will be considered in the sequel.

For values $r_2 \leqslant r_2^*$ all the relations presented above remain in force. To compute the flow characteristics for $r_2 > r_2^*$ the quantity r_2^* and the constants k_ν and r^* must be chosen.

5.4. *Formulas for the entire range of distances in the case of plane and cylindrical waves.* In the case of plane waves we choose $\nu = 1$, since only under this condition does the dependence $v_2(r_2)$ for small distances go over continuously into $v_2(r_2)$ for large distances.

Thus, proceeding from our basic assumptions made at the beginning of the chapter, for plane waves it is not possible to obtain formulas for computing the flow characteristics at large distances other than the formulas of subsection 5.3. Using (4.70) with $\nu = 1$ and the conditions on the shock wave, for the maximal overpressure we obtain the simple formula

$$\frac{p_2}{p_1} - 1 = \frac{4\gamma}{(\gamma+1)[-1+\sqrt{1+9\gamma\alpha R_2}]}. \tag{4.76}$$

Graphs of the law of motion of the shock wave and the flow characteristics on the front of the shock wave for $\gamma = 1.2, 1.4, 3$ in the case of plane symmetry are shown in Figures 56–59.

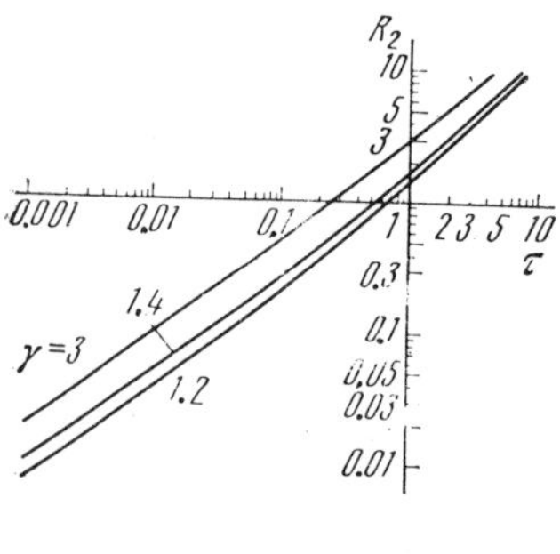

FIGURE 56. Dependence of R_2 on τ for various γ with $\nu = 1$

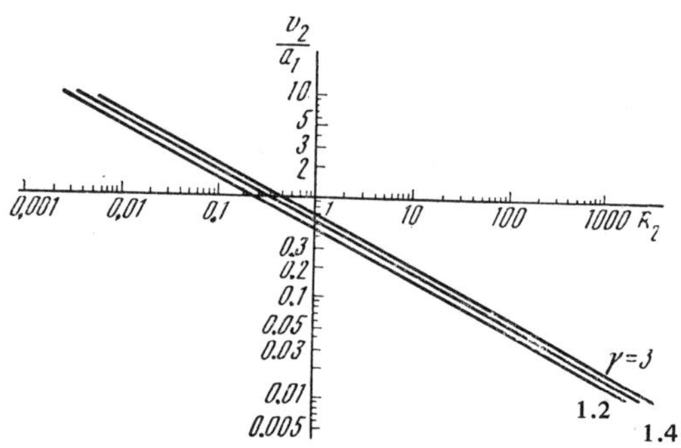

FIGURE 58. Dependence of the dimensionless velocity on R_2 for $\nu = 1$

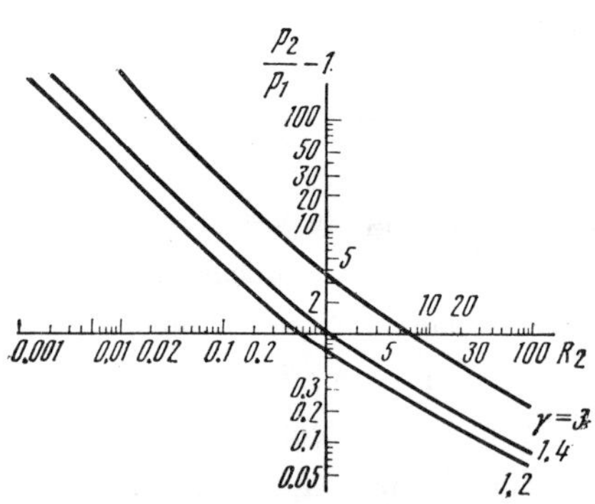

FIGURE 57. Maximal overpressures for $\nu = 1$

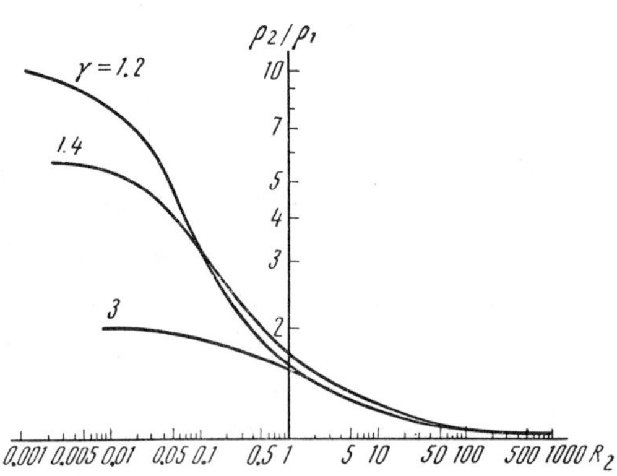

FIGURE 59. Variation of the density on the shock wave for $\nu = 1$

According to (4.65) and (4.67), for cylindrical symmetry in dimensionless variables it is possible to write

$$\frac{v_2}{a_1}=\begin{cases}\dfrac{(\gamma\alpha)^{-1/2}}{\gamma+1}R_2^{-1}, & 0\leqslant R_2\leqslant R_2^*,\\ \left[(\gamma\alpha)^{1/2}(\gamma+1)\right]^{-1}\bar{k}_2R_2^{-3/4}, & R_2^*\leqslant R_2<\infty,\end{cases}\qquad(4.77)$$

where $\bar{k}_2 = k_2(E_0/p_1)^{1/8}$.

The requirement of continuity of v_2/a_1 at the point $R_2 = R_2^*$ leads us to the following value of $\bar{k}_2$: $\bar{k}_2 = (R_2^*)^{-1/4}$. We choose R_2^* such that the transition from the formulas corresponding to (4.67) to the asymptotic formulas for the damping of shock waves should occur at values of a_1/D approximately equal to 0.9. To this value of a_1/D there corresponds an overpressure $p_2/p_1 - 1 = 0.27$, i.e. the shock wave is relatively weak, and use of the asymptotic formulas to compute the parameters of the shock wave does not introduce large errors.

From (1.62) and the function v_2/a_1 for $R_2 < R_2^*$ we find that the quantity $a_1/D \approx 0.9$ corresponds to $R^* = 2$. Thus on the basis of these arguments we take $R_2^* = 2$.

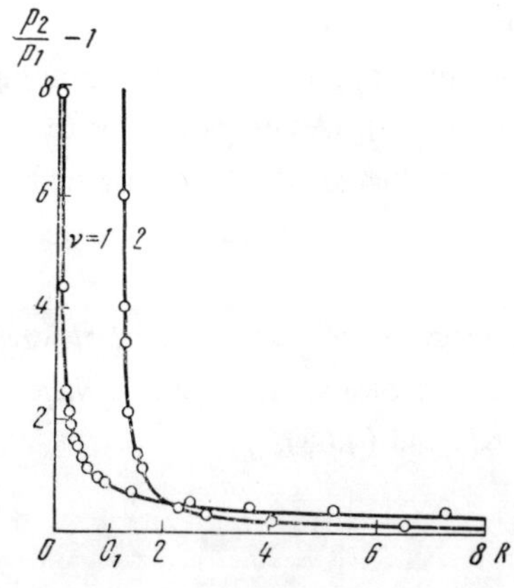

FIGURE 60. Comparison of calculated and approximate values of overpressures on the front of the shock wave for $\gamma = 1.4$

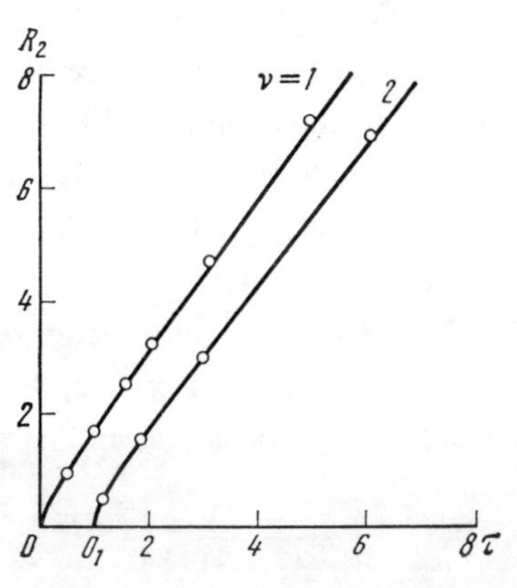

FIGURE 61. Comparison of calculated and approximate laws of motion of the shock wave for $\gamma = 1.4$

Using (4.77), (1.62), and the values of $\bar{k}_2$ and R_2^* obtained, we find formulas for the maximal overpressures:

$$\begin{aligned}\frac{p_2}{p_1}-1&=\frac{4\gamma}{(\gamma+1)\left(-1+\sqrt{16\gamma\alpha R_2^2+1}\right)}, && 0<R_2\leqslant 2,\\ \frac{p_2}{p_1}-1&=\frac{4\gamma}{(\gamma+1)\left(-1+\sqrt{16\sqrt{2}\,\gamma\alpha R_2^{3/2}+1}\right)}, && \infty>R_2\geqslant 2.\end{aligned}\qquad(4.78)$$

We further present formulas for finding the dependence $R_2(\tau)$ corresponding to the law (4.77). For $R_2 \leqslant 2$ the $R_2(\tau)$ dependence is given by (4.74). For $R_2 \geqslant 2$ we take for the motion of the shock wave the asymptotic law

$$a_1(t-t_*)=r_2\left[1-\left(\frac{E_0}{\gamma\alpha p_1}\right)^{1/2}k_2 r_2^{-3/4}+\ldots\right].$$

Passing to dimensionless variables τ and R_2 and substituting the value of k_2, we obtain

$$\tau-\tau^*=\gamma^{-1/2}R_2\left[1-\sqrt{\frac{1}{\gamma\alpha}}\,2^{-1/4}R_2^{-3/4}+\ldots\right], \qquad (4.79)$$

where τ^* is a constant. We choose τ^* from the condition that the values of τ found from (4.74) and (4.79) coincide for $R_2 = 2$. The following values were obtained from computations: $\tau^* = 0.17$ for $\gamma = 1.2$, $\tau^* = 0.25$ for $\gamma = 1.4$, $\tau^* = 0.41$ for $\gamma = 3$.

The relations (4.74), (4.76), (4.77), and (1.62) enable us to compute all the parameters of the front of the shock wave in the cylindrical case.

Graphs of the functions $R_2(\tau)$, $(p_2/p_1)(R_2) - 1$, $(v_2/a_1)(R_2)$ and $(\rho_2/\rho_1)(R_2)$ for various γ with $\nu = 2$ are given in [5] and will not be reproduced here.

Figure 60 shows the dependence of $\Delta p/p_1$ on R_2 obtained by computing according to the method considered in §3 and by (4.78). A comparison of the computed and approximate values of $R_2(\tau)$ is shown in Figure 61. Here data taken from the numerical solution are noted by circles. The comparison indicates good accuracy of the approximate formulas.

5.5. *Formulas for the entire range of distances in the case of spherical waves. Comparison of pressures for* $\nu = 1, 2, 3$. In the case of a spherical shock wave we have in dimensionless variables, according to (4.67) and (4.68),

$$\frac{v_2}{a_1}=\begin{cases}\dfrac{4}{5(\gamma+1)}\dfrac{1}{\sqrt{\alpha\gamma}}R_2^{-3/2}, & 0<R_2<R_2^*,\\[2ex] \dfrac{4}{5(\gamma+1)}\dfrac{1}{\sqrt{\gamma\alpha}}\dfrac{\bar{k}_3}{R_2\sqrt{\ln R_2-\ln l^*}}, & \infty>R_2\geqslant R_2^*.\end{cases} \qquad (4.80)$$

Here $\bar{k}_3$ and l^* are dimensionless constants related to k_3 and r^* by $k_3 = \bar{k}_3(p_1/E_0)^{1/6}$ and $r^* = l^*(E_0/p_1)^{1/3}$. For the dependence of v_2/a_1 on R_2 we require the continuity of the function $(v_2/a_1)(R_2)$ and its first derivative at $R_2 = R_2^*$. This requirement leads to the relations $\bar{k}_3 = (R_2^*)^{-1/2}$ and $\ln l^* = \ln R_2'{}^* - 1$. Just as in the cylindrical case, we take $R_2^* = 2$. This choice of R_2 is natural, since for $R_2 \geqslant 2$ the values of v_2/a_1 found from the asymptotic formula (4.12) give sufficiently good agreement with the values of v_2/a_1 obtained as a result of solving by numerical methods; this is demonstrated in Figure 55. We can also arrive at the choice of approximately the same R_2^* by arguments analogous to those used in choosing R_2^* in the case $\nu = 2$.

For the values of R_2^* and $\bar{k}_3$ chosen it is easy to obtain from (4.80) and (4.22) formulas for the overpressures:

$$\frac{p_2}{p_1}-1=\begin{cases}\dfrac{4}{\gamma+1}\dfrac{1}{-1+\sqrt{1+25\gamma\alpha R_2^3}}, & 0<R_2\leqslant 2,\\[2ex] \dfrac{4}{\gamma+1}\dfrac{1}{-1+\sqrt{1+50\gamma R_2^3\left(\ln\dfrac{R_2}{2}+1\right)}}, & \infty>R_2\geqslant 2.\end{cases}$$

The law of motion of the shock wave for values of $R_2 \leqslant 2$ is defined by (4.75). For $R_2 > 2$ we use the asymptotic formula valid for large distances to determine $R_2(\tau)$, as in the case of cylindrical symmetry. In dimensionless variables this formula has the form

$$\tau=\tau^*+\frac{R_2}{\sqrt{\gamma}}-\frac{0,4}{\gamma\sqrt{\alpha}}\bar{k}_3\sqrt{\ln(R_2/l^*)}. \tag{4.81}$$

Using (4.80) and the conditions (4.22) on the shock wave, we can also find the dependence $\rho_2/\rho_1(R_2)$ for various values of the adiabatic exponent γ. In [5] graphs are presented of the dimensionless overpressure for particles directly behind the front of the shock wave for $\gamma = 1.2, 1.4, 3$ and $\nu = 3$. Equation (4.75) was here integrated with initial conditions $\tau_0 = 10^{-5}$ and $R_{20} = \alpha^{-1/5}\cdot 10^{-2}$.

For the value $R_2 = 2$ transition was made from computing on the basis of (4.75) to computing according to (4.81). As in the cylindrical case, the constant τ^* was determined from the condition that the values of τ for $R_2 = 2$ should coincide and was found to be $\tau^* = -0.28$ for $\gamma = 1.2$, $\tau^* = -0.22$ for $\gamma = 1.4$ and $\tau^* = -0.19$ for $\gamma = 3$.

The magnitude of overpressure and the law $R_2(\tau)$ found from the approximate formulas were compared with computed data in order to determine the accuracy of computing on the basis of the approximate formulas proposed above. The comparison showed that in the spherical case for $p_2/p_1 - 1 \geqslant 0.05$ the error in determining the hydrodynamic characteristics of the wave front does not exceed 5% if the computed data are taken to be "exact".

For moments of time corresponding to $p_2/p_1 - 1 \geqslant 0.05$ the error in determining $R_2(\tau)$ does not exceed 3% [5].

To compare the damping laws of shock waves with different symmetries, in Figure 62 graphs are given of $p_2/p_1 - 1$ as a function of R_2 for $\nu = 1, 2, 3$ and $\gamma = 1.4$ which were constructed by the approximate formulas. It is evident from these graphs that the spherical shock wave dies out much faster than the cylindrical and plane shock waves.

We point out that the formulas obtained in this section are quite simple and possess high accuracy (they can be sharpened by using the damping laws (4.14) and (4.15)). These formulas have various applications (cf., for example, [45], [46]). The formulas presented in this section can also be generalized to include dissociation and ionization effects. The constants in the asymptotic relations (4.10)–(4.13) can be found on the basis of computations of the corresponding problems. These constants are given in [43] for the case of an explosion in air with spherical symmetry.

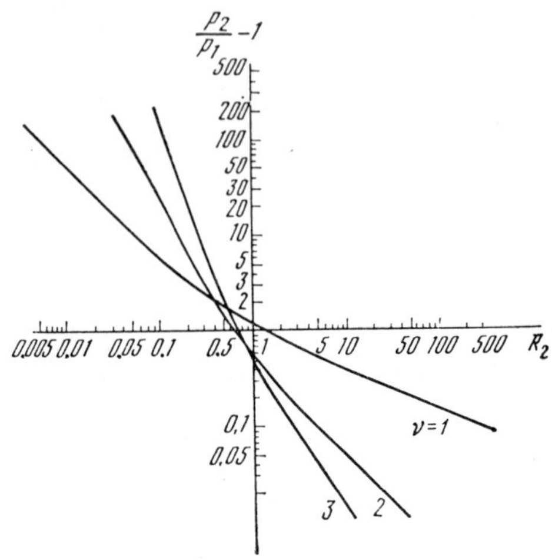

FIGURE 62. Graph of overpressure as a function of the radius for $\gamma = 1.4$. Comparison for different ν on the basis of the approximate theory

We have carried out work aimed at obtaining approximate formulas based on the approximation of the dependence $v_2(r_2)$ described for a perfect gas. Here good results are obtained if a more precise value of the energy parameter α is taken (taking into account the effective value of the adiabatic exponent γ_0, for example) and γ is assumed to be variable, depending on the density and pressure.

According to (4.64), for a strong explosion we have

$$v_{20} = \frac{4}{(\gamma+1)(\nu+2)} \sqrt{\frac{E_0}{\alpha(\gamma_0)\rho_1}}\, r_2^{-\nu/2}.$$

Substituting this dependence in the conditions on the shock wave and assuming that $\gamma = \gamma(p, \rho)$, we obtain implicit relations for $p_2(r_2)$, $\rho_2(r_2)$ and $D(r_2)$.

In a simplified version one may take $\gamma = 1.4$ and adjust only the value of α. This approach does not lead to large errors (especially for the function $r_2(t)$), since $\gamma + 1$ is weakly varying (from 2.1 to 2.4).

Comparison of our computations with the numerical solution of H. Brode [47] indicated that there is satisfactory agreement with regard to the dependence of $\Delta p/p_1$ on R_2 (of course, the agreement for the law of motion of the shock wave will also be good). We remark that in the case of approximation of $v_2(r_2)$ to compute the pressures p_2 it is convenient to use the formulas (2.69)–(2.71) and the approximating tables of N. M. Kuznecov (cf. Chapter 2).

§6. On conversion of dimensionless quantities to dimensional quantities. The similarity law(1)

As already noted, it is convenient to carry out the numerical solution of point

(1) Here we follow §6, Chapter IV, of the book [5] which was written by the author in collaboration with E. V. Rjazanov and N. S. Mel'nikova.

explosion problems in dimensionless parameters. For simplicity we shall assume that the gas is perfect with constant γ. As a result of computations carried out for any fixed values of ν and γ, we obtain the dependence of the unknown dimensionless functions on the dimensionless variables r/r_2 and τ; this makes it easy to find all the quantities characterizing the flow (for the same values of ν and γ) for any values of E_0, ρ_1 and p_1. For different values of ν and γ the computation must be carried out anew.

We consider the question of converting from dimensionless quantities to dimensional quantities. Suppose that we have the independent variables $\lambda = r/r_2$ and $\tau = t/t^0$ and $t^0 = r^0(\rho_1/p_1)^{1/2} = (E_0/p_1)^{1/\nu}\gamma^{1/2}/a_1$, where r^0 is a dynamic linear dimension and t^0 is the dynamic time.

The energy envolved during the explosion can be expressed in thermal or mechanical units (calories, ergs, kilogram-meters). For greater descriptiveness the energy may be characterized by the weight of some particular explosive. Often TNT is taken for a standard. (It is known that during the explosion of a so-called nominal atomic bomb over a short time interval approximately the same amount of energy is released as in the explosion of 20,000 tons of TNT.)

Recalling that

$$r = r_2\lambda, \qquad t = t^0\tau, \tag{4.82}$$

it is possible by knowing the function $R_2(\tau)$ from computation and assuming particular values for ρ_1, p_1 and E_0 to convert the dimensionless quantities λ and τ to the dimensional quantities r and t on the basis of (4.82).

In a similar manner it is possible to go over to the dimensional quantities of pressure, density, velocity, etc. As an example, we consider the following dimensional variables:

$$\begin{gathered} P(\sigma, \tau) = \frac{p}{p_1}, \quad G(\sigma, \tau) = \frac{\rho}{\rho_1}, \quad V(\sigma, \tau) = \frac{v}{a_1}, \\ R(\sigma, \tau) = \frac{r}{r^0}, \quad R_2(\sigma, \tau) = \frac{r_2}{r^0}, \end{gathered} \tag{4.83}$$

where σ is the dimensionless Lagrange coordinate, $\sigma = \xi/r^0$ and $\tau = t/t^0$. Conversion to the dimensional variables is accomplished via the formulas

$$\begin{gathered} p(\xi, t) = p_1 P, \quad \rho(\xi, t) = \rho_1 G, \quad v(\xi, t) = a_1 V, \\ r(\xi, t) = r^0 R, \quad r_2(t) = r^0 R_2, \quad \xi = r^0\sigma, \quad t = t^0\tau. \end{gathered} \tag{4.84}$$

It is sometimes required to convert from dimensional quantities corresponding to specific values of the dimensional initial parameters E_0, p_1 and ρ_1 to dimensionless variables (for the same values of γ and ν). We consider an example: the problem has been computed for certain values of γ and ν and definite values of E_{01}, p_{11} and ρ_{11}. It is required to find the solution corresponding to the same values of ν and γ but to different initial values of the energy, density, and pressure E_{02}, p_{12} and ρ_{12}. In this case we proceed as follows: using (4.83) and the known values E_{01}, p_{11} and ρ_{11}, we first go over to dimensionless variables, and then on the basis of (4.84) and the given E_{02}, p_{12} and ρ_{12} we find the relations of interest to us for the new system of dimensional variables. Thus, to convert the time we find first τ and then $t_{(2)}$:

$$\tau = \frac{t_{(1)}}{t_1^0}, \qquad t_{(2)} = t_2^0 \tau = \frac{t_2^0}{t_1^0} t_{(1)},$$

where

$$t_i^0 = \left(\frac{E_{0i}}{p_{1i}}\right)^{1/\nu} \left(\frac{\rho_{1i}}{p_{1i}}\right)^{1/2} \quad (i = 1,\ 2).$$

For conversion of the distance (the Euler coordinate of the particle) we have

$$R = \frac{r_{(1)}}{r_1^0}, \qquad r_{(2)} = r_2^0 R = \frac{r_2^0}{r_1^0} r_{(1)}, \qquad r_i^0 = \left(\frac{E_{0i}}{p_{1i}}\right)^{1/\nu}, \qquad (i = 1,\ 2),$$

where $t_{(1)}$ and $r_{(1)}$ are the dimensional time and Euler coordinate known from the computation carried out for the parameters E_{01}, p_{11} and ρ_{11}; $t_{(2)}$ and $r_{(2)}$ are the dimensional time and Euler coordinate converted to parameters.

Analogous formulas can be written down for converting the velocity, density, and other quantities. Here we assume that all quantities with indices 1 and 2 are given in the same system of units of measurement. From these formulas for $p_{11} = p_{12}$ and $\rho_{11} = \rho_{12}$ it is easy to find the relations

$$\frac{r_{(2)}}{r_{(1)}} = \left(\frac{E_{02}}{E_{01}}\right)^{1/\nu}, \qquad \frac{t_{(2)}}{t_{(1)}} = \left(\frac{E_{02}}{E_{01}}\right)^{1/\nu},$$

which express the so-called similarity law. For the case of a spherical explosion this law is often written as follows:

$$\frac{r_{(2)}}{r_{(1)}} = \left(\frac{w_1}{w_2}\right)^{1/3}, \qquad \frac{t_{(2)}}{t_{(1)}} = \left(\frac{w_1}{w_2}\right)^{1/3},$$

where w is the weight of the charge (the TNT equivalent for atomic explosions).

With the ratios of distances from the center of the explosion and of times thus chosen the magnitudes of the pressures will be the same. This similarity law is valid for a broad class of compressible media.

We consider further as a concrete example the conversion from the variables adopted in the work of Goldstine and von Neumann [4] to the dimensionless variables (4.83). In [4] the problem of a point spherical explosion ($\nu = 3$) was computed using the following initial data taken from the solution of the self-similar problem:(2)

$$\tau_{1N} = 0.0182575; \quad P_{2N} = 100; \quad G_{1N} = 1; \quad R_{2N} = 0.5 \quad (\gamma = 1.4), \quad (4.85)$$

where P_{2N} is the pressure on the front of the shock wave, G_{1N} is the density of the unperturbed medium, R_{2N} is the radius of the shock wave, and τ_N is the time. We shall assume here that all the quantities are dimensionless, being related to the dimen-

(2) In what follows the index N will indicate that the corresponding quantities are taken from [4].

sional quantities by (4.83):

$$P_N = \frac{p}{p_1}, \quad G_N = \frac{\rho}{\rho_1}, \quad \tau_N = \frac{t}{t_N^0}, \quad R_{2N} = \frac{r_2}{r_N^0}, \quad R_N = \frac{r}{r_N^0}, \quad \sigma_N = \frac{\xi}{r_N^0}, \tag{4.86}$$

where t_N^0 is a constant with the dimension of time, r_N^0 is a constant with the dimension of distance, r is the Euler coordinate, and ξ is the Lagrange coordinate.

In the self-similar problem on strong explosion the following relation is valid on the shock wave:

$$p_2 = \frac{8\rho_1}{25(\gamma+1)}\left(\frac{r_2}{t}\right)^2. \tag{4.87}$$

For the radius of the self-similar shock wave we have

$$r_2 = \left(\frac{E_0}{\alpha\rho_1}\right)^{1/5} t^{2/5}. \tag{4.88}$$

Substituting the expression for r_2 from (4.88) into (4.87) and using (4.86), we obtain

$$t_N^0 = \left(\frac{E_0}{\alpha p_1}\right)^{1/3}\left(\frac{\rho_1}{p_1}\right)^{1/2}\left[\frac{8}{25(\gamma+1)P_{2N}}\right]^{5/6}\frac{1}{\tau_N}.$$

For $\nu = 3$ and $\gamma = 1.4$ we have $\alpha = 0.851$. Hence

$$t_N^0 = \left(\frac{E_0}{0.851 p_1}\right)^{1/3}\left(\frac{\rho_1}{p_1}\right)^{1/2}\left[\frac{2}{15P_{2N}}\right]^{5/6}\frac{1}{\tau_{1N}} = \left(\frac{E_0}{p_1}\right)^{1/3}\left(\frac{\rho_1}{p_1}\right)^{1/2} k_N.$$

Recalling (4.85), we find the numerical values of the constant $k_N = 0.23229$. Thus $\tau_N = t/t_N^0 = t/k_N t^0$. Comparing this expression for the dimensionless time with (4.82), we have $k_N\tau_N = \tau$. Further, it is possible to obtain $R_2 = k_N R_{2N}$, $R = k_N R_N$, $\sigma = k_N\sigma_N$, $P = P_N$, $G = G_N$ and $V = \gamma^{-1}(\partial R_N/\partial\tau_N)$. These formulas give the relation between the dimensionless variables (4.86) used in [4] and the dimensionless variables (4.83).

§7. The use of analogy between an explosion and the flow around slender, blunt bodies

During the motion in a gas of slender bodies such as projectiles and rockets with hypersonic velocity, i.e. with a velocity several times exceeding the speed of propagation of acoustic waves in the surrounding medium, slender pointed bodies, subjected to the action of the medium, become blunted in their leading portion. Moving bodies may acquire a blunted shape in their leading part also in manufacture.

Thus, during hypersonic flights moving bodies are practically always blunted, and the solution of problems on the motion of blunt slender bodies is of great practical and theoretical importance. Motions of blunt bodies with hypersonic velocity are accompanied by the appearance in front of the body of a so-called detached shock wave which is not in contact with the surface of the body (Figures 63a and 64a). Methods

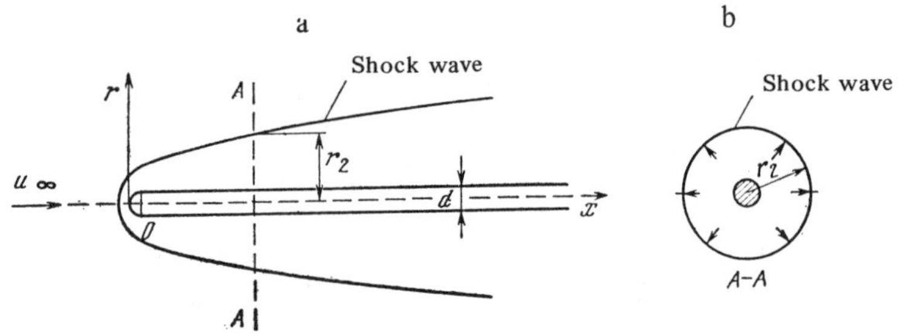

42 FIGURE 63. Diagram of flow around a cylinder with a spherical head (a) and the analogy with a cylindrical explosion (b)

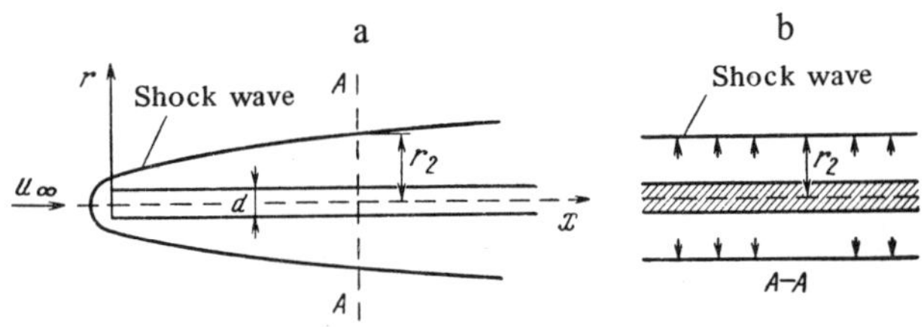

FIGURE 64. Diagram of flow around a blunt plate (a) and the analogy with the explosion of a plane charge (b)

of computing the flow behind a detached shock wave have been developed by O. M. Belocerkovskiĭ, V. V. Lunev, G. F. Telenin, P. I. Čuškin, and other authors [27], [28], [48]–[51].

There is a well-known analogy between the transient problem of explosion and the problem of stationary flow around bodies with hypersonic velocity which has been formulated in the works of G. G. Černyĭ [48] and other authors and is based on the use of the principle of plane sections.

The foundations of this analogy are also expounded in [5], [49]. Studies have shown that in planes perpendicular to the direction of the unperturbed motion of the gas the picture of the gas flow around a plate or cylinder must be essentially the same as during the propagation of a blast wave from a point explosion of plane and cylindrical charges (Figures 63b and 64b). The force of resistance X of the blunt body is approximately equal to the energy E_0, since the energy acquired by a layer of gas of unit thickness in surmounting the bluntness is $E_0 = X \cdot 1$.

Using the theory of explosions, one can obtain basic data on the flow around a thin blunt plate or a slender blunt cylinder (direct analogy). On the other hand, if we have the solution of the problem of hypersonic flow around a blunt plate or cylinder, we can determine approximately the parameters of plane and cylindrical explosions (inverse analogy). After the aforementioned principles of analogy had been formulated, the author's attention was attracted to the use of the theory of strong explosion and the linearized solution including counterpressure (cf., for example, [5], [48]). This is explained by the fact that the corresponding non-self-similar nonlinear problems had been little studied. In the work of M. A. Cikulin the principle of analogy was used not for a strong explosion but with application of experimental data for the parameters of the front based on explosions of cylindrical explosive charges [37]. A more complete investigation of direct analogy became possible only after obtaining the solution of the explosion problems using numerical and approximate methods.

For the use of the direct analogy the parameters of the transient problem are replaced by stationary parameters, where t and E_0 respectively are replaced by xu_∞^{-1} and $\frac{1}{2}c_x\rho_\infty u_\infty^2(\pi/4)^{\nu-1}d^\nu$.

Here the notation is as follows: x is the coordinate along the axis of the body measured from its foremost point; d is a characteristic cross-sectional dimension of the bluntness, u_∞ is the velocity of the incident gas flow directed along the x axis and c_x is the coefficient of resistance of the bluntness computed per unit area of cross section of the body and referred to velocity head.

In this correspondence the relation between the dimensionless quantities x/d, r/d and τ, R is given by

$$\frac{x}{d}=\left(\frac{\pi}{4}\right)^{(\nu-1)/\nu}\left(\frac{c_x\gamma^{(\nu+2)/2}}{2}\right)^{1/\nu}\mathrm{M}_\infty^{(2+\nu)/\nu}\tau,\qquad \frac{r}{d}=\left(\frac{\pi}{4}\right)^{(\nu-1)/\nu}\left(\frac{c_x\gamma}{2}\right)^{1/\nu}\mathrm{M}_\infty^{2/\nu}R, \tag{4.89}$$

where $\mathrm{M}_\infty = u_\infty/a_\infty$ is the Mach number of the oncoming flow.

Results obtained by the direct analogy are presented briefly below.

First of all, the approximate formulas for the parameters of the front of the shock wave considered in [5] make it possible to determine the form of the shock wave and write down its analytic expression. This was pointed out in [5] and considered in our work [20], where the following result was also obtained. The tangent dr/dx of the slope angle of the shock wave for large x tends in the stationary case to the corresponding quantity for the slope of the characteristics, i.e.

$$\frac{dr}{dx}\to(\mathrm{M}_\infty^2-1)^{-1/2}.$$

The speed of the shock wave in the nonstationary case tends to the speed of sound a_∞. Going over to the variables τ and R, we find for large τ

$$\frac{dR}{d\tau}\to\sqrt{\gamma}\,\frac{\mathrm{M}_\infty}{\sqrt{\mathrm{M}_\infty^2-1}},\qquad \frac{dR}{d\tau}\to\sqrt{\gamma}$$

for the cases of flow and explosion respectively. Thus for the flow the form of the shock wave at large distances will be the closer to the form obtained from the theory of explosion, the larger the number M_∞. This imposes a restriction on translation of data of the nonstationary problem to the flow problem (and conversely).

In [20] a comparison was made of the pressure distributions on a cylinder and of the forms of the leading shock waves obtained from approximate computation of a cylindrical explosion and by computation of the flow around bodies carried out by the method of characteristics [50]. This comparison indicated satisfactory agreement of results obtained by taking over data from explosion to flow outside the region of bluntness. The computations of higher accuracy considered in §3 made it possible to draw better-founded conclusions [34] on the accuracy of the analogy. It was found that for $M_\infty \sim 10$ both for $\nu = 2$ and $\nu = 1$ the gas-dynamic functions of explosion and flow are close to one another (in the sense of the analogy formulated) outside some neighborhood of the bluntness $(x/d > 10)$.

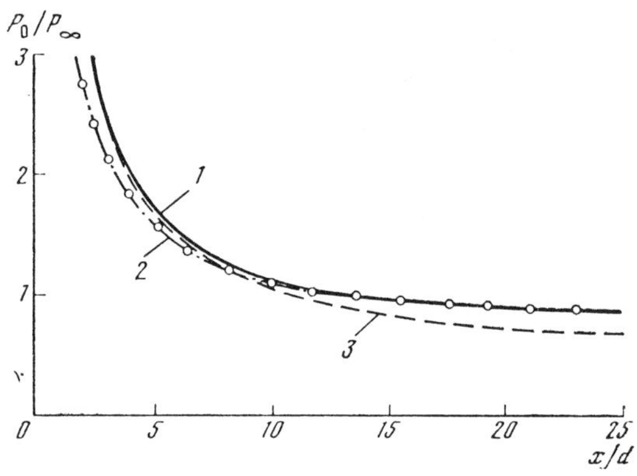

FIGURE 65. Comparison of the pressure distributions on a blunt cylinder computed from the theory of explosion and by the method of characteristics ($M_\infty = 10$; $\nu = 2$)

For illustration Figure 65 shows the pressure distribution on a cylinder with spherical blunting for $M_\infty = 10$ obtained from explosion computations (curve 1). Curve 2 corresponds to computations of the flow by the method of characteristics [50], while curve 3 refers to the linearized theory of explosions. As is evident from the graphs, the first two curves practically coincide sufficiently far away from the bluntness.

In recent years the problem of computing hypersonic flows has made strong advances. In this connection the question arises of using the universe analogy. Indeed, it has been shown that in the variables R and τ the dimensionless functions coincide outside a certain zone of bluntness, i.e. for sufficiently large τ and for inappreciable p_2/p_1. For $\nu = 2, M_\infty = 10$ and $p_2/p_1 < 3$ the agreement is rather good. Therefore for approximate estimates of the explosion parameters it is possible to use results of flow computations for $M_\infty \geqslant 10$. For this purpose one can use, for example, the recently published tables [51] of gas-dynamic functions of hypersonic flow around blunt cylinders with large numbers M_∞. The data given there make it possible to obtain pressure and density fields not only for a perfect gas but with account of the effects of dissociation and ionization as well. Naturally, the zone near the explosion (for small t) must be computed separately (for a perfect gas, for example, by using the effective value of γ).

§8. Explosion in a stationary translational flow of gas

In applications cases may arise of computing a point explosion in a stationary flow of gas moving with constant velocity U. The case of a strong spherical explosion has been discussed in [52]. If the process is viewed in a stationary coordinate system, then the solution must be converted using the Newton-Galilean transformation (cf. §1 of Chapter 1). Since the equations of gas dynamics are invariant under this transformation, the solution goes over into a new solution.

Suppose that some stationary system of rectangular Cartesian coordinates $\hat{x}^1, \hat{x}^2, \hat{x}^3$ has been chosen with origin at the center of the explosion at $t = 0$. We denote by U^j, $j = 1, 2, 3$, the projections of the vector of the initial velocity U on the coordinate axes and by v^j the corresponding projections of the vector of gas velocity after the explosion.

In considering the gas flow in the stationary coordinate system its parameters can then be computed by the transformation formulas $\hat{x}^j = x^j + U^j t$, $\hat{v}^j = v^j + U^j$, $\hat{p} = p$ and $\hat{\rho} = \rho$. Here v^j, p and ρ denote the velocity components, pressures, and density in the coordinate system x^j coupled to the advancing gas. We assume that the solution of the problem in this moving system is known. It can be found by the numerical method considered above or taken from tables of [35]. The problem thus reduces to a simple conversion of known data. In carrying out such a conversion in the case of cylindrical and spherical explosions one must consider the geometric relations following from the connection between Cartesian and cylindrical coordinates or between Cartesian and spherical coordinates.

In the spherical case we have

$$x^1 = r \sin\theta \cos\varphi, \qquad x^2 = \sin\theta \sin\varphi, \qquad x^3 = r \cos\theta,$$

$$v^1 = v \sin\theta \cos\varphi, \qquad v^2 = v \sin\theta \sin\varphi, \qquad v^3 = v \cos\theta,$$

where r, θ, φ are the spherical coordinates (of the moving system). It is clear from physical considerations that we have here the phenomenon of simple drift of the flow caused by a "steady wind". Unfortunately, during an explosion in the atmosphere the wind is seldom constant in magnitude or direction. Therefore the drift of the perturbation zone is more complicated. However, the formulas given are sometimes useful in determining the time of arrival of the shock wave at a given point of space if the vector U varies slowly in space and time. For large values of U the drift of the zone of perturbed motion drastically alters (for a stationary observer) the picture of the propagation of the shock wave.

§9. On reflection of the shock waves from a point explosion

9.1. *The initial stage of reflection of a plane wave from a wall parallel to it and of cylindrical and spherical waves from concentric cylindrical and spherical walls respectively.* Questions of the reflection of shock waves from a point explosion have not yet been worked out to any great extent. This is explained primarily by the relative complexity of the problems arising here. Naturally, to successfully solve reflection problems it is necessary to study the laws of propagation of shock waves in infinite space. In order to construct a more complete theory of point explosions, solution of the questions of shock-wave reflection is very essential.

We consider the question of determining the pressure, density, and velocity of the gas behind the front of a reflected shock wave for moments of time near the moment t_* at which the shock wave arrives at the wall. Because of the assumed geometry of the wall, the reflection of the shock wave from it will be normal reflection, and the flow behind the front of the reflected wave will be one-dimensional with plane, cylindrical, or spherical symmetry. In this case the pressure p_*, density ρ_*, and temperature T_* in the reflected wave are found from formulas well known in gas dynamics:

$$\frac{p_2}{p_1} = 2\left(\frac{p_2}{p_1} - 1\right) + \left[\left(\frac{p_2}{p_1} - 1\right)^2 \Big/ \frac{\gamma - 1}{\gamma + 1}\frac{p_2}{p_1} + 1\right], \tag{4.90}$$

$$\frac{\rho_*}{\rho_1} = \frac{\gamma p_2}{p_1}\frac{\rho_2}{\rho_1}\Big/\left[(\gamma - 1)\frac{p_2}{p_1} + 1\right], \qquad \frac{T_*}{T_1} = \frac{p_*}{p_1}\frac{\rho_1}{\rho_*}. \tag{4.91}$$

For the magnitude of the initial velocity of the reflected wave we have

$$\frac{D_*}{a_1} = \sqrt{\frac{2}{\gamma}}\left[(\gamma - 1)\frac{p_2}{p_1} + 1\right]\left[(\gamma + 1)\frac{p_2}{p_1} + (\gamma - 1)\right]^{-1/2}. \tag{4.92}$$

Since the wall is stationary, because of the boundary condition on it the velocity behind the front of the reflected shock wave v_* will be small for moments of time close to t_*.

If the distance r_* from the site of the explosion to the wall and the value of the characteristic length r^0 are given, then, knowing the dimensionless coordinate R_*, we find on the basis of the quantity $R_n = R_*$ from the tables of [35] or approximate formulas the dimensionless time τ_* of arrival of the wave at the wall, and from it the time t_* and also the corresponding values of q, p_2/p_1, and ρ_2/ρ_1. The basic parameters of the reflected shock wave for moments of time close to t_* are then computed from (4.90) and (4.91).

We point out that (4.90) and (4.91) for the problem in question are, strictly speaking, valid only at the moment of immediate reflection of the shock wave from the wall, since they are obtained for a homogeneous flow following the incident wave. However, these relations will be approximately satisfied for those moments of time where the variation of the flow parameters behind the incident wave can be neglected.

We shall present some considerations regarding the character of the subsequent flow. In the case of reflection of a wave from a strong explosion we actually have, from a physical point of view, a process of collision of a thin layer of gas with a stationary barrier and subsequent decay of any discontinuity. Here it is possible to construct various approximate models of the ensuing flow. We remark that certain questions of reflection for a strong explosion (in its initial stage) have been considered in [53], [54]. To resolve the question of the reflection of waves of moderate intensity it is necessary to use numerical methods. The flow of gas for the reflection being considered will be one-dimensional, but a reflected shock wave will propagate through the moving gas toward the center of the explosion. To solve this problem it is possible to apply the method of integral relations, where interpolation is carried out with respect to the spatial coordinate and the approximating system of differential equations is computed in time. The solution obtained for moments of time close to t_* can be used to prescribe the initial data. We mention that from the point of view of applications (explosions in air pockets in the ground, cylindrical explosions in xenon tubes to obtain light pulses, etc.) of greatest interest is the knowledge of the flow parameters in the immediate proximity of the wall. This should be considered in constructing approximate theories.

9.2. *The initial stage of regular reflection of plane, cylindrical, or spherical blast waves from a plane surface.* We first consider the plane case. Suppose that a plane blast wave impinges on an absolutely rigid plane wall so that the angle between the plane of the shock wave and the wall is nonzero and equal to α. Then for moments of time close to the collision of the wave with the wall it is possible to obtain asymptotic relations between the parameters of the incident and reflected waves from relations on the shock wave and the boundary condition on the wall for the velocity of the gas (cf. [55]).

We denote by α_*, p_* and ρ_* respectively the angle of reflection, pressure, and density behind the reflected wave at a moment of time immediately following reflection, and we introduce the following parameters:

$$\pi_1 = \frac{p_1}{p_2}, \qquad \pi_* = \frac{p_*}{p_2}, \qquad \omega = \tan\alpha, \qquad \omega_* = \tan\alpha_*.$$

If the angle α and parameter π_1 are given at the moment of encounter of the wave with the plane of reflection, then for determining ω_* we have the quadratic equation

$$m\left[(1-\mu)^2 - (\omega-\omega_*)^2 - (\mu+\omega\omega_*)^2\right] + m^2(1-\mu)^2(\omega-\omega_*) + \omega_* - \omega = 0, \tag{4.93}$$

where

$$\mu = \frac{\gamma-1}{\gamma+1}, \qquad m = \frac{(1-\pi_1)\,\omega}{1+\mu\pi_1+(\mu+\pi_1)\,\omega^2}.$$

The value of the relative pressure π_* behind the reflected wave is found as follows:

$$\pi_* = \frac{m\,(1+\mu\omega_*^2)+\omega_*}{\omega_* - m\,(\mu+\omega_*^2)}. \tag{4.94}$$

The ratios of the densities ρ_*/ρ_2 and temperatures T_*/T_2 can be found by means of (4.91). Since the point of intersection of the incident and reflected waves moves along the plane, to determine the velocity of the reflected shock wave D_* at the point of reflection we have the relation

$$D_* = D \frac{\sin \alpha_*}{\sin \alpha}. \tag{4.95}$$

Using the quantities D_* and ρ_*/ρ_2 and the law of mass conservation on passing through a surface of discontinuity, one can determine the normal component of gas velocity behind the reflected wave, v_*. The tangential component is found from the condition that it be continuous on passing through the discontinuity. We mention that the condition that (4.93) have no real roots gives values of the angles for which non-regular (Mach) reflection sets in.

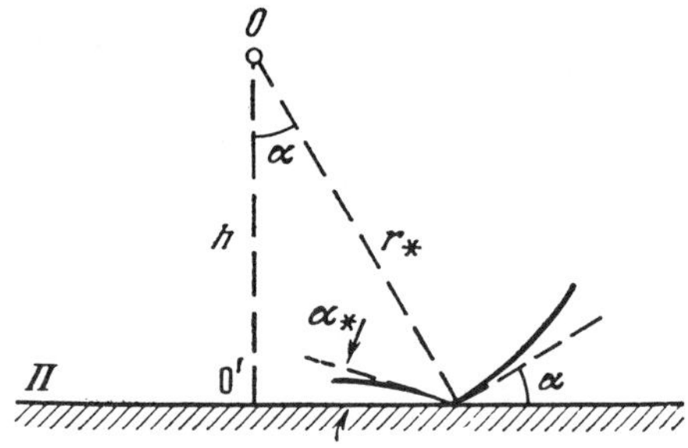

FIGURE 66. Diagram of regular reflection of a shock wave

We now consider the case of reflection of a cylindrical or spherical wave. Suppose that the explosion occurred at a distance h from the plane of reflection Π (in the cylindrical case we assume for simplicity that the line of the explosion is parallel to Π). We denote by α the angle at which the cylindrical or spherical wave arrives at Π at some moment of time t_* after the explosion. It is obvious from geometric considerations that α is equal to the angle between the perpendicular to Π and the radius vector drawn from the center of the explosion O to the point O' of reflection of the wave at the moment of time t^* (Figure 66). The gas flow during reflection of the shock wave will possess axial symmetry with axis passing through the point of the explosion and perpendicular to Π. The solution will depend on the parameters h, E_0, p_1, ρ_1 and γ, and for the dimensionless function p/p_1 we will have

$$\frac{p}{p_1} = P\left(\frac{t}{t^0}, \frac{r}{r^0}, \frac{h}{h^0}, \theta, \gamma\right),$$

where θ is the angular coordinate. For different ratios h/r^0 the problem must be solved anew (we assume that the value of γ is the same). The similarity law obtained in §6 is, strictly speaking, not satisfied here. If we change r^0 by varying the energy E_0, for example, then to convert the data to other h we must change r^0 in such a manner that the ratio $(h/r^0) = H_*$ remains constant; in other words, conversion is possible if

$$E_0 = \left(\frac{h}{H_*}\right)^\nu p_1. \tag{4.96}$$

For arbitrary E_0 conversion by ordinary similarity may lead to errors. This must always be taken into account. Here it can be mentioned that in [57] these facts are not always reckoned with and are stipulated in considering a spherical explosion; this can lead to misunderstandings. Of course, the similarity law may be applied in an approximate way for energies which do not deviate strongly from (4.96). As we have mentioned, the reflection of the shock wave for certain values of the angle of incidence α has a nonregular character [39], [40], [41], [50]–[54], [57]. For given Δp there is nonregular (Mach) reflection if the angle of incidence α is greater than a certain limiting angle α_0. In nonregular reflection there is a substantial increase of pressure (on the plane) as compared with the pressure that would be expected from regular reflection. Generally speaking, the value of the limiting angle α_0 increases with decreasing Δp. The lack of one-dimensionality of the problem, the restricted applicability of the similarity law, and the appearance of Mach reflection greatly complicate the theoretical solution of the problem. Consideration of this problem in its full generality is the subject of a special investigation. We shall consider only certain specific questions of this problem which refer basically to the stage of regular reflection.

Applying to the reflection of an element of a cylindrical or spherical wave the results of the solution to the problem on the reflection of a plane wave, one can determine the parameters of the gas directly at the moment of reflection of the wave element from the surface. Denoting by H_* the dimensionless quantity h/r^0, we have the equation

$$\cos \alpha = \frac{h}{r_*} = \frac{H_*}{R_*}. \tag{4.97}$$

Suppose that we are interested in the parameters of the reflected wave at a point O' of the plane Π at a distance r_* from the center of the explosion. If the parameter r^0 is known, then, knowing $R_* = r_*/r^0$, it is possible to compute the angle α from (4.97). By means of the tables of [35] or the formulas of §5, on the basis of the value $R_2 = R_*$ we find the quantities q_*, τ_* and the ratios p_2/p_1 and ρ_2/ρ_1. If for the ratio p_2/p_1 the angle α corresponds to the regime of regular reflection, then, using the notation introduced, we compute the parameters of the reflected wave at moments of time near t_* from (4.91) and (4.93)–(4.95).

Further, suppose that the relation between p_2/p_1 and the limiting angle of regular reflection is known. Such data follow from (4.93)–(4.95) on the basis of a theoretical analysis; they are presented, for example, in [55]. Then for a given value of h it is possible to find the boundary of the zone of regular reflection in the plane Π. We mention that for a spherical explosion with $\gamma = 1.4$ the problem of the initial stage of regular reflection has been solved by M. M. Vasil'ev [56]. A cylindrical explosion is apparently considered here for the first time (experimental data are given in [37]). Qualitative features of nonregular reflection are described in many places

[39], [41], [55], [57]–[60].

The question of determining the pressure on the plane at moments of time near the moment of reflection for the stage of nonregular reflection could be solved if formulas were known relating the pressures behind the incident and reflected waves for the triple-shock Mach configuration. Unfortunately, this question has not yet found a sufficiently precise solution. For the time being, therefore, our considerations can have only a qualitative and approximate quantitative interpretation.

Since the pressure on the plane does not vary monotonically with increasing α (even in the stage of regular reflection), the well-known problem arises of determining the most favorable (for given E_0) choice of explosion height h to obtain pressures $p_* \geqslant p_*^0$ on some surface Σ, where p_*^0 is a given pressure. Suppose that the explosion is spherical. Here we can also indicate the following problem: find the value of h for which the total force

$$F = \int_{\Sigma_0} p_* d\Sigma$$

is a maximum (for all t) for given Σ_0. Questions of the optimum choice of height of a spherical explosion are considered in a number of places (cf. [39], [40], [57]). The complete theoretical solution of this problem is not known to us either for spherical or for cylindrical waves.

BIBLIOGRAPHY

1. L. I. Sedov, *On the general theory of one-dimensional motions of a gas,* Dokl. Akad. Nauk SSSR **85** (1952), 723–726; addendum, ibid. **87** (1952), 4. (Russian) MR **14**, 425.

2. L. I. Sedov, *Similarity and dimensional methods in mechanics,* 4th ed., GITTL, Moscow, 1957; English transl., Academic Press, New York, 1959. MR **19**, 898; **21** #6840.

3. D. E. Ohocimskiĭ et al., *Computation of a point explosion taking back pressure into consideration,* Trudy Mat. Inst. Steklov. **50** (1957). (Russian) MR **20** #1422.

4. H. Goldstine and J. von Neumann, *Blast wave calculation,* Comm. Pure. Appl. Math. **8** (1955), 327–353. MR **17**, 314.

5. V. P. Korobeĭnikov, N. S. Mel′nikova and E. V. Rjazanov, *Theory of point explosions,* Fizmatgiz, Moscow, 1961. (Russian) MR **29** #5494.

6. Ja. B. Zel′dovič, *Theory of shock waves, and an introduction to gas dynamics,* Izdat. Akad. Nauk SSSR, Moscow, 1946. (Russian)

7. L. D. Landau, *Shock waves at large distances from their source,* Prikl. Mat. Meh. **9** (1945), 286–292. (Russian)

8. Ju. L. Jakimov, *On asymptotic solutions of the equations of one-dimensional nonstationary motion of an ideal gas, and on the asymptotic laws of damping of shock waves,* Prikl. Mat. Meh. **19** (1955), 681–692. (Russian)

9. G. M. Šefter, *An asymptotic solution of equations for one-dimensional unsteady motion of an ideal gas with cylindrical symmetry,* Dokl. Akad. Nauk SSSR **116** (1957), 572–575. (Russian) MR **20** #6908.

10. H. L. Brode, *Numerical solutions of spherical blast waves,* J. Appl. Phys. **26** (1955), 766–775. MR **16**, 971.

11. D. E. Ohocimskiĭ and Z. P. Vlasova, *The behavior of shock waves at large distances from the point of explosion,* Ž. Vyčisl. Mat. i Mat. Fiz. **2** (1962), 107–124 = USSR Comput. Math. and Math. Phys. **2** (1962), 107–127.

12. Pei Chi Chou, Robert R. Karpp and Shih Lien Huang, *Numerical calculation of blast waves by the method of characteristics,* AIAA J **5** (1967), 618–623; Russian transl., Raket Tehn. i Kosmonavtika **5** (1967), no. 4.

13. A. I. Žukov, *Application of the method of characteristics to the numerical solution of one-dimensional problems of gas dynamics,* Trudy Mat. Inst. Steklov. **58** (1960). (Russian) MR **23** #B727.

14. P. I. Čuškin, *The method of characteristics for spatial hypersonic flows,* Akad. Nauk SSSR Trudy Vyčisl. Centr **1968**. (Russian)

15. V. V. Rusanov, *Characteristics of the general equation of gas dynamics,* Ž. Vyčisl. Mat. i Mat. Fiz. **3** (1963), 508–527 = USSR Comput. Math. and Math. Phys. **3** (1963), 674–698.

16. G. G. Černyĭ, *The problem of point explosion,* Dokl. Akad. Nauk SSSR **112** (1957), 213–216. (Russian) MR **19**, 353.

17. ———, *Application of integral relationships in problems of propagation of a strong shock wave,* Prikl. Mat. Meh. **24** (1960), 121–125 = J. Appl. Math. Mech. **24** (1960), 159–165. MR **22** #6305.

18. N. S. Burnova (Mel'nikova), *Investigation of the point explosion problem,* Candidate's Dissertation, Moscow State Univ., Moscow, 1953. (Russian)

19. V. P. Korobeĭnikov, *On the analogy between cylindrical explosion and the hypersonic flow of gas around bodies,* PMTF Ž. Prikl. Meh. i.Tehn. Fiz. **1962**, no. 6, 45–49. (Russian)

20. N. S. Mel'nikova and T. M. Salamakin, *On the computation of point explosions in various media,* PMTF Ž. Prikl. Meh. i Tehn. Fiz. **1964**, no. 4, 155–160.

21. N. S. Mel'nikova, *On a point explosion in a medium with variable initial density,* Trudy Mat. Inst. Steklov. **87** (1966), 66–85 = Proc. Steklov Inst. Math. **87** (1967), 81–107.

22. T. H. Lee, R. Knystautas and G. G. Bach, *Theory of explosions,* MERL Rep. 69–10, McGill Univ., Montreal, 1969.

23. A. Sakurai, *On the propagation of cylindrical shock waves,* Exploding Wires, vol. 1, Plenum Press, New York, 1959, pp. 264–270; Russian transl., IL, Moscow, 1963.

24. V. V. Aduškin and I. V. Nemčinov, *Approximate determination of the gas parameters behind the front of a shock wave on the basis of the law of motion of the front,* PMTF Ž. Prikl. Meh. i Tehn. Fiz. **1963**, no. 4, 58–67. (Russian)

25. V. P. Andreev, *The method of successive approximations for one-dimensional transient problems of gas dynamics,* Izv. Akad. Nauk SSSR Meh. Židkost. i Gaza **1967**, no. 6, 31–36 = Fluid Dynamics **2** (1967), no. 6, 19–22.

26. A. A. Dorodnicyn, *One method for the numerical solution of certain nonlinear problems of aerodynamics,* Proc. Third All-Union Math. Congr. (Moscow, 1956), vol. 3, Izdat. Akad. Nauk SSSR, Moscow, 1958, pp. 447–453. (Russian)

27. O. M. Belocerkovskiĭ and P. I. Čuškin, *A numerical method of integral relations,* Ž. Vyčisl. Mat. i Mat. Fiz. **2** (1962), 731–759 = USSR Comput. Math. and Math. Phys. **2** (1962), 823–858. MR **27** #3090.

28. ———, *The numerical solution of problems in gas dynamics,* Basic Developments in Fluid Dynamics. Vol. 1, edited by M. Holt, Academic Press, New York, 1965, pp. 1–126.

29. V. P. Karlikov, V. P. Korobeĭnikov and E. V. Rjazanov, *An approximate method of solving problems of explosion in some ideal compressible media,* PMTF Ž. Prikl. Meh. i Tehn. Fiz **1963**, no. 2, 132–134. (Russian)

30. V. P. Korobeĭnikov and P. I. Čuškin, *A method of computing a point explosion in a gas*, Dokl. Akad. Nauk SSSR **154** (1964), 549–552 = Soviet Phys. Dokl. **9** (1964/65), 28–30.

31. ———, *On a method of computing point explosions in a gas with consideration of back pressure*, Proc. Fifth Session of the Scientific Council on Explosions, "Ilim", Frunze, 1965, pp. 74–82. (Russian) RŽMeh. **1966** #12Б133.

32. V. P. Karlikov, V. P. Korobeĭnikov and P. I. Čuškin, *Computing explosions in compressible media*, Second All-Union Congr. Theoret. Appl. Mech., Abstracts of Reports, "Nauka", Moscow, 1964, p. 103. (Russian)

33. V. P. Korobeĭnikov and P. I. Čuškin, *Plane, cylindrical, and spherical explosions in a gas with back pressure*, Trudy Mat. Inst. Steklov. **87** (1966), 3–34 = Proc. Steklov Inst. Math. **87** (1966), 1–40.

34. V. P. Korobeĭnikov, P. I. Čuškin and K. V. Šarovatova, *Tables of gas-dynamical functions for the initial stages of a point explosion*, Soobšč. po Vyčisl. Mat., vyp. 2, Izdat. Vyčisl. Centr Akad. Nauk SSSR, Moscow, 1963. (Russian). MR **32** #3408.

35. ———, *Gas-dynamical functions of a point explosion*, Akad. Nauk SSSR Trudy Vyčisl. Centr **1969**. (Russian)

36. M. A. Sadovskiĭ, *The mechanical action of shock waves in air on the basis of experimental data*, Fizika Vzryva, no. 1, Izdat. Akad. Nauk SSSR, Moscow, 1952. (Russian)

37. M. A. Cikulin, *Shock waves during the passage of coarse meteorites through the atmosphere*, "Nauka", Moscow, 1969. (Russian)

38. Hubert Schardin, *Measurement of spherical shock waves*, Comm. Pure Appl. Math. **7** (1954), 223–243.

39. *Effects of atomic weapons*, Los Alamos Scientific Laboratory, N. M., 1950.

40. Ju. S. Jakovlev, *The hydrodynamics of explosion*, Sudpromgiz, Leningrad, 1961. (Russian)

41. S. A. Lovlja et al., *Explosion phenomena*, Nedra, Moscow, 1966. (Russian)

42. A. S. Fonarev and S. Ju. Černjavskiĭ, *Computation of shock waves upon the detonation of spherical explosive charges in air*, Izv. Akad. Nauk SSSR Meh. Židkost. i Gaza **1968**, no. 5, 169–173 = Fluid Dynamics **3** (1968), no. 5, 121–125.

43. J. W. Miles, *Decay of spherical blast waves*, Phys. Fluids **10** (1967), 2706–2708.

44. V. P. Korobeĭnikov, *Approximate formulas for computing the characteristics of a shock wave front during point explosion in a gas*, Dokl. Akad. Nauk SSSR **111** (1956), 557–560. (Russian)

45. R. F. Fletcher, Dal Gerneth and Clark Goodman, *Explosion of propellants*, AIAA J. **4** (1966), 755–757.

46a. A. V. Zolotov, *The possibility of "thermal" explosion and the structure of the Tungus meteorite*, Dokl. Akad. Nauk SSSR **172** (1967), 805–808 = Soviet Phys. Dokl. **12** (1967/68), 101–104.

46b. ———, *Estimation of the parameters of the Tungus meteorite based on new data*, Dokl. Akad. Nauk SSSR **172** (1967), 1049–1052 = Soviet Phys. Dokl. **12** (1967/68), 108–111.

47. H. L. Brode, *Blast wave from a spherical charge*, Phys. Fluids **2** (1959), 217–229.

48. G. G. Černyĭ, *Gas flow with large hypersonic velocity*, Fizmatgiz, Moscow, 1959; English transl., *Introduction to hypersonic flow*, Academic Press, New York, 1961.

49. W. D. Hayes and R. F. Probstein, *Hypersonic flow theory*, Academic Press, New York, 1959; Russian transl., IL, Moscow, 1962.

50. P. I. Čuškin and N. P. Šulišnina, *Tables of hypersonic flow past blunt cones*, Vyčisl. Centr Akad. Nauk SSSR, Moscow, 1961. (Russian)

51. V. V. Lunev, K. M. Magomedov and V. G. Pavlov, *Hypersonic flow past blunt cones with consideration of the equilibrium physical-chemical transitions*, Vyčisl. Centr Akad. Nauk SSSR, Moscow, 1968; English transl., NASA Tech. Transl. TT F543, NASA, Washington, D. C., 1970.

52. È. I. Andriankin, *A perturbation method for the problem of strong explosions,* Izv. Akad. Nauk SSSR Otd. Tehn. Nauk **1958**, no. 12, 5–14. (Russian)

53. Tien Sun Chang and Otto Laporte, *Reflection of strong blast waves,* Phys. Fluids 7 (1964), 1225–1232.

54. F. Demmig, *Reflexion einer sphärischen Stosswelle an einer konzentrischen Kugelfläche,* Ann. Physik (7) **19** (1967), 113–127.

55. R. Courant and K. Friedrichs, *Supersonic flow and shock waves,* Interscience, New York, 1948; Russian transl., IL, Moscow, 1950. MR **10**, 637.

56. M. M. Vasil'ev, *On the reflection of a spherical shock wave from a plane,* Vyčisl. Mat., Sb. 6 (1960), 87–99. (Russian)

57. *The effect of nuclear weapons,* U.S. Atomic Energy Commission, Washington, D. C., 1957; Russian transl., Voenizdat, Moscow, 1960.

58. K. E. Gubkin, *Investigation of the reflection of shock waves by means of half-tone photographs,* Fizika Vzryva, no. 3, Izdat. Akad. Nauk SSSR, Moscow, 1955.

59. O. S. Ryžov and S. A. Hristianovič, *On nonlinear reflection of weak shock waves,* Prikl. Mat. Meh. **22** (1958), 586–599 = J. Appl. Math. Mech. **22** (1958), 826–843. MR **21** #6894.

60. T. Ginsburg, *Ausbreitung und Reflexion von starken Explosionswellen,* Sympos. Numerical Treatment of Partial Differential Equations with Real Characteristics (Rome, 1959), Liberia Eredi Virgilio Veschi, Rome, 1959, pp. 145–156. MR **22** #449.

61. N. A. Arhangel'skiĭ, *An algorithm for the numerical solution by the mesh method of the cylindrical blast problem, allowing for back pressure,* Ž. Vyčisl. Mat. i Mat. Fiz. **11** (1971), 222–236 = USSR Comput. Math. and Math. Phys. **11** (1971), no. 1, 289–307.

CHAPTER 5

PROBLEMS OF THE THEORY OF POINT EXPLOSION IN INHOMOGENEOUS MEDIA AND WITH NONSYMMETRIC ENERGY RELEASE

§ 1. An exact solution for a special $\rho_1(r)$ dependence

We consider the point explosion problem (PEP) where the initial pressure p_1 is constant, and the initial density of the gas is variable and changes with distance from the center of the explosion according to the law

$$\rho_1(r) = \frac{a}{\gamma}(\gamma-1)^2\left(\frac{\gamma+1}{2}\right)^{1-\beta}\left(\frac{r}{r^0}\right)^{-\omega}\left[\left(\frac{r}{r^0}\right)^{\nu} + \frac{\nu(\gamma^2-1)}{2\sigma_\nu\gamma}\right]^{-\beta}, \tag{5.1}$$

where γ is the ratio of specific heat capacities; a is an arbitrary positive constant;

$$\omega = [\nu(3-\gamma) + 2\gamma - 2]/[\gamma+1], \qquad \beta = [3\nu\gamma + 4 - \nu]/[\nu(\gamma+1)],$$

and $r^0 = (E_0/p_1)^{1/\nu}$ is the dynamic length; $\nu = 3, 2, 1$. It is evident from (5.1) that ρ_1 depends on γ and the dynamic length r^0 as parameters.

The one-dimensional adiabatic motions of the gas behind the wave are described by the system of equations (1.18)–(1.20) or (3.5). It is required to determine the dependence of the velocity, pressure, and density of the gas on the linear coordinate r and time t, and also the dependence of the radius of the shock wave r_2 on time.

The problem reduces to finding a solution of the system (3.5) with the initial conditions indicated above, boundary condition at the center of symmetry $v(0,t) = 0$, and conditions on the front of the blast wave which can be written in the form (1.62). It can be seen by direct verification that the solution of the problem posed is given by the formulas [1]

$$v = \frac{r}{kt}, \qquad k = \frac{\nu(\gamma-1)+2}{2}, \tag{5.2}$$

$$\rho = \frac{2p_1[kt]^{-(\nu-1)/k}}{\nu(\gamma^2-1)r}\frac{d}{dx}\left\{\frac{4\gamma}{b(\gamma+1)}[f(x)]^{-(\gamma-1)/2} - (\gamma-1)[f(x)]^{-\gamma}\right\}, \tag{5.3}$$

$$p = \frac{p_1[kt]^{-\gamma\nu/k}}{\gamma+1}\left\{\frac{4\gamma}{b(\gamma+1)}[f(x)]^{-(\gamma-1)/2} - (\gamma-1)[f(x)]^{-\gamma}\right\}, \tag{5.4}$$

$$\left(\frac{r_2}{r^0}\right)^{\nu} = \frac{(\gamma^2-1)}{2\gamma\sigma_\nu\left\{\frac{2}{(\gamma+1)b}[kt]^{-\nu(\gamma+1)/2k} - 1\right\}}, \tag{5.5}$$

where $x = r[kt]^{-1/k}$, $b = [\nu^2 p_1/(\sigma_\nu r^0)^2 a]^{1/(\beta-1)}$, and $f(x) \geqslant 0$ is a function not assuming negative values, defined by

$$\left(\frac{x}{r^0}\right)^\nu + \frac{(\gamma^2-1)}{2\sigma_\nu\gamma} f - \frac{2}{(\gamma+1)\,b}\left(\frac{x}{r^0}\right)^\nu f^{(\nu+1)/2} = 0. \tag{5.6}$$

Variation of the pressure directly behind the front of the shock wave is described by the formula

$$p_2 = p_1\left[1 + \frac{\nu(\gamma-1)}{\sigma_\nu}\left(\frac{r^0}{r_2}\right)^\nu\right].$$

We obtained the solution indicated from an exact solution of L. I. Sedov [2]. The method of constructing discontinuous solutions for this exact solution was worked out in [3].

From our solution in the particular case where $p_1 \to 0$ and $b \sim p_1 \to 0$ we obtain the known solution of the self-similar PEP where the initial density is distributed according to the law $\rho_1 = Ar_1^{-\omega}$ (cf. (2.28)).

§ 2. Explosion on the boundary separating two media

2.1. *The problem of strong plane, cylindrical or spherical explosions on the surface separating two ideal compressible media.* Suppose that there are two half spaces filled with ideal compressible media the internal energy of which depends on p and ρ according to the formula

$$\varepsilon = pf\left(\frac{\rho}{\rho_0}\right) + \text{const.} \tag{5.7}$$

If a point explosion occurs on the boundary of these two half spaces, then in the planar case the flow will be one-dimensional, while in the cylindrical and spherical cases the problem will depend on two spatial coordinates and the time. If the counterpressure is assumed equal to zero in both half spaces, the system of defining parameters will be the following: $E_0, \rho_0, \rho_{1-}, \rho_{1+}, r, t,$ where ρ_{1-} and ρ_{1+} denote the constant densities in the different half spaces. We call these the lower and upper half spaces. Since the equations of gas dynamics and the functions $f_-(\rho/\rho_0)$ and $f_+(\rho/\rho_0)$ do not contain dimensional constants, the problem formulated for a strong explosion is self-similar and in the general case will depend on two dimensionless variables of the form x^j/bt^δ or $\lambda = r/bt^\delta$ and θ, where θ is the polar angle. The choice of coordinate system depends on the method of solving the problems in question.

The boundary conditions here are the conditions on the shock waves in the upper and lower half spaces, the condition of equality of pressures on the boundary separating the two media, and the equality of the normal velocities of the gas on the boundary. All the equations needed can easily be obtained from the equations and conditions on surfaces of strong discontinuities presented in Chapter 1 and will not be considered here in detail.

For the case of a plane explosion on the boundary separating two perfect gases the self-similar problem has an exact analytic solution [4], [5]. The solution of the problem is essentially related to finding the distribution of the energy E_0 between the two half spaces from the condition that the pressures be equal on the boundaries of the half spaces.

For cylindrical and spherical explosions the problem has not yet been solved in detail. The simplest variants of the problem are the case of two perfect gases and the case where the medium in one of the half spaces is incompressible. By means of the example of a spherical explosion we shall consider the approximate solution of the question of the energy distribution between two half spaces filled with different perfect gases.

We shall follow the paper [6] which was written by the author in collaboration with G. A. Ostroumov. For the radius of the front of the shock wave $r_2(t)$ in the one-dimensional case we have the relation

$$r_2 = \left(\frac{E_0}{\alpha\rho_1}\right)^{1/5} t^{2/5}.$$

For the force F acting on the boundary of the half space for spherically symmetric flow we have for any γ

$$F = 2\pi p_2 \int_0^{r_2} \left(\frac{p}{p_2}\right) r\,dr = \frac{4\pi}{\gamma+1} \rho_1 D^2 r_2^2 K.$$

Substituting into this formula

$$Dr_2 = \frac{2}{5}\left(\frac{E_0}{\alpha\rho_1}\right) t^{-1/5},$$

we obtain

$$F = \frac{16\pi}{25}\frac{1}{\gamma+1}\left(\frac{E_0}{\alpha}\right)^{4/5} \rho_1^{1/5} K t^{-2/5}. \tag{5.8}$$

In (5.8) we have used the notation

$$\int_0^1 \left[\frac{p}{p_2}(\lambda)\right] \lambda\,d\lambda = K, \quad \lambda = \frac{r}{r_2}.$$

We now assume that the total forces acting in the upper and lower half spaces are approximately equal to the forces computed for the respective spherically symmetric flows [7]. Then from this hypothesis on the constant equality of the forces F_- and F_+ acting in both media, we obtain the relation

$$\frac{E_+^0}{E_-^0}=\left(\frac{\gamma_+ +1}{\gamma_- +1}\right)^{5/4}\frac{\alpha_+}{\alpha_-}\left(\frac{\rho_{1-}}{\rho_{1+}}\right)^{1/4}\left(\frac{K_-}{K_+}\right)^{5/4}. \tag{5.9}$$

Suppose, for example, that $\gamma_- = 7, \rho_{1-} = 1, \gamma_+ = 4$ and $\rho_{1+} = 2.65$; then, using the results of Chapter 2, we obtain $K_- = 1/5, K_+ = 0.195, \alpha_- = 0.0279$ and $\alpha_+ = 0.0766$; E_+^0 is the energy for the upper half space; E_-^0 is the energy for the lower half space (we remark that the parameters chosen correspond approximately to water and quartz). For these media we find $E_+^0 = 1.24E_-^0$ and $E_+^0 = 0.55E_0$, where $E_0 = E_+^0 + E_-^0$, i.e. the energy is about equally distributed between the half spaces. An estimate of the energy distribution between such media as water and air indicates that the principal part of the energy E_0 passes into the air; this corresponds to the physical picture of the phenomenon. We remark further that in the planar case ($\nu = 1$) the formula of the type (5.9) is exact. Computations indicate that the quantity K is weakly varying with increasing γ. For the values $\nu = 3$ and $3 \leqslant \gamma \leqslant 7$ we find that K is close to 0.2.

Approximate solutions of the problem of strong spherical and cylindrical explosions under other assumptions and in certain special cases have been considered by Ju. P. Raĭzer [8] and B. N. Rumjancev [9], [10], and also by L. V. Šuršalov [11]. Questions closely related to these problems of the impact and penetration of bodies into incompressible and compressible half spaces are considered in the works of L. I. Sedov [12], L. A. Galin [13], N. N. Moiseev [14], A. Ja. Sagomonjan [15], their students, and other authors. Explosion on the surface separating water and air has been studied experimentally by A. A. Deribas [16] and V. F. Minin [17]. Computation of a strong explosion on the surface of soft ground has been carried out in the work of S. S. Grigorjan and M. M. Martirosjan [18].

2.2. *On determining the flow parameters during explosion on a plane surface separating a gas from a liquid or solid medium.* We suppose that the upper half space is filled with gas and the lower with some other medium. We shall consider first the idealized case in which the lower half space if filled by an absolutely solid body. To determine, for example, by means of the tables [19], [20] the flow parameters on explosion of a charge on the boundary of a gas with an absolutely solid plane, the quantity E_0 is everywhere to be replaced by $\frac{1}{2}E_0$.

The problem of a spherical explosion at the vertex of an absolutely solid cone with solid angle φ is solved similarly. Here it is only necessary to note that the energy does not go to the entire solid angle 4π but to the angle $4\pi - \varphi$. It is also possible to consider the case of an explosion in a conical cavity in an absolutely solid body. For a cylindrical charge the appropriate wedge-shaped region is to be taken.

Suppose now that the lower half space is filled with a deformable medium, but the flow is of such a nature that processes occurring in the lower half space have a very weak effect on the gas flow in the upper half space. We moreover assume that the time of arrival of the shock wave at some fixed point of the upper half space is known (a time measured by experiment, for example). For such a point it is convenient to take some point located on a line perpendicular to the plane of the boundary and passing through the center of the explosion in the cylindrical and spherical cases.

It is then possible to consider in an approximate way those portions of the total energy E_0 of the explosion which passed to the upper and lower half spaces and to find the flow parameters of the gas in the upper half space. Indeed, on the basis of the coordinate of the fixed point r_* and the time of arrival of the shock wave t_* it is possible to determine the energy E^0_+ released into the upper half space by recalling the formulas $r_* = R_2 r^0$ and $t_* = \tau t^0$ and using the dependence of R_2 on τ obtained from the tables. It is obvious that the remainder of the energy E^0_- passes into the lower half space, since

$$E^0_+ + E^0_- = E_0. \tag{5.10}$$

Knowing the energy E^0_+, it is possible to obtain from the tables all the gas-dynamic parameters of interest to us and describe in an approximate way the explosion in the upper half space.

We remark that an analogous, but purely experimental approach to finding the energy distribution between two half spaces has been considered in [21].

2.3. *A plane explosion on the boundary separating two identical gases having the same initial pressures but different initial densities.* We shall discuss the question of using the results of computing a point explosion to solve the problem of determining the gas-dynamic parameters in a plane explosion on the boundary separating gases with initial parameters γ, p_1, ρ_{1+} and γ, p_1, ρ_{1-}.

Let us suppose that the Lagrange coordinate of the contact surface is $\eta_* = 0$. From the condition that the pressures be equal on this surface we have

$$P_1(0, \tau) = P_2(0, \tau), \tag{5.11}$$

where P_1 and P_2 are dimensionless pressures $(P = p/p_1)$ in the first and second half spaces respectively. We now take for the functions P_1 and P_2 the relations obtained for a homogeneous medium. Using the formula $\tau = t/t^0$ we can obtain from this the connection between the fraction of energy E^0_+ released into the first half space and the fraction E^0_- going to the second half space, namely,

$$E^0_+ = E^0_- \sqrt{\frac{\rho_{1-}}{\rho_{1+}}}. \tag{5.12}$$

Of course, formula (5.10) holds here also.

Thus, if a plane explosion with energy E_0 occurs on the boundary separating two identical gases, the physical characteristics of the flow in both half spaces can be determined from the tables of [13] using the relations (5.10)–(5.12). We mention that in the case of a plane explosion on the boundary for arbitrary ρ_1, p_1 and different γ the problem can be solved numerically using a generalization of the method presented in §3 of the preceding chapter.

§3. Approximate methods of determining the parameters of shock waves during explosion in a stratified inhomogeneous atmosphere

3.1. *Formulation of the problem and derivations following from dimensional analysis.* Investigation of the PEP in a quiescent gas with an initial density distribution having a layered character is of major interest. This type of density distribution occurs in the atmosphere of the earth and other planets, where the air density decreases with height. The variation of density with height affects the motion of the gas behind the front of the shock wave and complicates the flow picture. The temperature and pressure in the earth's atmosphere also change with height.

We suppose that the initial density and pressure of the gas vary according to the law

$$\rho = \rho_0 \Omega_1\left(\frac{z}{H}\right), \quad \Omega_1(0) = 1, \quad p = p_0 \Omega_2\left(\frac{z}{H}\right), \quad \Omega_2(0) = 1, \tag{5.13}$$

where z is the coordinate along which the density and pressure vary, ρ_0 is the density at $z = 0$, p_0 is the pressure at $z = 0$, and H is a constant with the dimension of length. For a spherical point explosion the motion of the gas will be two-dimensional having axial symmetry. The solution of this PEP is possible only numerically and is beset with considerable mathematical difficulties. For the case of an inhomogeneous atmosphere examples of a numerical solution are given in the works of K. I. Babenko, V. V. Rusanov, and A. M. Molčanov [22]–[24]. Computation of the initial stage of the explosion is considered in [25]. We give below several approaches to the solution of this problem.

For a cylindrical explosion the motion will be two-dimensional, transient motion if the line of explosion is parallel to the z axis or perpendicular to it. For other configurations the motion will be three-dimensional, and the basic functions depend on x^j and t. In the planar case the motion will be one-dimensional if the plane of the explosion is perpendicular to the z axis. For other configurations the flow will be two- or three-dimensional.

We consider the case of a spherical explosion. The system of equations of gas dynamics for axial symmetry in spherical coordinates is given in Chapter 1 (cf. (1.17)). The conditions on the shock wave can be written as follows:

$$\rho_1 D = \rho_2 (D - v_2) = m, \qquad v_2 m = p_2 - p_1,$$
$$\varepsilon_2 - \varepsilon_1 = \frac{1}{2}(p_2 + p_1)\left(\frac{1}{\rho_1} - \frac{1}{\rho_2}\right). \tag{5.14}$$

If we denote by $r_2 = r_2(t, \theta)$ the law of variation of the front of the shock wave, then in correspondence with the definition of the velocity D (cf. Chapter 1, §2) on the basis of (1.45) we have

$$D = \frac{\partial r_2}{\partial t}\left[1 + \left(\frac{1}{r_2}\frac{\partial r_2}{\partial \theta}\right)^2\right]^{-1/2}. \tag{5.15}$$

In this problem the unknown quantities are v_t, v_θ, p and ρ; they depend on r, θ and t. For the case of a perfect gas the defining parameters of the problem are

$$r,\ \theta,\ t,\ E_0,\ \rho_0,\ p_0,\ H,\ \gamma, \tag{5.16}$$

where r is the length of the radius vector and θ is the angle measured from the z axis. From the parameters (5.16) it is possible to form the following dimensionless combinations:

$$R = \frac{r}{r^0},\quad \theta,\quad \tau = \frac{t}{t^0},\quad h = \frac{H}{r^0},\quad \gamma, \tag{5.17}$$

where $r^0 = (E_0/p_0)^{1/3}$ is the local dynamic length and $t^0 = r^0(\rho_0/p_0)^{1/2}$ is the local dynamic time. On the basis of the π theorem we conclude that the unknown dimensionless quantities, the pressure p/p_0 for example, will depend on the dimensionless combinations indicated in (5.17) [26], [27]:

$$\frac{p}{p_0} = P(R, \tau, \theta, h, \gamma). \tag{5.18}$$

It follows from (5.18) that a computation made for a single energy E_0 and fixed r^0, t^0, h and γ cannot be used for other heights of the explosion unless H, p_0 and ρ_0 are changed in such a way that the quantity h remains constant. If the energy of the explosion is changed for fixed p_0 and ρ_0, then in order to use the computation made for some fixed E_0 we must vary the parameter H in such a way that h does not change. From these considerations it follows that the solution of problems on explosion in an inhomogeneous medium is much more complicated, since it is necessary to carry out computations for a series of parameters γ and h, with each solution two-dimensional. The solution of problems is complicated still further if the medium is not a perfect gas with constant heat capacities (for example, air with dissociation and ionization).

If one intends to solve the problem numerically, it is clear that the computation of even a single variant of the problem is quite a laborious task [23]–[25]. The necessity therefore arises of finding various approximate methods of solving the problem globally and approximate means of determining the parameters of the blast waves. These methods include the method of linearization, the method of approximating the quantities on the wave front, and the method of using exact solutions and the equations of hydrodynamics with approximate fulfillment of the conditions on the shock wave (here it is possible to use the solution of L. V. Ovsjannikov [28]). For this problem we shall consider in detail only the second method.

We note further that, as was pointed out by L. I. Sedov, in the case of the strong explosion of a spherically symmetric point charge ($p_1 = 0$) the problem will be self-similar if for the initial density distribution we have

$$\rho_1 = A_0 \frac{\chi(\theta)}{r^\omega},$$

where χ depends only on the polar angle θ and A_0 is a constant with the dimension $[A_0] = ML^{3-\omega}$. There can be self-similar solutions also in the cases of cylindrical and plane explosions for two-dimensional problems with variable initial density.

3.2. *Determination of the form and parameters of the front of a shock wave for explosion at a point.* Let $\rho_1 = \rho_0 e^{-z/H}$, and consider the explosion of a spherical point charge. For a strong explosion approximate methods of dealing with the variability of the density have been proposed in a number of works (cf. [29]–[35]). For sufficiently weak shock waves asymptotic methods of nonlinear geometric acoustics have been developed [36]–[39], and asymptotic laws of the damping of shock waves have been obtained which are more general than those considered in §2 of Chapter 4.

The development of the explosion from its strong initial stage to a stage near to the degeneration into acoustic waves is of practical interest. We mention that K. E. Gubkin [37] gives an approximate method of determining the pressure at large distances for an explosion in an inhomogeneous atmosphere, based on using asymptotic laws for the damping of shock waves and experimental data. Below approximate methods of determining the parameters of blast waves will be given [26], [27] which are based on results of solving the equations and problems of gas dynamics for strong and weak shock waves.

We note first of all that if for a fixed medium one of the quantities p_2, ρ_2, v_2, D is given as a function of r_2 and θ, then all the remaining quantities can be found from (5.15). Suppose that we are interested in determining the wave form and its law of variation in time. We introduce the dimensionless quantity $W = D/a_1$, where a_1 is the speed of sound in the quiescent medium. Passing to dimensionless variables in (5.16), we find

$$\beta \frac{\partial l}{\partial \tau} = W(l, \theta)\left[\left(\frac{1}{l}\frac{\partial l}{\partial \theta}\right)^2 + 1\right]^{1/2}, \quad l = \frac{r_2}{r^0}, \quad \beta = \left(\frac{p_0}{\rho_0}\right)^{1/2} \frac{1}{a_1}. \qquad (5.19)$$

If the function $W(l, \theta)$ is known from theoretical considerations (or from experiment), then (5.19) can be interpreted as an equation for determining $l(\tau, \theta)$.

Equation (5.19) is a nonlinear partial differential equation of first order. If at some moment $t = t_0$ ($\tau = \tau_0$) we are given

$$l = l_0(\theta),$$

then by solving the Cauchy problem for (5.19) we find the law of variation of $l = l(\tau, \theta)$.

We shall consider the question of theoretically determining the parameters of the wave front on the basis of data on the strong stage of the blast wave and asymptotic laws for the damping of waves at large distances. Here we follow mainly our joint work with V. P. Karlikov [26], [27].

We assume the explosion to be a point explosion and the gas to be perfect with $\gamma = 1.4$. In this case from (5.15) it is possible to find $v_2(q)$ and $\rho_2(q)$, where $q = a_1^2 D^2$. These relations are given by (1.63). Let

$$\Omega_1 = \Omega_2 = \exp\left(-\frac{z}{H}\right), \tag{5.20}$$

i.e. we consider the case of an isothermic atmosphere. For small values of z/H we have

$$\Omega_1 = 1 - \frac{r}{H}\cos\theta. \tag{5.21}$$

For a strong shock wave the PEP with density varying according to (5.13), (5.21) has been solved by Karlikov using the method of linearization (cf. [29], [30]). According to the results of these papers, the law of variation of the shock wave is given by the formula

$$l = \alpha^{-1/5}\tau^{2/5}\left(1 + \varkappa\alpha^{-1/5}\tau^{2/5}\right), \qquad \kappa = \frac{0.1605}{h}\cos\theta, \qquad \alpha = 0.851, \qquad \gamma = 1.4. \tag{5.22}$$

For the speed of gas particles behind the front $v_2(r_2, \theta)$ for small κ we have

$$v_{2L} = \frac{1}{3}\sqrt{\frac{E_0}{\alpha\rho_0}}\, r_2^{-3/2}(1 + 2\varkappa l)(1 + \kappa l)^{3/2}. \tag{5.23}$$

Instead of (5.23), in the same approximation (for small κ) it is possible to write

$$v_{2L} = \frac{1}{3}\sqrt{\frac{E_0}{\alpha\rho_0}}\, r_2^{-3/2}\left(1 + \frac{7}{2}\kappa l\right). \tag{5.24}$$

It follows from the foregoing that (5.23) and (5.24) give the function $v_2(r_2, \theta)$ for small values of r during an explosion in an atmosphere with density which varies according to (5.20).

It follows from the results of [38] that for large distances from the site of the explosion the following asymptotic formula is valid:

$$v_{1\mathrm{as}} = \frac{C_1 e^{ml}}{\rho_0 a_1 r_2}\left(\int_{r^*}^{r_2} \exp\left(\frac{r\cos\theta}{2H}\right)\frac{dr}{r}\right)^{-1/2}, \qquad m = \frac{\cos\theta}{2h}, \tag{5.25}$$

where C_1 and r^* are constants depending on the form of the wave. Generalizing the method considered in Chapter 4, we suppose that for the approximate determination of $v_2(r_2, \theta)$ up to considerable distances the following approximation may be used. We assume that up to some $r_*(\theta)$ formula (5.24) (or (5.23) with $h > 1$) holds, while for $r > r_*$ formula (5.25) holds. We choose the quantities C_1, r^* and r_* from the conditions of matching (5.24) and (5.25). It was shown in Chapter 4 that an analogous approximation in the one-dimensional case gives good results. Since sufficiently precise numerical or analytic solutions of the problem with counterpressure for an isothermic atmosphere are not yet known to us, it has not been possible to establish the accuracy of the proposed approximation by the corresponding comparisons. We remark that for small m we have from (5.25)

$$v_{2\text{as}} \sim \frac{1 + ml}{r_2^{3/2} \sqrt{m\left(1 - \frac{r^*}{r_2}\right) + \frac{1}{r_2} \ln \frac{r_2}{r^*}}},$$

which indicates the closeness of the quantitative behavior of the functions (5.24) and (5.25) for small ml. This also affirms the possibility of rough approximation of $v_2(r_2, \theta)$ by (5.24) and (5.25).

In dimensionless variables (5.24) and (5.25) take the form

$$V_1 = \omega l^{-3/2}\left(1 + \frac{7}{2}\kappa l\right) \quad \left(V = \frac{v_2}{a_1}\right), \tag{5.26}$$

$$V_{\text{as}} = \sigma\omega e^{ml} l^{-1} I^{-1/2} \quad \left(\omega = \frac{1}{3\sqrt{\alpha}}\right), \tag{5.27}$$

$$I = \int_{l^*}^{l} e^{m\xi} \frac{d\xi}{\xi} = E_i(ml) - E_i(ml^*), \tag{5.28}$$

where $E_i(x)$ is the integral exponential function and $l^* = r^*/r^0$. The quantity $l_*(\theta)$ can be chosen from the condition that q be close to one $(0.5 < q < 1)$. We shall assume that the transition to the asymptotic formulas takes place at $q = q_*$. Since $V = 2(1 - q)/(\gamma + 1)\sqrt{q}$, if q is known it is easy to find V and then to determine $l_*(\theta)$ by (5.26).

By analogy with the one-dimensional case (Chapter 4), we shall seek the quantities σ and l^* from the condition of matching the functions V defined by (5.26) and (5.27) and their derivatives with respect to l at $l = l_*$. For determining l^* and σ we find the formulas

$$\sigma = l_*^{-1/2} e^{-ml_*} I_*^{1/2}\left(1 + \frac{7}{2}\kappa l_*\right), \tag{5.29}$$

$$I_* = 0.5 e^{ml_*}\left[-\frac{1}{2} - ml_* + \frac{1}{1 + \frac{7}{2}\kappa l_*}\right]^{-1},$$

where $I_* = E_i(ml_*) - E_i(ml^*)$. We also indicate the relation between W and V:

$$W = \frac{\gamma+1}{4} V + \sqrt{\frac{(\gamma+1)^2}{16} V^2 + 1}. \tag{5.30}$$

The parameters and shape of blast waves in an isothermic atmosphere were computed for the indicated approximation of $V(l, \theta)$ and hence also $W(l, \theta)$.

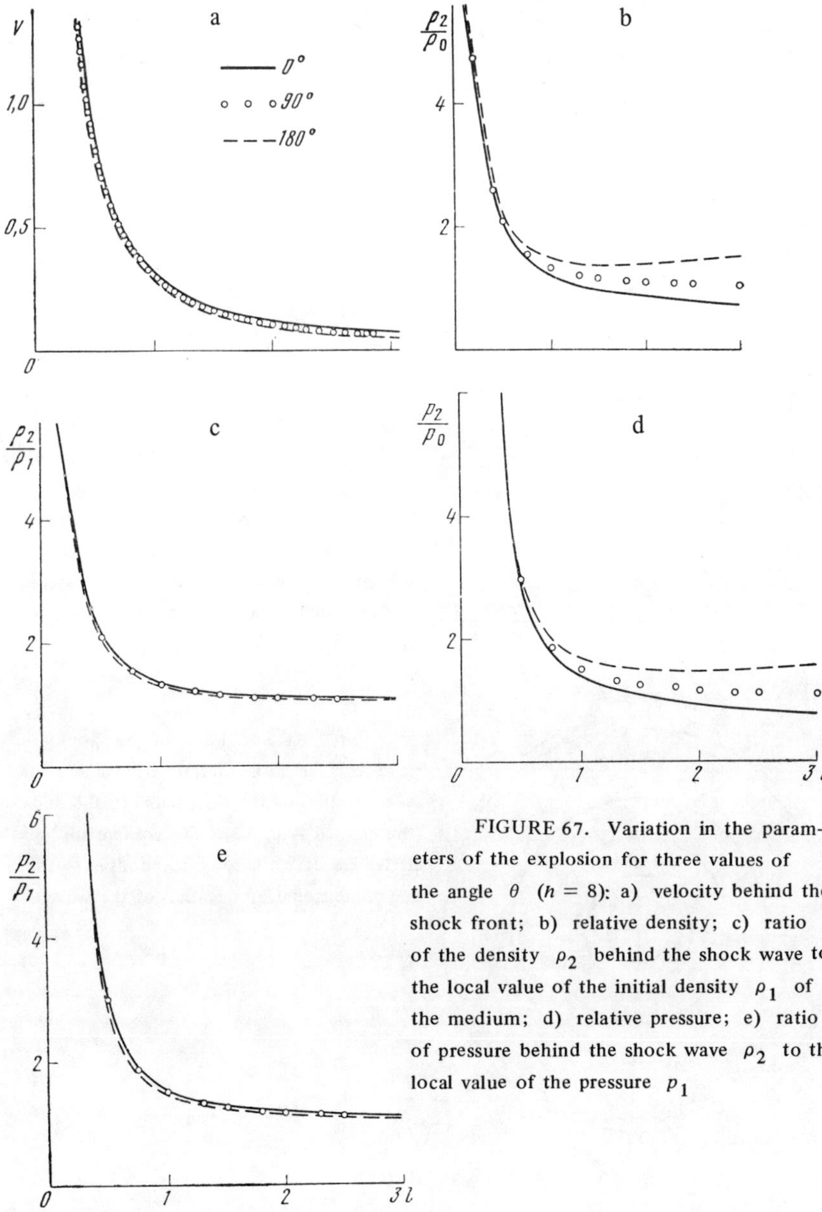

FIGURE 67. Variation in the parameters of the explosion for three values of the angle θ ($h = 8$): a) velocity behind the shock front; b) relative density; c) ratio of the density ρ_2 behind the shock wave to the local value of the initial density ρ_1 of the medium; d) relative pressure; e) ratio of pressure behind the shock wave p_2 to the local value of the pressure p_1

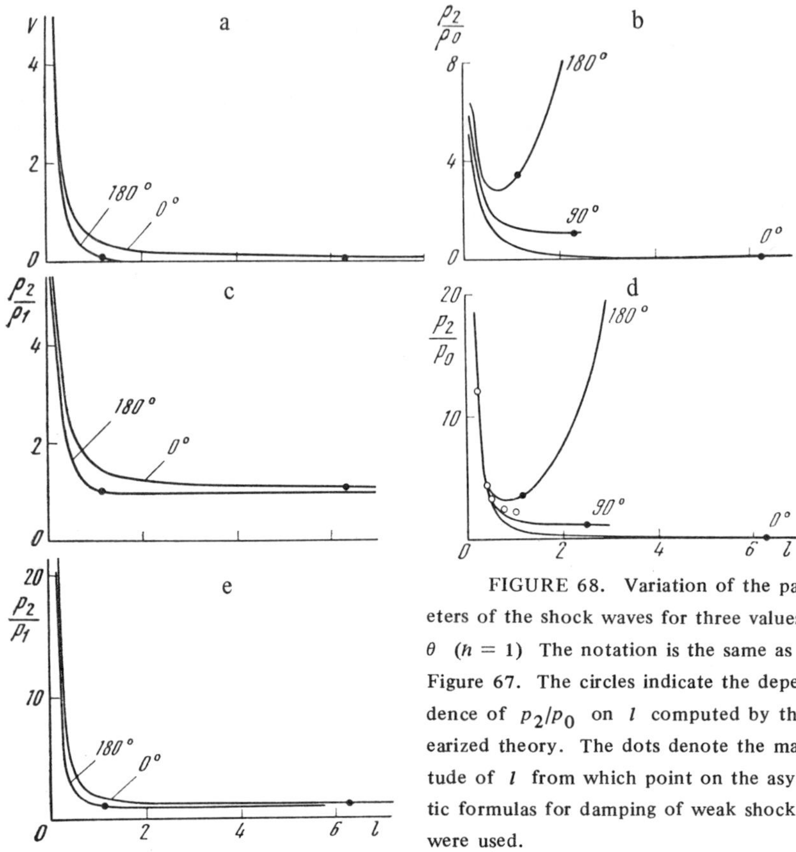

FIGURE 68. Variation of the parameters of the shock waves for three values of θ ($h = 1$) The notation is the same as in Figure 67. The circles indicate the dependence of p_2/p_0 on l computed by the linearized theory. The dots denote the magnitude of l from which point on the asymptotic formulas for damping of weak shock waves were used.

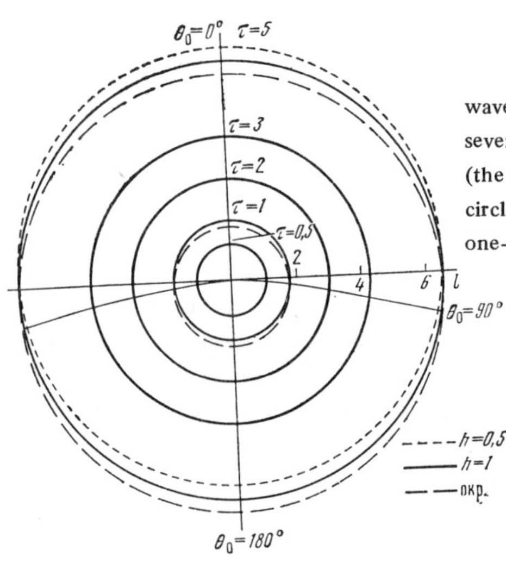

FIGURE 69. Form of the shock wave $l(\theta)$ in an isothermic atmosphere for several values of the time between 0.5 and 5 (the case h = 1; 0.5). For comparison, circles are drawn here corresponding to the one-dimensional propagation of the wave.

The pressure and density are found from the conditions on the shock wave in terms of the function $V(l, \theta)$ which is known from the approximation. The variation of the form of the shock wave is determined from (5.19) by numerically solving the Cauchy problem by the method of characteristics. The initial data for l are found from the relation (5.22) with $\tau_0 = 0.004$. Separate computations for τ not too large were also made from the approximation of V by (5.23). Here for $h > 1$ and $l < 2$ the computed forms of the shock waves differ from one another using the approximations (5.23) and (5.24). This indicates a certain correctness of these approximations.

Computations were carried out by the method described for various values of $h = H/r^0$ $(h = 0.5, 1, 5, 8)$. As an example we note that for the earth's atmosphere ($H = 8$ km) when $h = 1$ and $p = 0.1$ atm (the pressure at about 16 km) the energy of explosion is $E_0 \sim 5 \times 10^{22}$ erg. Computations indicated that the variability of the density has a substantial effect only for values of the parameter $h < 10$. Results of computations for $h = 8$ and $h = 1$ are shown in Figures 67–69 where graphs of $V(l, \theta), (\rho_2/\rho_0)(l, \theta), (\rho_2/\rho_1)(l, \theta), (p_2/p_0)(l, \theta), (p_2/p_1)(l, \theta)$ and the form of the shock wave are shown. Figure 69 shows the form of the shock wave for various τ with $h = 1$ (for $h = 8$ the form of the shock wave is close to a sphere). Circles with radius equal to the distance from the center to the point of intersection of the trajectories of the wave element (ray) with initial angle $\theta_0 = \pi/2$ with the shock wave are indicated by a dashed line. The shape of the wave for $h = 0.5$ is shown here for the value $\tau = 5$ (line with small dashes). From the graphs it is possible to see the extent to which the shape of the wave differs from a sphere.

In Figure 68e the circles denote the dependence of p_2/p_0 on l according to the linearized theory taking into account the variable density and constant counterpressure. In the method of linearization the formulas for p_2/p_0 and $l(\tau, \theta)$ can be obtained by using the principle of superposition of linear corrections (cf. Chapter 3). They have the form

$$\frac{p_2}{p_0} = \frac{0.157}{l^3}\left[1 - 0.36\frac{l}{h}\cos\theta\right] + 2.07,$$

$$l = \alpha^{-1/5}\tau^{2/5}\left[1 + \kappa\alpha^{-1/5}\tau^{2/5} + \frac{0.7 \cdot 25}{6} A_1 \alpha^{2/5}\tau^{6/5}\right], \tag{5.31}$$

$$A_1 = \text{const} = 1.92 \qquad (\gamma = 1.4).$$

The investigation of the problem and the results presented lead to the following conclusions.

1. For $h < 1$ the form of the shock wave differs from a sphere and has an "egg-shaped" appearance, being somewhat elongated at the top $(\theta = 0)$ and flattened at the bottom $(\theta = \pi)$. This is explained by the effects of decrease in the density with increasing z. For $h > 5$ the variation in the shape of the shock wave is

inappreciable. This conclusion agrees with data from the complete computation of the explosion for initial parameters corresponding to the standard atmosphere [23], [24].

2. The relative pressure p/p_1 varies rather weakly with θ, while the dimensionless pressure p/p_0 varies strongly as the polar angle θ changes. This implies a rule for approximately determining the pressure p_2 which can be used in practice: in order to determine the pressure in a stratified atmosphere it is necessary to use the curve $\widetilde{P}(l) = (p_2/p_1)(l)$ for the one-dimensional case and then, taking it as the graph of p_2/p_1 in the case of variable density and pressure, to find the desired pressure by the formula $p_2 = \widetilde{P}(l)p_1(r, \theta)$. This rule does not contradict conclusions following from an analysis of the system of equations of gas dynamics (1.17).

The cases of cylindrical and plane explosions can be studied in a similar mannar.(1) If the medium is weakly inhomogeneous (for example, the distance H over which the density of the medium varies appreciably is considerably less than the characteristic dynamic length r^0), then for approximate computation of the parameters of the shock waves one can use the hypothesis of local one-dimensional flows (the analogue of the well-known hypothesis of "plane sections"), neglecting local variation of the form of the wave along the axis or plane of the explosion.

§4. Nonsymmetric release of energy

It was noted in Chapter 2 that in studying the action of a laser beam on a medium we encounter questions of explosion when the energy is released along a line, while the specific energy E_0 is variable: $E_0 = E_0(s)$ if s is the coordinate along the line of the explosion. If the distribution of $E_0(s)$ is such that dE_0/ds is small, then by making use of the hypothesis of plane sections it is possible to compute the gas flow on the basis of the locally homogeneous theory. It would be possible to find explicit corrections to the solution by the method of linearization if $E_0 = E_0^0 + \mu f(s)$, where μ is a small parameter. An analogous remark holds also for the plane explosion if the energy E_0 varies along the plane $E_0 = E_0(x^2, x^3)$ (the plane of the explosion is prependicular to the x^1 axis). Explosion of a semi-infinite line or semi-infinite plane is a classical example of nonsymmetric energy release. For a perfect gas with $p_1 = 0$ the corresponding problems are self-similar. The problem of the semi-infinite line was formulated by L. I. Sedov [40]; one of the possible methods of solving it was considered in [41]. Another case of nonsymmetric evolution of energy is that of an explosion along some curve or along a smooth surface. We shall consider the case of explosion along a certain curve. Suppose that the curvature of the curve is small, and the specific energy E_0 is constant; then for distances $r \ll R$, where R is the radius of curvature of the line of the explosion, it is possible to apply

(1) *Added in proof.* Approximate methods of determining the parameters of spherical and cylindrical shock waves have been developed in the recent publications [42] and [43].

(approximately) the theory of explosion along a line. Thus the semi-infinite line may have a variable specific energy $E_0(s)$. These cases are realized in practice (the laser beam; cosmic bodies in the earth's atmosphere, for example, the explosion of the Tungus meteorite).(2)

For linear equations such as the heat equation or wave equation, problems of this type can be solved by using the principle of superposition of elementary solutions considered in Chapter 1. For the nonlinear case we do not know the exact solution of these problems.

BIBLIOGRAPHY

1. V. P. Korobeĭnikov, *The exact solution of a nonlinear problem involving an explosion in a gas of variable initial density*, Dokl. Akad. Nauk SSSR 117(1957), 947–948. (Russian) MR 20 #5001.

2. L. I. Sedov, *Integration of the equations of one-dimensional motion of a gas*, Dokl. Akad. Nauk SSSR 90 (1953), 735. (Russian) MR 15, 175.

3. V. P. Korobeĭnikov and E. V. Rjazanov, *Construction of exact discontinuous solutions of the equations of one-dimensional gas dynamics, and their applications*, Prikl. Mat. Meh. 22 (1958), 265–268 = J. Appl. Math. Mech. 22 (1958), 362–367. MR 20 #6896.

4. Ja. B. Zel'dovič, *Motion of a gas under the action of a momentary pressure (shock)*, Akust. Ž. 2 (1956), 28–38 = Soviet Phys. Acoustics 2 (1956/57), 25–35. MR 18, 620.

5. R. I. Nigmatullin, *A strong plane explosion on the boundary of two ideal, calorically perfect gases*, Vestnik Moskov. Univ. Ser. I Mat. Meh. 1965, no. 1, 83–87. (Russian)

6. V. P. Korobeĭnikov and G. A. Ostroumov, *More on cavitation destruction*, Akust. Ž. 11 (1965), 458–462 = Soviet Phys. Acoustics 11 (1965/1966), 394–397.

7. G. A. Ostroumov, *On the mechanism of cavitation destruction*, Akust. Ž. 9 (1963), 198–204 = Soviet Phys. Acoustics 9 (1963/64), 160–164.

8. Ju. P. Raĭzer, *The notion of a gas under the action of a concentrated shock on its surface (an explosion on the surface)*, PMTF Ž. Prikl. Meh. i Tehn. Fiz. 1963, no.1, 57–66. (Russian)

9. B. N. Rumjancev, *On a limiting case of the propagation of strong blast waves in an inhomogeneous medium*, PMFT Ž. Prikl. Meh. i Tehn. Fiz. 1963, no. 1, 127–129. (Russian)

10. ———, *On the theory of explosions in inhomogeneous media with adiabatic index close to* 1, PMTF Ž Prikl. Meh. i Tehn. Fiz. 1963, no. 4, 118–120. (Russian)

11. L. V. Šuršalov, *On the problem of a strong explosion at the boundary of a half-space containing a perfect gas*, Prikl. Mat. Meh. 33 (1969), 358–363 = J. Appl. Math. Mech. 33 (1969), 346–351.

12. L. V. Sedov, *Two-dimensional problems in hydrodynamics and aerodynamics*, GITTL, Moscow, 1950; English transl., Interscience, New York, 1965.

13. L. A. Galin, *Impact of a body on the surface of a compressible fluid*, First All-Union Congr. Theoret. Appl. Mech; Abstracts of Reports, Izdat. Akad. Nauk SSSR, Moscow, 1960, p. 57. (Russian)

14. È. P.Borisova, P. P. Korjakov and N. N. Moiseev, *Plane and axially symmetric automodel (similarity) problems of penetration and of stream impact*, Prikl. Mat. Meh. 23 (1959), 347–360 = J. Appl. Math. Mech. 23 (1959), 490–507. MR 22 #5257.

15. A. Ja. Sagomonjan, *Spatial problems of the transient motion of a compressible fluid*, Izdat. Moskov. Gos. Univ., Moscow, 1962. (Russian)

16. A. A. Deribas and S. I. Pohožaev, *The problem of an explosion on the surface of a liquid*, Dokl. Akad. Nauk SSSR 144 (1962), 524–526 = Soviet Phys. Dokl. 7 (1962/63), 383–384.

---

(2) *Added in proof.* Numerical solution of the problem of explosion of a semi-infinite charge is given in [44].

17. V. F. Minin, *Explosion on the surface of a fluid*, PMTF Ž. Prikl. Meh. i Tehn. Fiz. **1964**, no. 3, 159–161. (Russian)

18. S. S. Grigorjan and M. M. Martirosjan, *Ground waves generated by surface explosives*, Astronautica Acta **15** (1969/70), 283–288.

19. V. P. Korobeĭnikov, P. I. Čuškin and K. V. Šarovatova, *Tables of gas-dynamical functions for the initial stages of a point explosion*, Soobšč. po Vycisl. Mat., vyp. 2, Izdat. Vycisl. Centr Akad. Nauk SSSR, Moscow, 1963. (Russian) MR **32** #3408.

20. V. P. Korobeĭnikov and P. I. Čuškin, *Gas-dynamical functions of a point explosion*, Akad. Nauk SSSR Trudy Vyčisl. Centr 1969. (Russian)

21. V. D. Alekseenko, *Experimental investigation of the energy distribution during contact explosion,* Fiz. Gorenija i Vzryva 3 (1967), no. 1, 152–155 = Combustion, Explosion and Shock Waves **3** (1967), no. 1, 97–98.

22. K. I. Babenko et al., *Methods of solving some two-dimensional problems*, Questions of Comput. Math. and Comp. Techniques, Mašgiz, Moscow, 1963, pp. 99–103. (Russian) RŽ Mat. 1963 #11B37.

23. V. V. Rusanov and È. È. Šnol′, *Difference methods in spatial problems of gas dynamics*, Report presented at the Fourth All-Union Math. Congr., Leningrad, 1961. (Russian)

24. V. V. Rusanov, *Computation and investigation of multidimensional gas flows by the method of finite differences*, Doctoral Dissertation, Institute of Problems of Mechanics, Moscow, 1968. (Russian)

25. H. S. Kestenboĭm, F. D. Tureckaja and L. A. Čudov, *Point explosion in an inhomogeneous atmosphere,* PMTF Ž. Prikl. Meh. i Tehn. Fiz. **1969**, no. 5, 25–28 = J. Appl. Mech. Tech. Phys. **10** (1969), 709–712.

26. V. P. Korobeĭnikov and V. P. Karlikov, *Determination of the shape and parameters of the shock front caused by an explosion in an inhomogeneous medium*, Dokl. Akad. Nauk SSSR **148** (1963), 1271–1274 = Soviet Phys. Dokl. **8** (1963/64), 137–140.

27. V. P. Korobeĭnikov, *Gas dynamics of explosions*, Annual Rev. Fluid Mech. **3** (1971), 317–346.

28. L. V. Ovsjannikov, *A new solution of the equations of hydrodynamics*, Dokl. Akad. Nauk SSSR **111** (1956), 47–49. (Russian) MR **19**, 89.

29. V. P. Karlikov, *Solution of the linearized, axially symmetric problem of a point explosion in a medium with variable density*, Dokl. Akad. Nauk SSSR **101** (1955), 1009–1012. (Russian) MR **16**, 1173.

30. ———, *The linearized problem of the propagation of a strong explosion in an inhomogeneous atmosphere*, Vestnik Moskov. Univ. Ser. Mat. Meh. Astr. Fiz. Him. **1959**, no. 4, 27–39. (Russian) MR **22** #10469.

31. ———, *A linearized solution of the strong explosion problem in a medium with linear density distribution*, Vestnik Moskov. Univ. Ser. I Mat. Meh. **1960**, no. 1, 60–65. (Russian) MR **27** #6469.

32. È. I. Andriankin, *The perturbation method for the strong explosion problem*, Izv. Akad. Nauk SSSR Otdel. Tehn. Nauk **1958**, no. 12, 6–14. (Russian)

33. A. S. Kompaneec, *A point explosion in an inhomogeneous atmosphere*, Dokl. Akad. Nauk SSSR **130** (1960), 1001–1003 = Soviet Phys. Dokl. **5** (1960/61), 46–48. MR **22** #10465.

34. È. I. Andriankin et al., *Propagation of a strong explosion in an inhomogeneous atmosphere*, PMTF Ž. Prikl. Meh. i Tehn. Fiz. **1962**, no. 6, 3–7. (Russian)

35. Dallas D. Laumbach and Ronald F. Probstein, *A point explosion in a cold exponential atmosphere*, J. Fluid Mech. **35** (1969), 53–75.

36. K. E. Gubkin, *Propagation of discontinuities in sound waves*, Prikl. Mat. Meh. **22** (1958), 561–564 = J. Appl. Math. Mech. **22** (1968), 787–793.

37. ———, *Nonlinear geometric acoustics and its applications*, Some Problems of Math. and Mech. (Dedicated to M. A. Lavrent′ev on his Sixtieth Birthday), Izdat. Sibirsk. Otdel. Akad. Nauk SSSR, Novosibirsk, 1961, pp. 69–76. (Russian)

38. O. S. Ryžov, *The damping of shock waves in inhomogeneous media*, PMTF Ž. Prikl. Meh. i Tehn. Fiz. **1961**, no. 2, 15–25. (Russian)

39. O. S. Ryžov and G. M. Šefter, *On the energy of acoustic waves propagating in moving media*, Prikl. Mat. Meh. **26** (1962), 854–866 = J. Appl. Math. Mech. **26** (1962), 1293–1309. MR **27** #4471.

40. L. I. Sedov, *Similarity methods in nonlinear continuum mechanics*, Proc. Third All-Union Math. Congr. (Moscow, 1956), vol. III, Izdat. Akad. Nauk SSSR, Moscow, 1958, pp. 480–490. (Russian)

41. V. P. Korobeinĭkov and P. I. Čuškin, *Plane, cylindrical, and spherical explosions in a gas with back pressure*, Trudy Mat. Inst. Steklov. 87 (1966), 3–34 = Proc. Steklov Inst. Math. 87 (1966), 1–40.

42. *H. S. Kestenboĭm and G. G. Rosljakov, *Numerical solution of one-dimensional problems of explosions*, Izdat. Moskov. Gos. Univ., Moscow, 1971. (Russian)

43. *V. A. Bronšten, *Propagations of spherical and cylindrical blast waves in an inhomogeneous atmosphere with regard for back pressure*, PMTF Ž. Prikl. Meh. i Tehn. Fiz. 1972, no. 3, 84–90 = J. Appl. Mech. Tech. Phys. 13 (1972), 334–339.

44. *V. P. Korobeĭnikov, P. I. Chushkin [Čuškin] and L. V. Shurshalov [Šuršalov], *Gas dynamics of the flight and explosion of meteorites*, Astronautica Acta 17 (1972), 339–348.

*Added in proof

CHAPTER 6

EXPLOSION IN A COMBUSTIBLE MIXTURE OF GASES

§1. Formulation of the problem

The present chapter is devoted to the fundamental problem of a quiescent gas in which exothermic chemical reactions are possible. Because of explosion a strong shock wave propagates through the gas which heats the gas to the point where combustion reactions are possible. Such problems arise in the excitation of detonation in combustible mixtures of gases and in thermonuclear detonation, and their study is of both theoretical and practical interest. Extensive investigation of these new problems has begun in recent years in the work of the author, R. I. Solouhin, G. G. Černyĭ, V. A. Levin. J. Lee, A. Oppenheim, and E. Bišimov.

The motion of the gas will be assumed to be one-dimensional. We shall consider the general picture of flow of the medium. It follows from physical considerations that during the first instants after explosion the gas will move according to the laws of an ordinary point explosion [1], [2], since the contribution of the energy of combustion to the total energy balance is still not great. It is possible to estimate those distances up to which it is expedient not to consider the effect of the energy of detonation on the motion of medium. Let Q be the calorific power per unit mass of the combustible mixture. As an example we consider the case of constant initial density ρ_1 and constant energy Q. If the width of the combustion zone is considered negligibly small as compared with the radius of the wave r_2, then the energy evolved during the burning of the gas will have the magnitude

$$U = \frac{1}{\nu}\sigma_\nu \rho_1 Q r_2^\nu, \qquad \sigma_\nu = 2(\nu - 1)\pi + (\nu - 2)(\nu - 3).$$

Assuming that $E_0 > U$, we arrive at a condition for weak effect of the energy of detonation on the flow:

$$r_2 < r_*, \qquad r_*^\nu = \frac{\nu E_0}{\sigma_\nu Q \rho_1}. \tag{6.1}$$

Condition (6.1) will be satisfied also in the case where the width of the reaction zone is considered or the combustible mixture is incompletely burned, since in these cases the combustion energy will be less than U.

If the charge has finite radius r_0, then in applying the ordinary theory of point explosions to describe the motion we must use the estimate

$$r_0 < r < r_*. \tag{6.2}$$

Conditions (6.1) and (6.2) impose a strong upper bound on the applicability of the laws of point explosion in an inert gas to flows of an exploding medium. However, if the energy E_0 is not great and is released in a small volume, then for $r_2 < r_*$ the flow will proceed basically as during an ordinary point explosion. On the other hand, for instants of time where $E_0 < U$ combustion processes play a major role, and the gas flow has the basic features of explosive combustion [1], [3].

For values of r_2 close to r_* the gas flow assumes a more complex, intermediate character.

The distinguishing features of the flow in question are that next to the front of the shock wave there is a zone where chemical reactions occur, and in a neighborhood of the center of the explosion there is a certain region in which (as also in ordinary explosions) the particles of the gas have a high entropy.

From a mathematical point of view the solution of this problem reduces to integrating the system of equations of gas dynamics and chemical kinetics with boundary conditions on surfaces of strong discontinuity and in a neighborhood of the center of the explosion. It must moreover be considered that at the initial moment $t = 0$ a finite energy E_0 is released at the center of symmetry and initial values of the velocity, density, pressure, and other parameters of the medium are prescribed. For a complete description of the motion of the gas it is necessary to consider the kinetics of all the basic chemical reactions of combustion and also processes of ionization and dissociation in zones of high temperatures.

In considering the kinetics of chemical reactions the solution of the problem formulated above involves considerable mathematical difficulties even in the case where viscosity and heat conduction are neglected. In addition to the quantity Q, the equations of gas dynamics will contain a number of other parameters (the energy of activation, coefficients in the expressions for the rates of reactions, etc.). The presence of a considerable number of constant parameters and large gradients of the functions in the zone of chemical reactions makes analytic and numerical solution of the problems very difficult. In a number of cases there are rather simple approaches to investigation of the problem in question. The following models of the gas flow can be introduced.

1. The model of a detonation wave where it is assumed that the substance burns in the immediate neighborhood of the front of the shock wave and the shock wave together with the zone of chemical reactions behind it is considered to be a single surface of strong discontinuity.

2. A simple model with kinetics, i.e. a model of flow of the gas with chemical reactions where the reaction process is described by a single variable quantity—the mass concentration of the unburned substance [3], [8].

3. A flow model with two fronts in which a shock wave propagates through the unperturbed gas behind which at some distance there is a combustion wave where

total evolution of heat occurs, i.e. an induction zone or zone of ignition delay is introduced (cf., for example [4]–[7], [9]).

4. A model for including chemical reaction with heat evolution in the flow behind the front of the shock wave after passage of an induction period. The reaction process is also described by a single variable quantity—the mass concentration of unburned matter; but here the reverse reaction of "recombination" can also be considered.

5. Complete account of the basic chemical reactions behind the front of the shock wave.

Our investigations have dealt mainly with only models 1–4.

In the case of the model of a detonation wave the problem in question has been studied in detail for a perfect gas with constant heat capacities. Here it should be mentioned that the solution of the problem for this model in certain special cases was first given by I. S. Šikin [10]. The parameters of the front of a strong detonation wave were determined approximately in the work of V. A. Levin [11]. In this work a condition analogous to (6.1) was also formulated.

§2. Solution of the problem for the model of a detonation wave

Because of the above-noted special features of the flow in the initial stage of the explosion, the denotation will not follow the Chapman-Jouguet rule (C-J). We shall consider this question in somewhat more detail for the case of a perfect gas with a constant ratio of specific heat capacities. We denote by D the speed of the shock wave, by v_2 the velocity of the gas behind the front of the shock wave, and by a_2 the corresponding speed of sound. Then from the conditions on the shock wave for a strong explosion we have

$$v_2 + a_2 = \frac{D}{\gamma+1}\left(2 + \sqrt{2\gamma(\gamma-1)}\right).$$

It follows from this expression that for $\gamma > 1$ we have $v_2 + a_2 > D$, i.e. in the initial stage of the process a case of overcompressed detonation occurs. For a C-J wave, as is known, $v_2 + a_2 = D$. During the propagation of the shock wave (in an exploding gas) the quantity $v_2 + a_2$ decreases, and here passage to the regime of C-J detonation is possible. During an explosion in a chemically inert gas a condition of C-J type is actually satisfied only asymptotically as $t \longrightarrow \infty$, $v_2 \longrightarrow 0$, $a_2 \longrightarrow a_1$ and $D \longrightarrow a_1$.

We shall consider the mathematical formulation of the problem for the model of a detonation wave. The original system of equations of gas dynamics for one-dimensional adiabatic motions of the gas we take in the form (1.18)–(1.20). For the solution of the PEP in a quiescent exploding medium it is required to find a solution of the system (1.18)–(1.20) with boundary conditions on the front of the shock wave, i.e. for $r = r_2$ (cf. §2, Chapter 1):

$$\rho_2 (D - v_2) = \rho_1 D, \qquad p_2 = p_1 + \rho_1 D v_2, \tag{6.3}$$
$$\varepsilon_1 + \frac{p_1}{\rho_1} + \frac{D^2}{2} + Q = \varepsilon_2 + \frac{p_2}{\rho_2} + \frac{1}{2}(D - v_2)^2.$$

In addition, we have a boundary condition at the center of symmetry:

$$v(0, t) = 0. \tag{6.4}$$

At the instant of time $t = 0$ at the center of symmetry a finite energy E_0 is released, and the following conditions are prescribed:

$$v(r, 0) = 0, \qquad p(r, 0) = p_1 = \text{const}, \qquad \rho(r, 0) = \rho_1(r)$$
$$(r > 0), \qquad r_2(0) = 0. \tag{6.5}$$

We shall consider the solution of this problem for the case of a perfect gas and assume that the initial density is variable: $\rho_1 = Ar^{-\omega}$. The system of defining parameters of the problem contains the constants $Q, A, p_1, E_0, \omega, \gamma$, and ν, and, in addition, the solution will depend on the coordinate r and time t. For simplicity we shall assume the detonation wave to be strong and neglect the initial pressure of the gas p_1. We introduce the effective pressure of the explosion by the formula

$$p_* = \frac{1}{\gamma} Q \rho_{10},$$

where ρ_{10} is a constant with the dimension of density which may be written as follows: $\rho_{10} = A(r^0)^{-\omega}$, where $r^0 = (E_0/p_*)^{1/\nu}$. The quantity p_* can be written in the more explicit form

$$p_* = \left(\frac{QA}{\gamma}\right)^{\nu/(\nu-\omega)} E_0^{-\omega/(\nu-\omega)}.$$

Once can include the parameter p_* in the system of defining parameters in place of Q.

The simplest considerations of dimension theory imply that for the problem in question in the case of a strong detonation wave there is a similarity law analogous to the similarity law for an ordinary explosion (cf. §6, Chapter 4). The quantity p_* here plays the role of the initial pressure of the gas.

From the system of defining parameters of the problem it follows that if we introduce the dimensionless functions

$$f = \frac{v}{D}, \qquad g = \frac{\rho}{\rho_1}, \qquad h = \frac{q}{p_*}\left(\frac{r_2}{r^0}\right)^{\omega} p, \tag{6.6}$$

then they will depend on certain constant parameters and two dimensionless variables for which we take

$$\lambda = \frac{r}{r_2}, \qquad q = \frac{Q}{D^2}. \tag{6.7}$$

We also introduce the dimensionless radius of the shock wave $R_2(q) = r_2/r_0$. After having transformed equation (1.20) of the original system taking into account the condition of adiabatic flow and having gone over to the dimensionless variables (6.6) and (6.7), we obtain the system

$$\begin{aligned}
&D_{\lambda,q}f + \frac{1}{\gamma g}\frac{\partial h}{\partial \lambda} - \eta\frac{f}{2q} = 0, \qquad \eta = R_2\frac{dq}{dR_2},\\
&D_{\lambda,q}g + g\frac{\partial f}{\partial \lambda} - \omega g + \frac{(\nu-1)fg}{\lambda} = 0,\\
&D_{\lambda,q}h + \gamma h\frac{\partial f}{\partial \lambda} - \omega h - h\left(\frac{h}{q} - \gamma(\nu-1)\frac{f}{\lambda}\right) = 0,\\
&D_{\lambda,q} = (f-\lambda)\frac{\partial}{\partial \lambda} + \eta\frac{\partial}{\partial q}.
\end{aligned} \tag{6.8}$$

From (6.3) on setting $p_1 = 0$ we find

$$(f_2 - 1)g_2 = -1, \qquad h_2 = \gamma f_2, \qquad \frac{1}{2} + q = \frac{1}{\gamma - 1}\frac{h_2}{g_2} + \frac{1}{2}(f_2 - 1)^2. \tag{6.9}$$

We note that for $q = 0$ this problem goes over into the self-similar problem on strong explosion (cf. Chapter 2). For $E_0 = 0$ we obtain the self-similar problem on the propagation of a detonation wave [1], [3] which is also appropriately included in the class of PEP. For a complete solution of the problem we must follow the transition from one self-similar regime of flow (strong point explosion in an inert gas) to the other self-similar regime (detonation wave or point explosion in a detonating gas with negligibly small energy of explosion). We note also that if in the equation of energy conditions (6.3) and (6.9) we place a "minus" sign in front of Q (assuming that $Q > 0$), we have the PEP with absorption of energy behind the wave front (as during vaporization, for example). Here there will be additional damping of the wave. The methods of solution developed below are also applicable in this case. The problem of explosion including the effect of the energy of detonation can be solved by applying numerical methods.

For the initial stage of the explosion when

$$\frac{(p_2 - p_*)}{p_*} > 1, \qquad q < q_j, \qquad r_2 < r_*,$$

a simpler path can be proposed based on linearization of the initial equations about the self-similar solution for strong explosion. Here q_j denotes the quantity q for the C-J wave. A solution of the linearized problem in the variables chosen can be constructed in analogy with the case of considering counterpressure treated in Chapter 3. We represent the unknown functions in the form

$$\begin{aligned} f(\lambda, q) &= f_0(\lambda) + qf_1(\lambda) + O(q^2), \\ g(\lambda, q) &= g_0(\lambda) + qg_1(\lambda) + O(q^2), \\ h(\lambda, q) &= h_0(\lambda) + qh_1(\lambda) + O(q^2), \\ \frac{R_3}{q}\frac{dq}{dR_2} &= \frac{i\nu - \omega}{1 + A_1 q + O(q^2)}, \end{aligned} \tag{6.10}$$

where $O(q^2)$ denotes terms of order q^2 for small q. The solution (6.10) is substituted into the system (6.8). Then, neglecting terms of order q^2 and higher, we obtain a system of linear ordinary differential equations for the linear corrections f_1, g_1 and h_1; the coefficients in this system depend on the self-similar solution f_0, g_0, h_0. In addition, it is necessary to find the constant A_1 determining the non-self-similar correction to the law of motion of the shock wave.

From (6.4) and (6.9) we find the boundary values of the functions f_1, g_1 and h_1:

$$\begin{aligned} f_1(1) &= -(\gamma - 1), & h_1(1) &= -\gamma(\gamma - 1), \\ g_1(1) &= -\frac{(\gamma + 1)^2}{\gamma - 1}, & f_1(0) &= 0. \end{aligned} \tag{6.11}$$

Suppose now that $\omega = 0$. We shall consider this case in more detail, following §3 of Chapter 3 where we considered the linearized problem taking account of counter-pressure. It is easily seen that because of the dimensionless variables chosen the system of linear equations for f_1, g_1 and h_1 coincides with the corresponding system of §5 of Chapter 3. This system is to be solved with the boundary conditions (6.11). In the case of arbitary ν and γ the problem can be solved by numerical integration. The computation has been carried out for a number of values γ for $\nu = 1, 2, 3$ according to the technique worked out in Chapter 3.

The values of the constant A_1 obtained are shown in Table 6.

TABLE 6

Values of the constant A_1

ν	γ						
	8/7	1.2	1.3	1.4	5/3	2	3
1	0.5091	0.7246	1.121	1.545	2.811	4.685	12.28
2	0.4695	0.6694	1.037	1.428	2.592	4.296	11.10
3	0,4551	0.6489	1.005	1.383	2.502	4.132	10.59

Curves of p/Qp_* (for $\nu = 3$) and the case $\gamma = 1.2$, $q = 0.15$, $\theta = 1$, and curves of $p/\theta p_1$ for the case $\gamma = 3$, $q = 0.04$, $\theta = 5$ are constructed in Figure 70; here the dashed lines correspond to explosion without considering detonation $(h = h_0)$.

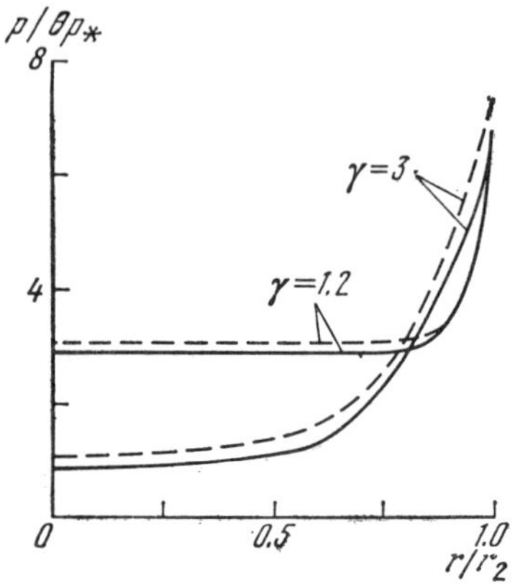

FIGURE 70. Spatial distribution of relative pressures for two values of γ. The dashed lines correspond to the case $q = 0.1$

We also present the formulas giving the law of motion of the shock wave $R_2(\tau)$ in parametric form (cf. (3.77)):

$$R_2^\nu = \frac{\delta^2}{\alpha\gamma} q \exp(A, q), \qquad \delta = \frac{2}{\nu + 2}, \qquad R_{20}^\nu = \frac{\delta^2}{\alpha\gamma} q,$$
$$r = \delta\left(\frac{q}{\gamma}\right)^{1/2} R_{20}\left[1 + \frac{\nu\delta + 2}{2\nu\delta(\nu\delta + 1)} A_1 q\right], \tag{6.12}$$
$$\tau = \frac{t}{t^0}, \qquad t^0 = r^0\left(\frac{\rho_1}{p_*}\right)^{1/2}, \qquad \alpha = \alpha(\gamma, \nu).$$

The functions $R_2(\tau)$ for the various ν and $\gamma = 1.4$ are shown in Figure 71a. We remark also that in the case $\nu = 3, \gamma = 7$ the linearized problem has an exact analytic solution in analogy with the situation of considering counterpressure (here $A_1 = 60$). In the case of variable initial density $(\omega \neq 0)$ the linearized problem can be studied analogously.

Since the applicability of the linear solution is limited to small values of q, numerical methods must be used to compute the problem for large q (and correspondingly for large times t) when the transition of the wave to the C-J regime occurs. To this end it is possible to use the method of computing problems on explosion developed in §3 of Chapter 4 [12]. In using this method the computing scheme is changed only at places corresponding to the computation of the parameters of the front of the shock wave. The initial data can be prescribed for some value of τ or $q < q_j$ using the results of solving the linearized problem considered above. Numerical solution for the case of constant initial density was carried out in this fashion. The computations were mainly carried out up to times when the parameters of the shock wave become close to the parameters of the C-J wave $(q_j/q - 1 < 0.01)$. In using the method of computation developed in Chapter 4 and based on the scheme of the numerical method of integral relations, the region of integration between the detonation wave and the center of the explosion was decomposed in space into an even number of strips n. Computations were made for the cases $n = 2$ and $n = 4$. The results presented below refer to the case $n = 4$, i.e. the unknown functions were approximated by polynomials of fourth degree (the problem was programmed by E. Bišimov).

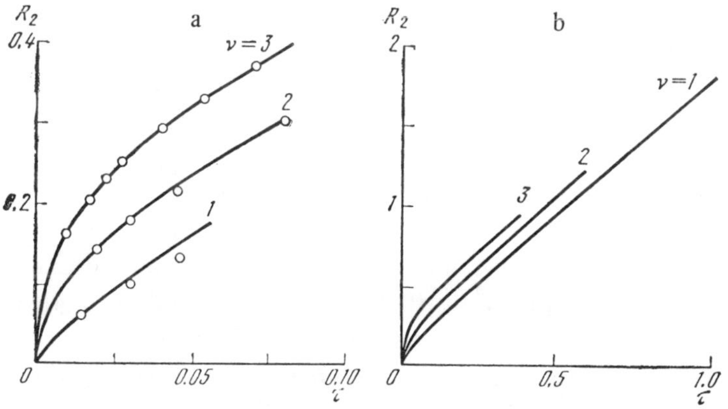

FIGURE 71. Dependence of the law of motion of an overcompressed detonation wave on time $R_2(\tau)$ a) according to the linearized solution (the circles denote results of the numerical solution of the nonlinear problem); b) by numerical solution (in the cases $\nu = 2, 3$ the end points of the curves correspond to transition to the Chapman-Jouguet regime)

As a result of the computations it was found that in the case of explosion in a perfect gas with constant γ spherical and cylindrical shock waves approach the parameters of the C-J wave rather quickly. In the planar case the approach to the C-J regime is realized after much greater time intervals (theoretically the transition to the C-J regime occurs after an infinite amount of time [13]). Using the results of computing a plane explosion, it is possible to determine the constants in the asymptotic formulas for the damping of overcompressed detonation waves obtained by G. G. Černyĭ [13] and find the analytic relations for the parameters of the front of the detonation wave as it goes over to the C-J regime. Thus for $\gamma = 1.4$ for the dependence of the coordinate of a strong plane detonation wave on time we have

$$\tau = \tau_0 + \frac{R_2}{k}\left(1 + \frac{m^2}{R_2^2}\right), \quad \tau_0 = -0.0835, \quad k = 1.64,$$
$$m = 0.0741, \quad t = E_0 p_*^{-3/2} \rho_1^{1/2} \tau, \quad R_2 = \frac{p_*}{E_0} r_2. \tag{6.13}$$

Comparison of the values of τ found by this formula with the numerical solution of the problem indicates that it possesses good accuracy even for rather small values of R_2 $(R_2 \sim 0.5)$.

Figure 71b shows the functions $R_2(\tau)$ for $\nu = 1, 2, 3$ on the basis of the numerical solution.

In the cases $\nu = 2, 3$ the curves are terminated at values close to the moments of transition to the C-J regime. The curve of $R_2(\tau)$ for the planar case can be continued for large times by (6.13).

If R_j denotes the distance to the point where the strong wave goes over to the C-J regime, in the spherical and cylindrical cases we have

$$R_j = \kappa \left(\frac{E_0}{p_*}\right)^{1/\nu}, \qquad p_* = \frac{\rho_1 Q}{\gamma}, \tag{6.14}$$

where the coefficient κ depends on γ and ν (for $\gamma = 1.4$ the coefficient κ is close to one).

It follows from (6.13) that during explosions in a combustible mixture with the same low initial temperature and equal E_0 but different pressures the quantity R_j will increase with a decrease of pressure and hence also of the initial density ρ_1. This conclusion is in qualitative agreement with the experimental results of [14], where the question of propagation of blast waves in mixtures of hydrogen and oxygen was considered. We remark that (6.14) reflects the similarity law previously mentioned.

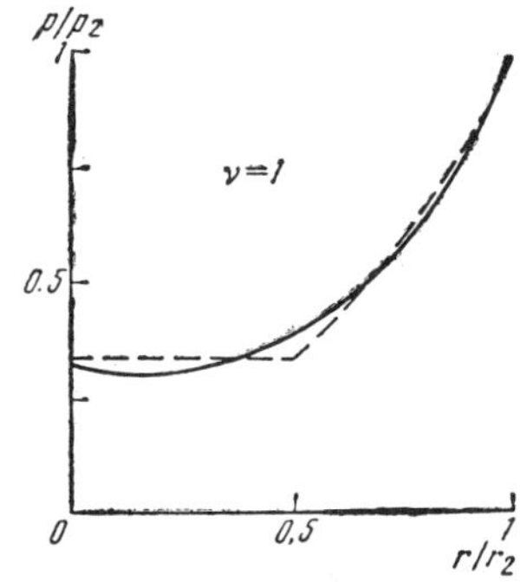

FIGURE 72. Comparison of the pressure profiles for the C-J wave and a slightly overcompressed shock wave ($\tau \sim 1$)

Figure 72 shows the spatial distribution of relative pressure p/p_2 for the case of a plane explosion in a gas ($\gamma = 1.4$) for large values of τ ($\tau = 0.65$). Here we also show a curve (the dashed line) constructed on the basis of the known exact solution of the problem on propagation of a strong self-similar plane C-J detonation wave (under the assumption that the initial energy of the explosion E_0 is negligible). Here it is evident that in a neighborhood of the wave front the gas flow is quite close to the regime of flow for a C-J explosion.

We mention that computation of the problem for large times in the case of spherical and cylindrical waves becomes very complicated in view of the presence of singularities in the conditions on the front of the shock wave near the C-J point.

Solution of the problem in question with consideration of counterpressure has been carried through by E. Bišimov. He also studied the case of variable evolution of heat $Q = Q(\xi_0)$, where ξ_0 is the initial coordinate of a particle of the gas [15], [16]. The results presented there and also data from investigations of the asymptotic behavior of shock waves [13], [18] afford the conclusion that in the model of a detonation wave the overcompressed wave in the cases of spherical and cylindrical explosions passes to the C-J regime at a finite distance from the center of the explosion, while for a plane explosion the passage to the C-J regime is realized only at considerable distances.

Thus, for explosive mixtures of gases (or of liquid and solid explosives in the perfect-gas approximation) to which model 1 is applicable it is possible to neglect at certain distances the energy of initiation of the cylindrical and spherical detonation waves. For plane waves these distances are considerably greater.

To conclude this section we remark that the results we have obtained are published in [17] and [18].

§3. On certain properties of gas flow with consideration of the kinetics of chemical reactions

Consideration of the kinetics of chemical reactions may in a number of cases strongly affect the picture of gas flow during a point explosion [18]–[20].

1. We shall consider the simplest model of such flows (model 2). We will neglect the effect of viscosity and heat conduction on the gas flow and assume that the course of chemical reaction can be described by a single chemical variable β, the mass concentration of nonreacted molecules. The chemical energy βQ per unit mass of the combustible mixture is then included. In the system of equations (1.18)–(1.20) the energy equation is changed; its right side is now not zero but equal to the quantity

$$-\rho r^{\nu-1}\frac{d}{dt}(\beta Q).$$

For a perfect gas with constant γ the energy equation can be transformed to the form [8]

$$\frac{dp}{dt}-\frac{\gamma p}{\rho}\frac{d\rho}{dt}+(\gamma-1)\rho\frac{d(\beta Q)}{dt}=0. \qquad (6.15)$$

We take the equation of chemical kinetics in the form [8], [18], [19]

$$\frac{d\beta}{dt}=-L\beta^m p^n \rho^l \exp\left(-\frac{u_0}{\mu}\frac{\rho}{p}\right). \qquad (6.16)$$

Here u_0 is the energy of activation, μ is the mean molecular weight, m is the order of the reaction, n and l are constants, L is a positive constant, $\beta = 1$ at the start of the reaction, and $\beta = 0$ at the end of the reaction.

Thus the complete system of equations contains (1.18), (1.19) and (6.15), (6.16). The boundary conditions are the usual conditions on the shock wave and condition (6.4).

In this problem it can be assumed that $\beta = 1$ directly behind the front of the shock wave. It follows from (6.16) that β can only decrease in a particle of the gas.

We also write down the integral law of energy conservation, which takes the form

$$E_0 = \sigma_\nu \int_0^{r_2} \left(\frac{\rho v^2}{2} + \varepsilon\rho - Q(1-\beta)\rho\right) r^{\nu-1} dr + \frac{\sigma_\nu}{\nu} \rho_1 \varepsilon_1 r_2^\nu. \tag{6.17}$$

This relation is very useful in considering a number of questions. For $\beta = 0$ and $\xi_1 = 0$ we obtain from (6.17) the integral law of energy conservation in the case of a strong detonation wave (§2).

The solution of the PEP for this flow model with chemical reactions can be obtained in the general case only by approximate or numerical methods. It is possible to try to find a class of self-similar solutions of the equations of gas dynamics and chemical kinetics. We shall assume that

$$\frac{u_0}{\mu} = B p^{\alpha_1} \rho^{\alpha_2}, \qquad \rho_1 = A r^{-\omega}. \tag{6.18}$$

If the energy E_0 is neglected, i.e. we consider the problem of the propagation of a strong shock wave which is self-sustaining due to chemical energy, then the system of defining parameters is the following: $Q, L, B, A, m, l, n, \omega, \alpha_1, \alpha_2, \gamma, \nu$. From dimensional considerations it is easy to show that for $\alpha_1 = \alpha_2 = 0$ the problem will be self-similar provided that $l + n = 1/\omega$.

We thus conclude that for a strong shock wave the problem of propagation of the detonation will possess self-similarity if the initial density varies according to the law $\rho_1 = Ar^{-1/(l+n)}$. If $E_0 \neq 0$, the problem of explosion in a chemically active gas will be self-similar only under the condition that Q depends on the gas parameters (or on the gas parameters, the coordinate, and time).

Thus, if under the condition (6.18) we have the dependence $Q = Q_0 p^{\alpha_1} \rho^{\alpha_2}$, then the problem on strong point explosion is self-similar under the conditions

$$2(1-\alpha_1)(\delta - 1) + \delta\omega(\alpha_1 + \alpha_2) = 0, \qquad 2 = \delta(\nu + 2 - \omega),$$
$$\delta[3l + n + (\nu - 1)(l + n)] - 1 - 2l = 0. \tag{6.19}$$

It is also possible to indicate classes of self-similar solutions [19] if it is assumed that L, B and Q_0 depend on certain powers of r and t. For self-similar motions the system of equations of gas dynamics and chemical kinetics reduces in special (dimensionless) variables to a system of four ordinary differential equations:

$$\begin{aligned} &\varphi(\lambda)\frac{dg}{d\lambda} + \frac{2}{\gamma+1} g \frac{df}{d\lambda} - \omega g + \frac{2(\nu-1) gf}{(\gamma+1)\lambda} = 0, \\ &g\left(\frac{\delta-1}{\delta} f + \varphi(\lambda)\frac{df}{d\lambda}\right) + \frac{\gamma-1}{\gamma+1}\frac{dh}{d\lambda} = 0, \\ &\left[2\frac{\delta-1}{\delta} + \omega(\gamma-1)\right] h + \varphi(\lambda)\left(\frac{dh}{d\lambda} - \frac{\gamma h}{g}\frac{dg}{d\lambda}\right) = \end{aligned} \tag{6.20}$$

$$= -\sigma_2 g\left[2\frac{\delta-1}{\delta}\beta h^{\alpha_1}g^{\alpha_2} + \varphi(\lambda)\frac{d}{d\lambda}(\beta h^{\alpha_1}g^{\alpha_2})\right],$$

$$\varphi(\lambda)\frac{d\beta}{d\lambda} = -\sigma_0 h^n g^l \beta^m \exp(-\sigma_1 n^{\alpha_1-1}g^{\alpha_2+1}),$$

where

$$\varphi(\lambda) = \frac{2f}{\gamma+1} - \lambda, \quad f = \frac{v}{v_2}, \quad h = \frac{p}{p_2}, \quad g = \frac{\rho}{\rho_2}, \quad \lambda = \frac{r}{r_2},$$

and v_2, p_2 and ρ_2 are the values of the velocity, pressure, and density behind the strong shock wave. For the self-similar case of a strong explosion the function $r_2(t)$ has the form

$$r_2 = \left(\frac{E_0'}{\alpha_* A}\right)^{\delta/2} t^{\delta}, \quad \delta = \frac{2}{2+\nu-\omega}.$$

Here α_* is a quantity depending on the dimensionless parameters of the problem defined (as in Chapter 2) by the law of energy conservation:

$$\alpha_* = 2\sigma_\nu \frac{\delta^2}{\gamma^2-1}\int_0^1 [gf^2 + h - \sigma_2 g^{\alpha_2+1}h^{\alpha_1}(1-\beta)]\lambda^{\nu-1}d\lambda. \tag{6.21}$$

It is necessary to find a solution of the system (6.20) with boundary conditions $f(1) = g(1) = h(1) = 1$ on the shock wave and condition (6.21) or the condition $f(0) = 0$ at the center of the explosion.

In contrast to the usual self-similar problems of gas dynamics [1], the consideration of chemical reactions greatly complicates the solution. Thus the system of self-similar equations for the cases considered has no classical adiabatic integral and does not reduce to the investigation of a single ordinary differential equation of first order. Complicated singularities may thus arise in a neighborhood of the center. It is therefore necessary to apply approximate and numerical methods even for studying the self-similar problems of the propagation of shock waves in a combustible mixture taking the reaction kinetics into account.

To control the accuracy of the computations one can make use of the integral law of mass conservation which in the dimensionless variables chosen has the form

$$\int_0^1 g\lambda^{\nu-1}d\lambda = \frac{\gamma-1}{\gamma+1}\frac{1}{\nu-\omega}.$$

If $\alpha_2 = 0$, from the self-similarity conditions for $\omega = 0$ we have $n = (\nu+2)/2\nu$.

Another interesting case is that where $\omega = \nu - 1$. As we pointed out in Chapter 3, this case corresponds to a point explosion or propagation of a detonation wave in a gas flow in a wedge-shaped ($\nu = 2$) or conical ($\nu = 3$) nozzle. In the case of explosion there is the relation $\omega\alpha_2 = \nu - \omega - \nu\alpha_1$. Examples of solutions for self-sustaining waves ($\alpha_* = 0$) with $\omega = 1$ and $\alpha_1 = \alpha_2 = 0$ are given in [19].(1)

2. For the further description of explosion phenomena in a chemically active gas mixture it is sometimes possible to use a flow model in which a surface of discontinuity is introduced—a combustion wave behind the front of the propagating shock wave (model 3).

In this model it is supposed that, after the front of the shock wave passes a particle of the gas, instantaneous combustion of the matter in it occurs after a certain time t_{ind} with release of energy Q on the front of a combustion wave. The time t_{ind} is usually called the time of ignition delay or the induction time. It is known [4]–[7], [9] that the induction time depends on the thermodynamic parameters of the gas. Suppose that for t_{ind} we have

$$t_{\text{ind}} = Lp^{-n_1}\rho^{-l_1}\exp\left(\frac{K}{T}\right), \tag{6.22}$$

where n_1, l_1, k and L are constants ($K > 0, L > 0$).

Thus for a stoichiometric mixture of hydrogen with air one can take [21], [22]

$$t_{\text{ind}} = 4.5 \cdot 10^{-9}\frac{1}{p}\exp\left(\frac{10^4}{T}\right), \tag{6.23}$$

where t_{ind} is expressed in seconds, the pressure p is in atmospheres, and T is the temperature (°K).

Deferring the discussion of the detailed characteristics of the gas flow behind the front of the shock wave, let us examine the rough qualitative variation of t_{ind} behind the front of the shock wave. It follows from (6.22) and (6.23) that t_{ind} is small for large p and T. As the shock wave dies out and p and T decrease in the gas particle, the quantity t_{ind} increases, and thus the distance between the front of the shock wave and the combustion wave also increases, and the combustion front becomes detached from the front of the shock wave. Thus from a certain time onward the shock wave and the zone of combustion reactions cannot be taken as a single surface of discontinuity—the detonation wave. For sufficiently large t the combustion reaction may proceed intensively only outside some neighborhood of the points on the front of the shock wave (in a zone of sufficiently high temperatures), and combustion will practically not occur in a neighborhood of the front of the shock wave if it is strongly damped. It is possible that this was the flow picture in the experiments

(1) In this work the case $\omega = 0$, $E_0 \neq 0$, $\alpha_1 = 1$, $\alpha_2 = 0$ is also considered. However, the extrapolation of the solution to the region $\lambda < 0.2$ performed there is not precise.

described in [14] during explosion of a charge in mixtures of hydrogen with oxygen in a rather long broad pipe.

The possibility of detachment of the detonation wave in the phenomenon of point explosion was first deduced by us in 1967 [18]. Experimental and theoretic data on the flow around bodies of a hypersonic stream of a combustible mixture also served as indirect corroboration of the flow picture described [22], [23]. By the inverse analogy between hypersonic flow around blunt bodies and the phenomenon of a cylindrical explosion (cf. Chapter 4, §7) we can qualitatively describe the phenomenon of cylindrical explosion using the picture of stationary flow. If use is made of this analogy and data is taken over from the flow around a blunt cylinder, then in the case of a cylindrical explosion in a mixture of hydrogen and air (or oxygen) it is natural to expect the appearance of a flow picture with two fronts: the front of the shock wave and a combustion front following it in accordance with (6.23). Simultaneously with us R. I. Solouhin, J. H. Lee, and A. K. Oppenheim [24] arrived at an analogous conclusion on the basis of experiments with a laser beam in a combustible mixture. The phenomenon of decay of a detonation wave has also been observed during diffraction of detonation waves [25].

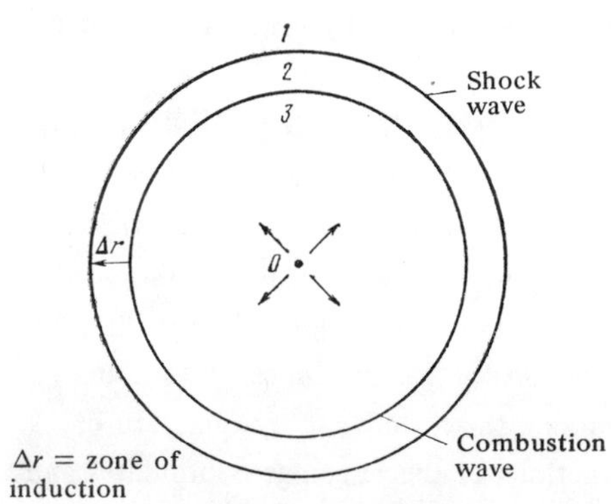

FIGURE 73. Scheme of gas flow in the model of two fronts

Work has recently been carried out [20] on the theoretical justification of the effect of detachment of shock waves during the initial stage of a point explosion. Some results of this investigation are presented below. Figure 73 shows the scheme of gas flow in the model of two fronts. In region 1 the gas is at rest, and in region 2 the gas moves behind the usual shock wave. The gas is ignited as the combustion wave passes, and region 3 is filled with combustion products. If the change in the adiabatic exponent on passage through a discontinuity is neglected and the phenomenon of heat conduction is not considered, then the equations for the motion of the gas in regions of continuity reduce to (1.18)–(1.20). The conditions on the shock wave have the usual form. The conditions on the combustion front can be written as follows:

$$\frac{\rho_*}{\rho_3}=\frac{N_1}{\gamma+1},\qquad v_3=D_*+\frac{v_*-D_*}{\gamma+1}N_1,\qquad p_3=\frac{p_*N}{(\gamma+1)\Omega},$$
$$N=\gamma+\Omega+\gamma\sqrt{(\Omega-1)^2-2(\gamma^2-1)Q(D_*-v_*)^{-2}},\tag{6.24}$$
$$\Omega=\frac{\gamma p_*}{\rho_*}\frac{1}{(D_*-v_*)^2},\qquad N_1=\gamma+\Omega+\frac{1}{\gamma}(\gamma+\Omega-N),$$

where p_*, ρ_*, v_* and p_3, ρ_3, v_3 represent the pressure, density and velocity in front of and behind the front of the combustion wave respectively. The relations (6.24) follow from the general conditions on strong discontinuities obtained in §2 of Chapter 1.

We now introduce the fraction of induction time dc according to the formula [7]: $dc = -dt/t_{\text{ind}}$. From this and (6.22) we have

$$\frac{dc}{dt}=-k_1p^{n_1}\rho^{l_1}e^{-E_1\rho/p}.\tag{6.25}$$

We shall assume that $c = 1$ on the front of the shock wave. Then the instant of time $\bar{t}$ at which $c = 0$ gives us the termination of the induction period for a gas particle for a variable field of pressures and densities. If t_* is the instant at which the shock wave crosses the particle, then $t_{\text{ind}} = \bar{t} - t_*$.

Thus in order to solve the problem we must integrate the system (1.18)–(1.20), (6.25) in region 2 and the system (1.18)–(1.20) in region 3 with the appropriate initial and boundary conditions. In (6.25) the total derivative appears on the right. From the point of view of computations it may be more efficacious to use the equations in the Lagrange variables. For the initial moments of time when (6.1) holds we can neglect the effect of the combustion wave on the gas dynamics. Equation (6.25) then affords us the possibility of determining the instant $\bar{t}$ by quadrature, since p and ρ may be considered known functions of the Lagrange coordinates and time from the gas-dynamical problem. For a strong explosion, this solution is given in Chapter 2. Computations carried out in this manner (see also the more complicated model of the reactions considered below) in the Lagrange variables completely support the aforementioned qualitative arguments in favor of uncoupling of the front of an overcompressed detonation wave (Figure 74).

Figure 75 shows the graphs of the lines $c = 1, R_2(\tau) = r_2/r^0$ (the shock wave)- and $c = 0$ (combustion front), $R_f = r_f/r^0$ for the case $\nu = 2, \gamma = 1.3$, $E_0 = 10^{10}$ erg/cm^2, $p_1 = 10^6$ dyne/cm^2, $\rho_1 = 10^{-3}$ g/cm^2 with constants $1/5E_1$ and k_1 close to those for a hydrogen-oxygen mixture ($Q \approx 7 \times 10^{10}$ erg/g).

It follows from the graphs of Figures 75 and 76 that the combustion front and the shock wave practically coincide for small τ ($\tau = t/t^0$), where r^0 and t^0 represent the dynamic length and time formed from E_0, ρ_1 and p_1. With increasing τ the detonation wave decays into a shock wave and a flame front. This situation occurs also in the case of other symmetries. The computations indicate that with increase in E_0 the moments of time t from which point on there is a substantial growth in the distances between r_2 and r_j increase.

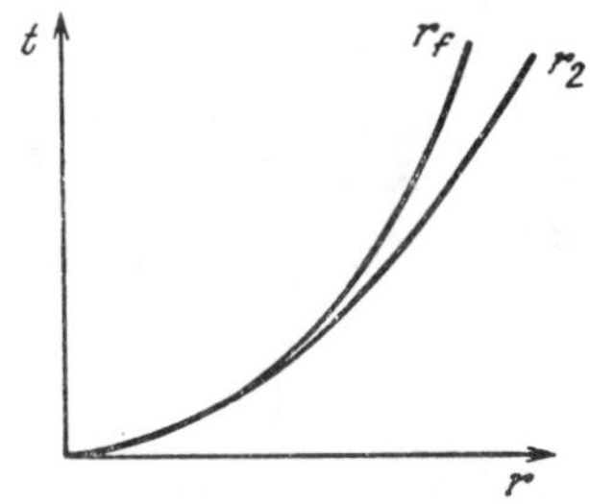

FIGURE 74. Scheme of uncoupling of an overcompressed detonation wave

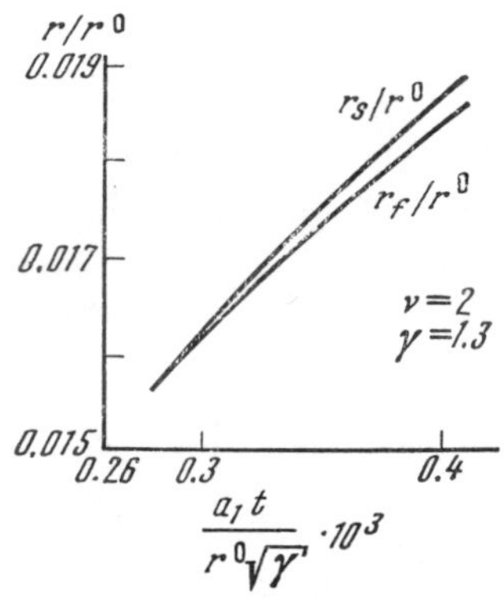

FIGURE 75. Graphs of the law of motion of the shock wave and the ignition front for a model mixture

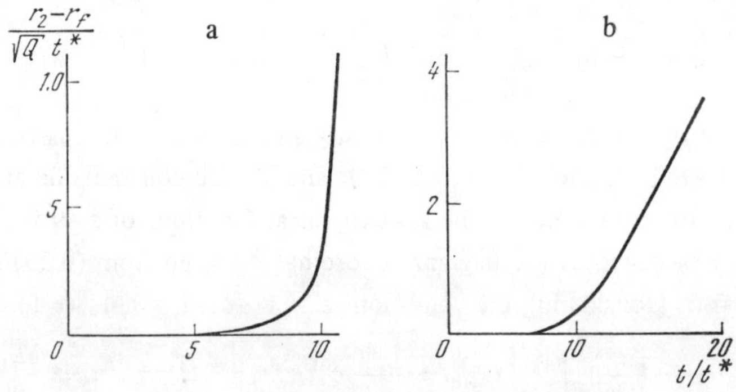

FIGURE 76. Growth of the width of the induction zone in time (model reaction, $\gamma = 1.4$, $\nu = 2$, $E_0 = 10^{10}$ erg/cm^2, $t^* = 10^{-7}$ sec)
a) for a gas particle behind the wave front in dependence on the time of termination of the induction period; b) in space

In carrying out the computations it was assumed that $n_1 = 1$ and $l_1 = 0$, i.e. that the factor in front of the exponential in (6.25) depended only on the pressure. R. I. Solouhin has pointed out to the author that for a hydrogen-oxygen mixture it is better to take this factor dependent on the density ρ. The computations subsequently performed by V. V. Markov for the case $n_1 = 0$ and $l_1 = 1$ indicate that the qualitative picture of uncoupling of the wave is preserved also for such values of n_1 and l_1. Only the numerical values of the functions $c(\tau)$ and $R_f(\tau)$ are changed. An example of computing the induction zones for $\nu = 1, 2, 3$ is shown in Figure 77 (a stoichiometric mixture of acetylene and oxygen, $\rho_1 = 0.5 \times 10^{-3}$ g/cm^3, $p_1 = 10^6$ dyne/cm^2, $\gamma = 1.4$, $E_0 = 3 \times 10^6$ erg/cm$^{3-\nu}$, $n_1 = 0$, $l_1 = 1$). The possibility of the appearance of combustion in a neighborhood of the center where the density is low was not considered here.

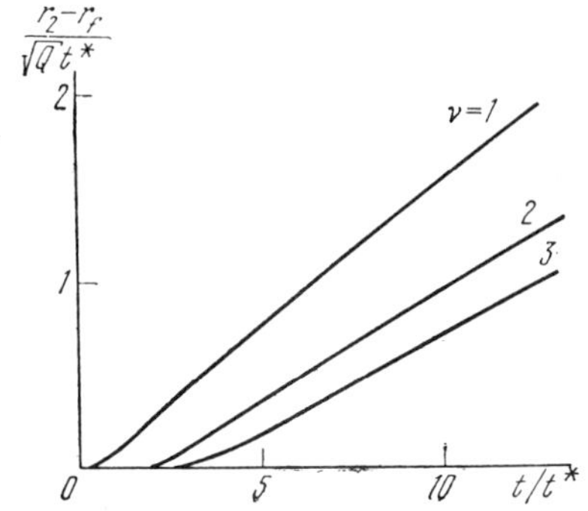

FIGURE 77. Variation of the width of the induction zone behind a strong wave for a mixture of acetylene and oxygen ($E_0 = 3 \times 10^6$ erg/cm$^{3-\nu}$; $\gamma = 1.4$)

A rough qualitative estimate of the induction time can be obtained by using the approximation of the self-similar solution proposed in [26]. Indeed, we assume that

$$\frac{1}{T} = \frac{1}{T_*} + \frac{a}{T_*}\ln\frac{t}{t_*}, \qquad \rho = \rho_*\left(\frac{t_*}{t}\right)^{2b}, \qquad p = p_*\left(\frac{t_*}{t}\right)^{2b\gamma}, \tag{6.26}$$

where T_*, ρ_* and p_* are the temperature, density and pressure in the particle passing across the shock wave at the instant t_*, and a and b are constants determined from the condition of approximating the gas-dynamical functions of a strong explosion (the time t_* may be taken as the Lagrange coordinate). Then from (6.25) for $n_1 = 1$ and $l_1 = 0$ (considering the condition $c = 1$ for $t = t_*$) we find

$$c = 1 - \frac{t_*}{t_{0\,\text{ind}}\Omega}\left[1 - \left(\frac{t_*}{t}\right)^{\Omega}\right], \qquad t_{0\,\text{ind}} = \frac{1}{k_1 p_*} e^{E_1\rho_{k}/p_*}, \qquad \Omega = 2b\gamma + \frac{aE_1\rho_*}{p_*} - 1.$$

From this we obtain for the time $\bar{t}$ of termination of the induction period

$$\bar{t} = t_*\left[1 - \frac{t_{0\,\text{ind}}}{t_*}\Omega\right]^{-1/\Omega}. \tag{6.27}$$

Using for the induction time the relation $t_{\text{ind}} = \bar{t} - t_*$, we have from (6.27)

$$t_{\text{ind}} = t_*\left[1 - \frac{t_{0\,\text{ind}}}{t_*}\Omega\right]^{-1/\Omega} - t_*. \tag{6.28}$$

In the spherical case for $\gamma = 1.3$ it was found in [26] that $a = 0.44$ and $b = 0.75$. Computations carried out at our request by T. Baĭteliev with the constants for an oxygen-hydrogen mixture indicate that (6.28) correctly describes the phenomenon of uncoupling of a detonation wave, although for small t it gives large errors in determining t_{ind}. We mention that approximate methods of estimating t_{ind} have also been developed in [27], [28]. The analysis above referred only to the stage of a strong shock wave. For large times the gas-dynamical functions will differ from those for a strong explosion, and the solution of the problem can be obtained either by the method of approximating the density in the Lagrange variables or numerically by the method of characteristics, by the method of integral relations or by the method of through calculation [31]. We remark also that the model of two fronts may have limited applicability due to the appearance of instabilities.

3. We now consider certain special features of the solution of the problems for the model of including chemical reaction with the evolution of heat in the flow behind the front of the shock wave after passage of an induction period (model 4). We shall be concerned only with the case of a strong explosion in a quiescent gas.

The equations describing the course of the chemical reactions we take in the form of the Arrhenius relations

$$\frac{dc}{dt} = -\frac{1}{t_{\text{ind}}} = -k_1 p^{n_1} \rho^{l_1} \exp\left(-\frac{E_1 \rho}{p}\right), \tag{6.29}$$

$$\frac{d\beta}{dt} = -k_2 \beta^{m_1} p^{n_2} \rho^{l_2} \exp\left(-\frac{E_2 \rho}{p}\right) + k_3 (1-\beta)^{m_2} p^{n_3} \rho^{l_3} \exp\left(-\frac{E_3 \rho}{p}\right). \tag{6.30}$$

Before the start of reaction (6.30) it is assumed that $\beta = 1$, and $c = 1$ on the front of the shock wave. When c becomes zero the induction period is at an end, and the reaction with evolution of heat begins. The reaction (6.29) proceeds with no evolution of heat.

We write the system of equations describing the one-dimensional motion of the gas in the form

$$\rho \frac{dv}{dt} + \frac{\partial p}{\partial r} = 0, \qquad \frac{\partial \rho}{\partial t} + \frac{\partial \rho v}{\partial r} + \frac{(\nu - 1)\rho v}{r} = 0,$$

$$\rho \frac{dH}{dt} - \frac{dp}{dt} = 0, \qquad H = \frac{\gamma}{\gamma - 1} \frac{p}{\rho} + \beta Q. \tag{6.31}$$

Here H is the enthalpy, β is the fraction of unreacted molecules, and Q is the total heat evolved per unit mass of the gas.

In studying the problem we must, as heretofore, consider the conditions on the shock wave, the condition of release of energy E_0, and the initial conditions.

For these models the conditions of self-similarity of the problem for various l_i, m_i and n_i and dependencies of E_i and the released heat Q on pressure and density (and also on the coordinate and time) will actually coincide with those considered for model 2. During the propagation of a strong blast wave in a quiescent gas of constant density in the self-similar case the absolute width of the induction zone grows in proportion to $t^{2/(\nu+2)}$. However, the relative width of this zone, equal to the ratio of the difference of the coordinates of the shock wave and the combustion wave to the coordinate of the shock wave, remains constant. We remark that for this self-similar motion the quantities E_i and Q are proportional to the pressure. Consideration of the finite rate of the chemical reaction here also leads to a substantial change in the qualititive picture of the gas flow.

If the quantities E_i and A are constant, then the PEP is not self-similar. We shall consider explosion in a one-dimension quiescent medium. At times close to the starting time the magnitude of the total energy evolved during combustion in the volume bounded by the shock wave is considerably less then the energy of the explosion, i.e.

$$E_0 > U_4 = \sigma_\nu \int_0^{r_2} Q(1-\beta)\rho r^{\nu-1} dr, \qquad \sigma_\nu = 2(\nu-1)\pi + (\nu-2)(\nu-3), \tag{6.32}$$

and therefore the effect of combustion on the gas-dynamical flow is small.

For the initial stage of the explosion when (6.32) is valid one can seek a solution using the method of linearization with respect to a small parameter μ proportional to the ratio $Q\rho_1/E_0$. Then for any unknown function f we have

$$f = f_{(0)} + \mu f_{(1)} + \sigma(\mu). \tag{6.33}$$

After substituting functions of the form (6.33) into (6.19)–(6.31) (and their linearizations) we obtain for f_0 and f_1 a system of partial differential equations one of which, a nonlinear equation, contains only leading terms of the expansion, i.e. the quantities $v_{(0)}, \rho_{(0)}$ and $p_{(0)}$, and the other, a linear equation, contains the quantities $v_{(1)}, \rho_{(1)}$ and $p_{(1)}$. Here the system for the leading terms itself decomposes into two systems. The solution of the gas-dynamical equations is formed on self-similar functions describing the flow from a strong point explosion considered in Chapter 2 (the effect of chemical reactions is negligibly small). The chemical reactions in this case proceed on the given flow field and are described by (6.29)–(6.30), in which each function is to be assigned the index (0). We shall assume that $m_1 = m_2 = 2$, $n_1 = 1$ and $n_2 = 2$.

We introduce the dimensionless quantities

$$G = \frac{\rho_{(0)}}{\rho_1}, \quad P = \frac{p_{(0)}}{\rho_1 Q}, \quad V = \frac{v_{(0)}}{\sqrt{Q}}, \quad \tau = \frac{t}{t^*}, \quad y = \frac{r}{\sqrt{Q}t^*},$$
$$\delta_0 = t^* k_1 Q \rho_1, \quad \delta_1 = \frac{E_1}{Q}, \quad \delta_2 = t^* k_2 Q^2 \rho_1^2, \quad \delta_3 = \frac{E_2}{Q},$$

t^* is the characteristic time. Then in the Euler variables the equations for determining c and β assume the form

$$\frac{\partial c}{\partial \tau} + V\frac{\partial c}{\partial y} = -\delta_0 P \exp\left(-\delta_1 \frac{G}{P}\right), \tag{6.34}$$

$$\frac{\partial\beta}{\partial t} + V\frac{\partial\beta}{\partial y} = -\delta_2 P^2\beta^2\left\{\exp\left(-\delta_3\frac{G}{P}\right) - (1-\beta)^2\exp\left[-(\delta_3+1)\frac{G}{P}\right]\right\}. \quad (6.35)$$

The system (6.34)–(6.35) is a model for the kinetics of all chemical reactions of second order $(m = 2)$. Since in this approximation the functions V, G and P are assumed given, the system consists of two independent first-order partial differential equations. The initial data for them are given at $r = r_2$, $t = t_*$ $(c = 1, \beta = 1)$ and at $r = r_f$, $t = \bar{t}$ $(c = 0, \beta = 1)$, while in the region between r_2 and r_j we have $\beta = \text{const}$.

Thus we have a Cauchy problem for (6.34) and (6.35), the solution of which can be found by the standard method of characteristics. We remark that if the induction time has already been found, for example, from a solution in the Lagrange variables described for model 3, then it remains only to determine β. To control the accuracy of determining t_{ind} the equation for c was integrated anew. The ordinary differential equations defining the variation of the functions along the characteristics were solved by the numerical Runge-Kutta method. We shall not go into the details of these computations. On the basis of a method worked out by E. Bišimov a program was written for the computer BÈSM-3 and the computations were carried out for the values of the parameters $t^* = 10^{-7}$ sec, $\delta_0 = 10^3$, $\delta_1 = 5.1$, $\delta_2 = 20$. These constants were chosen so that the kinetic equations should correspond to the induction times and the resulting reaction for mixtures of hydrogen-oxygen type. The data obtained on the induction time and width of the induction zone coincided with the corresponding quantities for model 3 (cf. Figures 75 and 76). We mention that Figure 76a shows the variation of the relative difference $(r_2 - r_f)/Q^{1/2}t^* = \Delta y$ along a trajectory of a particle as a function of the relative time of termination of the induction period $\bar{t}/t^*$. These computations also indicated that consideration of the finiteness of the rates of the chemical reactions leads to a qualitatively new picture of development of the flow as compared with the model of an infinitely thin detonation wave. Due to the existence of large negative gradients for the temperature, pressure, and density of the gas behind the wave, in the region of abrupt expansion of the flow the ignition zone separates from the shock wave. The ignition delay time increases strongly during propagation of the wave in spite of the fact that the blast wave is still rather strong: $p_2 > 50p_1$. This leads to the decay of the detonation wave into an ordinary condensation jump and a flame front. Computations indicate that for this model of the direct and reverse reaction (6.30) a major role is played by the reverse reaction which is often neglected in model description of gas motions behind detonation waves. The spatial distribution of the concentration β is given in Figure 78. Due to the high temperature, in a neighborhood of the center of the explosion the mixture is not completely burned, and the thermal energy is only partially released. The ratio of the energy evolved in model 4 to that which would be evolved in model 3 with complete combustion is approximately equal to one half for $\tau = 10$. By this time the ignition zone has already parted from the blast wave, but the magnitude of the energy evolved during combustion in the entire perturbed region is still negligibly small as compared with the energy of the explosion ($U_4/E_0 \sim 0.02$). (This circumstance bespeaks the rather high accuracy of the approximation adopted.)

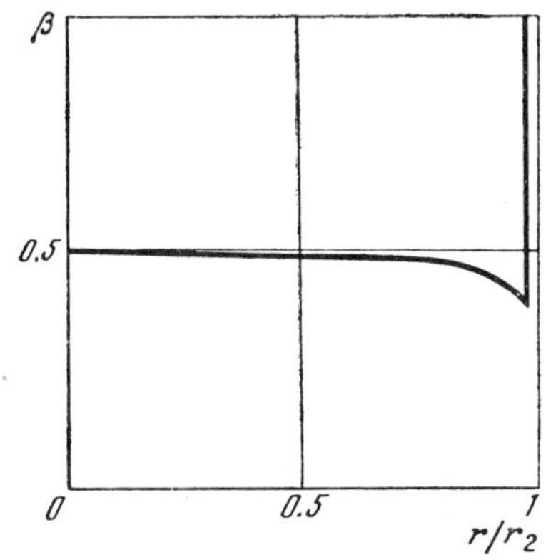

FIGURE 78. Variation of concentration β behind a strong shock wave (model reaction, $\gamma = 1.4$, $\nu = 2$, $E_0 = 10^{10}$ erg/cm)

In this approximation there are in principle no difficulties in solving the problem taking into account the complete kinetics of the chemical reactions. Questions of the kinetics of the chemical reactions of combustion and explosion have at present been extensively worked out by scientists of N. N. Semenov's school (for a survey of this work see [29]) and other authors. Using these results it is possible to compute the processes of chemical reactions during strong explosion in a gas.(2)

In conclusion we mention that an analogous analysis could be carried through also for reactions of thermonuclear detonation. In principle there are here no differences in the sense of the role of the kinetics of the reactions [29], [30]. However, as was pointed out in [30], in investigating the structure of the thermonuclear detonation wave a substantial role will be played by radiation effects, since thermonuclear reactions take place at extremely high temperatures (see also the recently published paper [32]). We remark that in a neighborhood of the shock wave instabilities may arise which destroy the one-dimensionality of the flow and lead to the regeneration of detonation combustion (this is often observed in experiments).

---

(2) *Added in proof.* For the cylindrical case a numerical solution of the complete problem with kinetics (6.29)–(6.30) is given in [33].

BIBLIOGRAPHY

1. L. I. Sedov, *Similarity and dimensional methods in mechanics.* 4th ed., GITTL, Moscow, 1957; English transl., Academic Press, New York, 1959. MR 19, 898; 21 #6840.

2. James Serrin, *Mathematical principles of classical fluid mechanics,* Handbuch der Physik (herausgegeben von S. Flügge), Band 8.1, Strömungsmechanik I, Springer-Verlag, Berlin, 1959, pp. 125–263. MR 21 #6836b.

3. Ja. B. Zel'dovič and A. S. Kompaneec, *Theory of detonations,* GITTL, Moscow, 1955; English transl., Academic Press, New York, 1960.

4. K. I. Ščelkin, *Two cases of unstable combustion,* Ž. Èksper. Teoret. Fiz. 36 (1959), 600–606 = Soviet. Phys. JETP 9 (1959), 416–421. MR 21 #6211.

5. R. M. Zaĭdel' and Ja. B. Zel'dovič, *One-dimensional instability and the damping of an explosion,* PMTF Ž. Prikl. Meh. i Tehn. Fiz. 1963, no. 6, 59–65. (Russian)

6. R. I. Solouhin, *On detonation in a gas heated by a shock wave,* PMTF Ž. Prikl. Meh. i Tehn. Fiz. 1964, no. 4, 42–48. (Russian)

7. R. B. Gilbert and R. A. Strehlow, *Theory of detonation initiation behind reflected shock waves,* AIAA J. 4(1966), 1777–1783.

8. V. V. Puhnačev, *On the instability of the Chapman-Jouguet detonation,* PMTF Ž. Prikl. Meh. i Tehn. Fiz. 1963, no. 6, 66–73. (Russian)

9. R. I. Solouhin, *The zone of the exothermic reaction in a one-dimensional shock wave in a gas,* Fiz. Gorenija i Vzryva 2 (1966), no. 3, 12–18 = Combustion, Explosion and Shock Waves 2 (1966), no. 3, 6–10.

10. I. S. Šikin, *On exact solution of equations with shock waves and detonation waves in one-dimensional gas dynamics,* Dokl. Akad. Nauk SSSR 122 (1958), 33–36 = Soviet Phys. Dokl. 3 (1958/59), 915–917. MR 20 #7472.

11. V. A. Levin, *Approximate solution of the problem of a strong point explosion in a combustible mixture,* Izv. Akad. Nauk SSSR Meh. Židkost. i Gaza 1967, no. 1, 122–124 = Fluid Dynamics 2 (1967), no. 1, 80–81.

12. V. P. Korobeĭnikov and P. I. Čuškin, *Plane, cylindrical, and spherical explosions in gas with back pressure,* Trudy Mat. Inst. Steklov. 87 (1966), 3–34 = Proc. Steklov Inst. Math. 87 (1966), 1–40.

13. G. G. Černyĭ, *Asymptotic law of propagation of a plane detonation wave,* Dokl. Akad. Nauk SSSR 172 (1967), 558–560 = Soviet Phys. Dokl. 12 (1967/68), 21–22.

14. A. J. Laderman, *Detonability of hydrogen-oxygen mixtures in large vessels at low initial pressures,* AIAA J. 4 (1966), 1784–1790.

15. E. Bišimov, *Numerical solution of the problem of a strong point explosion in a detonable gas,* Some Questions of Differential Equations, "Nauka", Alma-Ata, 1969, pp. 94–103. (Russian) RŽ Mat. 1969 #9Б554.

16. E. Bišimov, *Plane, cylindrical and spherical explosions with back pressure in a detonable gas,* Izv. Akad. Nauk Kazah. SSR Ser. Fiz.-Mat. 1970, no. 1, 43–48. (Russian)

17. V. P. Korobeĭnkov, *Point explosion in a detonating gas,* Dokl. Akad. Nauk SSSR 177 (1967), 295–298 = Soviet Phys. Dokl. 12 (1967/68), 1003–1005.

18. ———, *The problem of point explosion in a detonating gas,* Astronautica Acta 14 (1968/69), 411–419.

19. E. Bišimov et al., *One dimensional nonstationary flows of a combustible mixture of gases taking into account the finite rate of chemical reactions,* Izv. Akad. Nauk SSSR Meh. Židkost. i Gaza 1968, no. 6, 7–19 = Fluid Dynamics 3 (1968), no. 6, 1–8.

20. V. P. Korobeĭnkov and V. A. Levin, *Strong explosion in a combustible mixture of gases,* Izv. Akad. Nauk SSSR Meh. Židkost. i Gaza 1969, no. 6, 48–51 = Fluid Dynamics 4 (1969), no. 6, 30–32.

21. A. Ferri, P. Libby and V. Zakkay, *Theoretical and experimental investigation of supersonic combustion,* Polytech. Inst. of Brooklyn and General Applied Science Lab., Inc., 1962.

22. M. P. Samozvancev, *On the stabilization of detonation waves by means of poorly streamlined bodies,* PMTF Ž. Prikl. Meh. i Tehn. Fiz. **1964**, no. 4, 126–129. (Russian)

23. G. G. Černyĭ, *Supersonic flow past bodies with formation of detonation and combustion fronts,* Problems of Hydrodynamics and Continuum Mechanics. Contributions in Honor of the Sixtieth Birthday of Academician L. I. Sedov, published simultaneously in Russian and English, "Nauka", Moscow; Soc. Indust. Appl. Math., Philadelphia, Pa., 1969, pp. 145–169.

24. J. H. Lee, R. I. Soloukhin and A. K. Oppenheim, *Current views on gaseous detonation,* Astronautica Acta **14** (1969), 565–584.

25. L. G. Gvozdeva, *Experimental investigation of the diffraction of detonation waves in a stoichiometric mixture of methane and oxygen,* PMTF Ž. Prikl. Meh. i Tehn. Fiz. **1961**, no. 5, 53–56. (Russian)

26. Ja. B. Zel'dovič and Ju. P. Raĭzer, *The physics of shock waves and high-temperature hydrodynamic phenomena,* 2nd ed., "Nauka", Moscow, 1966; English transl., vols. 1, 2, Academic Press, New York, 1966, 1967.

27. R. I. Soloukhin, *Nonstationary phenomena in gaseous detonation,* Twelfth Sympos. (Internat.) on Combustion (Poitiers, 1968), The Combustion Institute, Pittsburgh, Pa.

28. E. A. Lundstrom and A. K. Oppenheim, *On the influence of nonsteadiness on the thickness of the detonation wave,* Twelfth Sympos. (Internat.) on Combustion (Poitiers, 1968), The Combustion Institute, Pittsburgh, Pa.; cf. Proc. Roy. Soc. London Ser. A **310** (1969), 463–478.

29. N. N. Semenov, *Development of the theory of chain reactions and thermal ignition,* Znanie, Moscow, 1969. (Russian)

30. Ann L. Fuller and Robert A. Gross, *Thermonuclear detonation wave structure,* Phys. Fluids **11** (1968), 534–544.

31. L. N. Busurina et al., *Numerical computation of detonation,* Ž. Vyčisl. Mat. i Mat. Fiz. **10** (1970), 239–243 = USSR Comput. Math. and Math. Phys. **10** (1970), no. 1, 333–339.

32. M. S. Chu, *Thermonuclear reaction waves at high densities,* Phys. Fluids **15** (1972), 413–422.

33. *V. P. Korobeĭnikov et al., *Propagation of blast waves in a combustible gas,* Astronautica Acta **17** (1971/72), 529–537.

* Added in proof.

CHAPTER 7

POINT EXPLOSION IN AN ELECTRICALLY CONDUCTING GAS INCLUDING THE EFFECT OF THE MAGNETIC FIELD

§1. Introduction

The problems considered in the preceding chapters can be extended to the case where the medium is an electrically conducting gas which is capable of interacting with applied magnetic fields. This generalization of the PEP is of interest for astrophysical applications, for the propagation of blast waves in the ionosphere, for problems arising in the focusing of a laser beam, in the detonation of an explosive substance or passage of electrical discharges in a high-temperature plasma under laboratory conditions, in studying devices creating impulse sources of current for explosion magnetohydrodynamic generators, etc. We shall consider the formulations of the problems and the basic equations describing the motion of the gas.

Suppose that in a quiescent electrically conducting gas with an initial magnetic field H_1 an instantaneous release of energy, i.e. an explosion, occurs at a point, along a line, or over a plane. A shock wave starts to propagate through the gas which perturbs the initial magnetic field, initial pressure p_1 and the initial density ρ_1.

We denote by E_0 the energy released during the explosion. If it is assumed that the ensuing motion is described by the system of equations of magnetohydrodynamics (MHD), we arrive at the problem of integrating the system of MHD equations with boundary conditions at the center of explosion, on the front of the blast wave, and at infinity.

Since the magnetic field introduces anisotropy into the motion, in contrast to the corresponding problem for the equations of gas dynamics, even for constant p_1 and ρ_1 in the spherical case the problem in its precise formulation does not reduce to a one-dimensional problem, while in the planar and cylindrical cases it is reducible to a one-dimensional problem only for special configurations of the magnetic field. The system of equations of MHD for the motions of a nonviscous perfect gas which does not conduct heat has the form (1.69)–(1.73). The electric field E and the current density j are computed from the formulas

$$\boldsymbol{E}=\frac{\boldsymbol{j}}{\sigma}-\frac{1}{c}\boldsymbol{v}\times\boldsymbol{H},\qquad \boldsymbol{j}=\frac{c}{4\pi}\operatorname{curl}\boldsymbol{H}, \tag{7.1}$$

where σ is the conductivity and c is the speed of light.

If the magnetic Reynolds numbers are high $(R_m \gg 1)$, then in the system of MHD equations one may neglect the dissipative terms; for small magnetic Reynolds numbers $(R_m < 1)$ for the motion of the gas one may neglect the variation of the initial electromagnetic field. For motions of the gas with finite electrical conductivity we shall assume that the conductivity of the gas depends on pressure and density in the following manner:

$$\sigma = \sigma_1 \rho^n p^m, \tag{7.2}$$

where σ_1, n and m are constants. Thus in solving problems with finite electrical conductivity there appear three additional parameters: σ_1, n and m. This also complicates the investigation.

We shall consider the case of infinite conductivity and gas motion with finite numbers R_m. The theory presented below was developed mainly by the author; we gave both the formulations of the basic problems and the methods of solving them. A number of questions were worked out jointly with V. P. Karlikov and E. V. Rjazanov.

§2. One-dimensional problems for the motions of an ideally conducting gas in which the velocity is perpendicular to the field vector

2.1. *Formulation of the problems.* We shall consider the formulation of problems for one-dimensional flows of an infinitely conducting gas. In this case it makes sense to study only cylindrical and plane point explosions. We assume that the vector H is directed perpendicular to the trajectories of the gas particles. In the cylindrical case the field vector H may, in particular, be directed along the axis of symmetry, along tangents to concentric circles with center on the symmetry axis, or in the general case, it may create a screw field with components H_φ and H_z (the magnetic force lines are the screw lines; see the diagram of gas motion in Figure 79). In the case of a plane explosion the flow may be one-dimensional even when the direction of the field vector is inclined to the plane of the explosion. We shall consider this case in the next section.

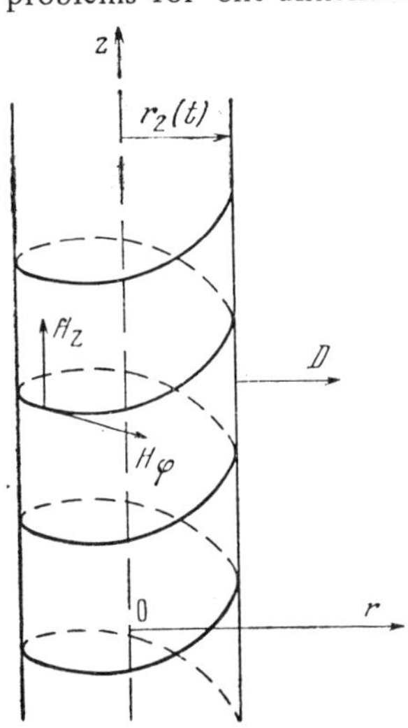

FIGURE 79. Scheme of the propagation of a cylindrical shock wave in a screw field

From the system of MHD equations one can obtain the equations for one-dimensional flows in the form

$$\begin{gathered}
\rho \frac{dv}{dt} = -\frac{\partial}{\partial r}(p+h) - \frac{2(\nu-1)}{r} h_\varphi, \\
\frac{d\rho}{dt} = -\rho\left(\frac{\partial v}{\partial r} + \frac{(\nu-1)v}{r}\right), \\
\frac{d}{dt}\left(\frac{p}{\rho^\gamma}\right) = 0, \quad \frac{d}{dt}\left(\frac{h_\varphi}{r^2\rho^2}\right) = 0, \quad \frac{d}{dt}\left(\frac{h_z}{\rho^2}\right) = 0, \\
\frac{d}{dt} = \frac{\partial}{\partial t} + v\frac{\partial}{\partial r}, \qquad h = h_z + (\nu-1)h_\varphi, \\
h_z = \frac{H_z^2}{8\pi}, \qquad h_\varphi = \frac{H_\varphi}{8\pi}.
\end{gathered} \tag{7.3}$$

We shall consider two types of shock waves: MHD waves and shock waves with a conductivity jump which ionize the gas. For MHD shock waves in a medium with infinite conductivity we have from (1.107)–(1.110)

$$\begin{gathered}
[\rho(v-D)] = 0, \qquad [\rho v(v-D) + p + h] = 0, \\
\left[(v-D)\left(\frac{p}{\gamma-1} + \frac{\rho v^2}{2} + h\right) + (p+h)v\right] = 0, \\
[h_\varphi(v-D)^2] = 0, \qquad [h_z(v-D)^2] = 0.
\end{gathered} \tag{7.4}$$

For shock waves with a conductivity jump we shall assume that H remains continuous in passing through the surface of discontinuity. This means that the magnetic viscosity in the shock layer is assumed to be greater than the other dissipative coefficients [1]. The discontinuities of the remaining quantities are subject to the gas-dynamical conditions

$$\begin{gathered}
[\rho(v-D)] = 0, \qquad [\rho v(v-D) + p] = 0, \\
\left[(v-D)\left(\frac{p}{\gamma-1} + \frac{1}{2}\rho v^2\right) + pv\right] = 0.
\end{gathered} \tag{7.5}$$

From the system (7.3) one can obtain the equation of the energy conservation law in the form

$$\frac{\partial}{\partial t}\left(\frac{\rho v^2}{2} + \frac{p}{\gamma-1} + \frac{H^2}{8\pi}\right) + r^{1-\nu}\frac{\partial}{\partial r}\left[r^{\nu-1}v\left(\frac{\rho v^2}{2} + \frac{\gamma p}{\gamma-1} + \frac{H^2}{4\pi}\right)\right] = 0. \tag{7.6}$$

We mention some well-known properties of the system (7.3).

PROPERTY 1. For $\gamma = 2$ the system (7.3) has the integral

$$p = \Phi_1(\xi) h_z, \tag{7.7}$$

where $\Phi_1(\xi)$ is an arbitrary function of the Lagrange coordinate ξ. This implies that if a solution of the system (7.3) for $\gamma = 2$ and $h_z = 0$ is known, then it is easy to find a solution of this system also for the case $\gamma = 2$ and $h_z \neq 0$. In particular, knowing a solution of the ordinary gas-dynamical equations for $\gamma = 2$, we obtain a solution of (7.3) for $h_\varphi = 0$ which contains an additional arbitrary function of ξ.

In the case of MHD shock waves and $\gamma = 2$ a substitution of the form $p^* = p + h_z$ reduces (7.4) to the form of conditions for $h_z = 0$ if the frozen-in condition for h_z is not considered.

PROPERTY 2. Since the system (7.3) contains no dimensional constants, in those problems for which the auxiliary conditions contain no more than one constant with dimensions independent of E_0 the condition for self-similarity of the motion will be satisfied. In this case it is possible to study the problem by reducing its equations to a system of ordinary differential equations.

PROPERTY 3. For $H_\varphi = 0$ the system of equations (7.3) and conditions (7.4) lead to the gas-dynamical conditions for gas motion with special thermodynamic properties (cf. also Chapter 1). Indeed, the system (7.3) and conditions (7.4) give the frozen-in integral $H_z = H_{z1}\rho/\rho_1$. If we introduce new thermodynamic quantities ρ^* and ϵ^* for the "mixture" of gas and field, we have the relations

$$p^*(S, \rho) = p + h_z = f(S)\rho^\gamma + \frac{1}{8\pi}\left(\frac{H_{z1}}{\rho_1}\right)^2 \rho^2, \tag{7.8}$$

$$\varepsilon^* = \frac{p}{\rho(\gamma-1)} + \frac{H^2}{8\pi\rho} = \frac{p^*}{\rho(\gamma-1)} + \frac{\gamma-2}{\gamma-1}\left(\frac{H_{z1}}{\rho_1}\right)^2 \frac{\rho}{8\pi}, \tag{7.9}$$

where S is the entropy and $f(S)$ is an arbitrary function of the entropy.

Formulas (7.8) and (7.9) define thermodynamic functions of the gas whose flow is considered in studying MHD problems. It also follows from Property 3 that for $\gamma = 2$ the problem reduces to the gas-dynamical problem with equations of state for an ideal gas in correspondence with Property 1. The properties of the equations and boundary conditions enumerated provide the foundation for alterations in the formulations of the PEP as compared with the case of gas dynamics. We proceed to concrete problems for equations (7.3)–(7.6).

2.2. *Constant initial magnetic field parallel to the front of the shock wave.* We consider the PEP with $H_1 = H_{z1}$ for the case of plane and cylindrical explosions. It follows from (7.9) that in this case even for a strong explosion ($p_1 = 0$) the problem is not self-similar. If the magnetic field is weak, i.e. $H_{z1}^2 < 8\pi p_1$, the initial distribution of the field H_z can be found from the frozen-in integral $H_z = \rho H_{z1}/\rho_1$.

To treat the problem of explosion for large times, including the effect of the field H_z and counterpressure, it is necessary to use numerical methods analogous to the methods developed for the problems of gas dynamics.

We shall consider here the method we published in [2] of treating the problem based on application of the method of integral relations. We take the initial equations in divergence form:

$$\frac{1}{\nu}\frac{\partial\rho}{\partial t}+\frac{\partial}{\partial\zeta}(\zeta\rho u)=0,\qquad \frac{1}{\nu}\frac{\partial\varepsilon}{\partial t}+\frac{\partial}{\partial\zeta}[\zeta u(\varepsilon+p^*)]=0,$$

$$\frac{1}{\nu}\frac{\partial p^{1/\gamma}}{\partial t}+\frac{\partial}{\partial\zeta}(\zeta u p^{1/\gamma})=0,\qquad \frac{\partial}{\partial t}\frac{\rho u}{r^{\nu-2}}+\nu\frac{\partial}{\partial\zeta}(\rho v^2+p^*)+\frac{\nu-1}{\zeta}\rho v^2=0,\qquad (7.10)$$

$$\frac{H_z}{H_{z\infty}}=\frac{\rho}{\rho_\infty},\quad \rho_\infty=\rho_1,\quad H_{z\infty}=H_{z1},\quad \varepsilon=\frac{\rho v^2}{2}+\frac{p}{\gamma-1}+\frac{H^2}{8\pi},\quad \zeta=r^\nu,\quad u=\frac{v}{r}.$$

We introduce new independent variables

$$h^*=\frac{p^*q}{p_\infty},\qquad G=\frac{H_z}{H_{z\infty}},\qquad \xi=\frac{\zeta}{\zeta_n},\qquad q=\frac{\gamma p_\infty}{\rho_\infty D^2},\qquad \psi=\left(\frac{q}{p_\infty}p\right)^{1/\gamma},$$

$$\varphi=\frac{ur_n}{D},\quad g=\frac{\rho}{\rho_\infty},\qquad e=\frac{\varepsilon q}{p_\infty}.\qquad (7.11)$$

In this section the index n will denote the quantities directly behind the wave front, and the index ∞, the initial parameters.

In the new variables, (7.10) can be written in the form

$$\begin{gathered}\frac{\partial g}{\partial q}+\mu\left[\frac{\partial}{\partial\xi}(g\varphi\xi)-\frac{\partial}{\partial\xi}(g\xi)+g\right]=0,\\ \frac{\partial e}{\partial q}+\mu\left[\frac{\partial}{\partial\xi}(e\varphi\xi)-\frac{\partial}{\partial\xi}(e\xi)+\frac{\partial}{\partial\xi}(h^*\varphi\xi)+e\right]-\frac{e}{q}=0,\\ \frac{\partial\psi}{\partial q}+\mu\left[\frac{\partial}{\partial\xi}(\psi\varphi\xi)-\frac{\partial}{\partial\xi}(\psi\xi)+\psi\right]-\frac{\psi}{\gamma q}=0,\qquad (7.12)\\ \frac{\partial m}{\partial q}+\frac{\mu}{\nu}\left[m+\nu\frac{\partial}{\partial\xi}\left(gf^2+\frac{h^*}{\gamma}-m\xi\right)+\frac{\nu-1}{\xi}gf^2\right]-\frac{m}{2q}=0,\qquad G=g.\end{gathered}$$

A relation between the dimensionless time $\tau=t/t^0$ and q follows from the kinematic condition $D=dr_n/dt$:

$$\tau=R_n\frac{\mu}{\nu}\sqrt{\frac{q}{\gamma}}.$$

Here and throughout this section the prime denotes differentiation with respect to q. In addition, we introduce the notation

$$\mu=\frac{\delta'}{\delta}+\frac{1}{q},\qquad m=\varphi g\xi^{(2-\nu)/\nu},\qquad h^*=(\psi^*)^\gamma,\qquad f=\frac{v}{D},\qquad R=\frac{r}{r^0},$$

$$\delta=\frac{\sigma_\nu R_n^\nu}{\nu q},\quad r^0=\left(\frac{E_0}{p_\infty}\right)^{1/\nu},\quad t^0=r^0\left(\frac{\rho_\infty}{p_\infty}\right)^{1/2},\quad \sigma_\nu=2(\nu-1)\pi+(\nu-2)(\nu-3).$$

We note that in (7.10) and (7.12) not all the equations are independent. One of the equations of these systems can be viewed as a consequence of the others by considering the relation between ϵ, v, ρ and H. The "extra" relations between the unknown functions can be used for additional control of the accuracy of the solution. Since we have the frozen-in integral $G=g$, to solve the problem it suffices to find only the three functions g, ψ and φ.

The conditions on the shock wave in the new variables take the form

$$g_n(\varphi_n - 1) = -1, \qquad g_n = G_n, \qquad m_n(\varphi_n - 1) + \frac{h_n^*}{\gamma} = \frac{1}{\gamma} q\left(1 + \frac{\gamma}{2}\beta\right),$$
$$\beta = \frac{H_\infty^2}{4\pi p_\infty}, \qquad (\varphi_n - 1)e_n + h_n^*\varphi_n = q\left(\frac{1}{\gamma-1} + \frac{\gamma}{2}\beta\right). \tag{7.13}$$

Thus we must find a solution of (7.12) with boundary conditions (7.13) and the symmetry condition at the center of the explosion.

Integrating (7.12) with respect to ξ from some $\xi_l = \xi_l(q)$ to $\xi_n = 1$ and noting (7.13), we obtain the system of integral relations

$$\mathcal{J}_{1l}' + g_l\xi_l' - \mu[1 + g_l\xi_l(\varphi_l - 1) - \mathcal{J}_{1l}] = 0,$$
$$\mathcal{J}_{2l}' + e_l\xi_l' - \mu\left[e_l\varphi_l\xi_l - e_l\xi_l + h_l^*\varphi_l\xi_l - \mathcal{J}_{2l} + q\left(\frac{1}{\gamma-1} + \frac{\gamma}{2}\beta\right)\right] - \frac{1}{q}\mathcal{J}_{2l} = 0,$$
$$\mathcal{J}_{3l}' + \psi_l\xi_l' - \mu\left[\psi_l\xi_l(\varphi_l - 1) - \mathcal{J}_{3l} + \frac{\psi_n}{g_n}\right] - \frac{1}{\gamma q}\mathcal{J}_{3l} = 0, \tag{7.14}$$
$$\mathcal{J}_{4l}' + m_l\xi_l' + \mu\left[\frac{1}{\nu}\mathcal{J}_{4l} + m_l\xi_l - \frac{h_l^*}{\gamma} - g_l f_l^2 + \frac{\nu-1}{\nu}\mathcal{J}_{5l} + \right.$$
$$\left. + \frac{q}{\gamma}\left(1 + \frac{\gamma}{2}\beta\right)\right] - \frac{1}{2q}\mathcal{J}_{4l} = 0,$$

where

$$e_l = \frac{\gamma}{2} g_l\varphi_l^2\xi_l^{2/\nu} + \frac{\psi_l^\gamma}{\gamma-1} + q\frac{\gamma\beta}{2} g_l^2, \qquad f_l = \varphi_l\xi_l^{1/\nu}, \qquad m_l = g_l\varphi_l\xi_l^{(2-\nu)/\nu},$$
$$h_l^* = \psi_l^\gamma + \gamma q\frac{\beta}{2} g^2. \tag{7.14a}$$

In the system (7.14) we have introduced the following notation:

$$\mathcal{J}_{1l} = \int_{\xi_l}^1 g\,d\xi, \qquad \mathcal{J}_{2l} = \int_{\xi_l}^1 e\,d\xi, \qquad \mathcal{J}_{3l} = \int_{\xi_l}^1 \psi\,d\xi, \qquad \mathcal{J}_{4l} = \int_{\xi_l}^1 m\,d\xi, \qquad \mathcal{J}_{5l} = \int_{\xi_l}^1 \frac{gf^2}{\xi}\,d\xi.$$

The system (7.14) together with (7.14a) may serve as basis for constructing computing schemes for the numerical solution of the problem. If $\beta < 1$ the treatment of the problem for any γ is analogous to the gas-dynamical case. For small q the effect of the field is small, and the gas-dynamical functions obtained numerically in Chapter 3 can be taken for initial data. In analogy to the results of Chapter 4, in the region of integration $0 \leqslant \xi \leqslant 1$ we introduce the central interval $\xi_0(q)$, having fixed the Lagrange coordinate on the basis of the initial data. As in the corresponding gas-dynamical problem (cf. Chapter 4), inside the central interval we perform the computation according to special asymptotic formulas.

We determine the solution in the region between the boundary of the central interval $\xi = \xi_0(q)$ and the shock wave $\xi = \xi_n = 1$ by the method of integral relations. At the nth approximation we decompose this region into n strips by the intervening lines:

$$\xi_i(q) = \frac{n-i}{n}\xi_0 + \frac{i}{n} \qquad (i = 1,\ 2,\ \ldots,\ n-1).$$

If we approximate the integrands by Lagrange interpolation polynomials with interpolation nodes on the lines $\xi = \xi_k$ $(k = 0, 1, \cdots, n)$, we can obtain an approximating system of ordinary differential equations.

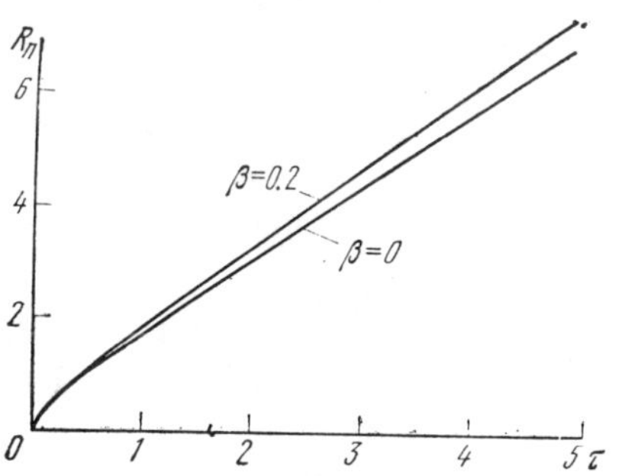

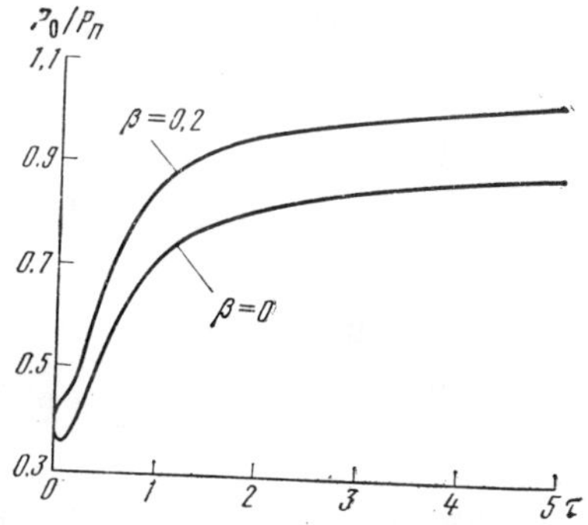

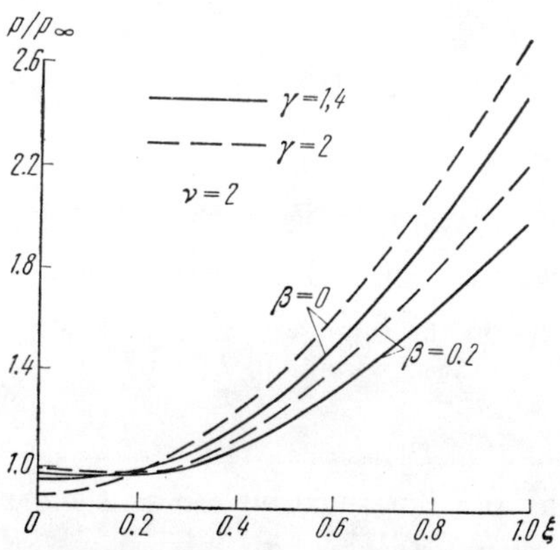

FIGURE 80. Law of motion of a cylindrical shock wave in a parallel field

FIGURE 81. Variation of the relative pressure at the center of explosion ($\nu = 2$; $\gamma = 1.4$)

FIGURE 82. Distribution of pressures with respect to ξ for two values of γ

This approximating system will differ from the analogous system of §3 of Chapter 4 only in the coefficients of the equations obtained from the second and last integral relations of (7.14). For the approximating system of equations we will have a Cauchy problem whose numerical solution enables us to determine all the unknown functions. General control of accuracy can be obtained by satisfying the integral laws of energy and mass conservation. As examples, computations of the problem were carried out for the cases $n = 4$, $\nu = 2$, $\gamma = 1.4$ or $\gamma = 5/3$, and various $\beta < 1$. For the plane case the computation was done with $\gamma = 1.4$ and $\beta = 0.2$. Figure 80 shows the computed dependence $R_n(\tau)$, $\nu = 2$. Here for comparison the corresponding dependence for $\beta = 0$ is given. Figure 81 shows the variation of the function p_0/p_n with respect to τ $(p_0 = p(0, t))$ for $\gamma = 1.4$, $\nu = 2$, $\beta = 0$ and $\beta = 0.2$. The dependence $P(\xi) = p/p_\infty$ for $q = 0.43$, $\beta = 0.2$ and $\beta = 0$ is shown in Figure 82. The scheme of the method of integral relations can also be used for values $\beta > 1$. In this case, however, it is desirable to solve the linearized problem of explosion with counterpressure and magnetic field exactly. Its approximate solution is considered in [3].

Since for $\gamma = 2$ the problem reduces to the corresponding gas-dynamical problem, for arbitrary β in order to obtain the solution in this case it suffices to use the results of the computation of the gas-dynamical PEP by the method considered in Chapter 4.

If we denote by g_2, P_2 and f_2 $(P = p/p_\infty)$ the unknown functions of the problem of gas dynamics, and by P, g and f the corresponding functions of the magnetohydrodynamical problem, we have

$$P^* = \frac{p^*}{p^*_\infty} = P_2, \quad g = G = g_2, \quad f = f_2, \quad P = P_2(1+\beta) - \beta g^2, \quad h_z = \Phi(\xi)\rho^2.$$

We shall give special consideration to the case of a strong wave.

Neglecting P^*_∞ in the conditions on the shock wave (7.4), we obtain

$$v_n = \frac{2}{3}D, \qquad \rho_n = 3\rho_\infty, \qquad p^*_n = \frac{2}{3}\rho_\infty D^2, \qquad h_{zn} = 9h_{z\infty}.$$

We shall assume that $h_{z\infty} = \text{const}$. The solution for $p^*(r, t), \rho(r, t)$ and $v(r, t)$ is known (cf. Chapter 2).

We shall find the arbitrary function $\Phi(\xi)$ entering in $h_z(r, t)$. From the conditions on the jump we obtain

$$\frac{h_{zn}}{\rho_n^2} = \frac{h_{z\infty}}{\rho_\infty^2}, \qquad \Phi(\xi) = \frac{h_{z\infty}}{\rho_\infty^2}.$$

From this it follows that we can satisfy the MHD equations and the conditions on the shock wave (7.4) if we take

$$h_z = \frac{h_{z\infty}}{\rho_\infty^2}\rho^2.$$

Recalling the form of the solution of the analogous problem in ordinary gas dynamics, we have

$$
\begin{aligned}
&r=r_n\mu^{-\delta}\left[2\left(\frac{3}{2}-\mu\right)\right]^{-\delta\nu/2}\left[4\left(\mu-\frac{3}{4}\right)\right]^{\delta/2}, \qquad v=v_n\lambda\mu,\\
&\rho=\rho_n\left[4\left(\mu-\frac{3}{4}\right)\right]^{\nu\delta/2}\left[2\left(\frac{3}{2}-\mu\right)^{(\nu-2)/(\nu+2)}\exp\left(-3\nu\delta\frac{1-\mu}{\frac{3}{2}-\mu}\right)\right],\\
&p=p_n\mu^{\delta\nu}\left[2\left(\frac{3}{2}-\mu\right)^{\delta(\nu-2)}\exp\left(-3\nu\delta\frac{1-\mu}{\frac{3}{2}-\mu}\right)\right]-h_z,\\
&h_z=\frac{h_{z\infty}}{\rho_\infty^2}\rho^2, \qquad \frac{3}{4}\leqslant\mu\leqslant 1.
\end{aligned}
\tag{7.15}
$$

Here r_n is the coordinate of the shock wave. The dependence $r_n(t)$ coincides with the gas-dynamical dependence.

For a strong explosion in a nonconducting medium with the conductivity jump $\sigma_\infty = 0$, $\sigma_n = \infty$ from the boundary condition $h(r_n, t) = h_{z\infty}$ we find in analogy to the foregoing

$$\Phi(\xi)=\frac{h_{z\infty}}{9\rho_\infty^2}.$$

The solution of the problem in this case is given by (7.15) if it is assumed that $h_z = h_{z\infty}\rho^2/9\rho_\infty^2$. It follows from (7.15) that if high temperatures are preserved at the center of explosion, the pressure in a neighborhood of the front of the shock wave is lower than the corresponding pressure during explosion with no magnetic field.

It also follows from the formula for h_z that the magnetic pressure during explosion in an infinitely conducting medium is nine times greater than the magnetic pressure during explosion in a nonconducting medium. We mention that for an explosion with a conductivity jump on the shock wave the effect of the change in the initial field H_∞ caused by the radiated electromagnetic wave was not considered [1]. This effect may be appreciable if the characteristic velocities of the motion of gas particles are high and $H_{z\infty}$ is not small.

The solution (7.15) was first published by us in [4], [5]. This solution was subsequently obtained by A. Sakurai [6].

To reckon with the counterpressure and magnetic field when $\gamma = 2$ we make use of the numerical solution of the problem considered in Chapter 4. Graphs of the pressure are given in Figures 82 and 83. From these graphs it is evident that for $q > 0.5$ the pressure distribution differs strongly from the gas-dynamical pressure especially near the center. This difference is not only of quantitative character but of qualitative character as well: there are substantial pressure gradients in a neighborhood of the center of the explosion.

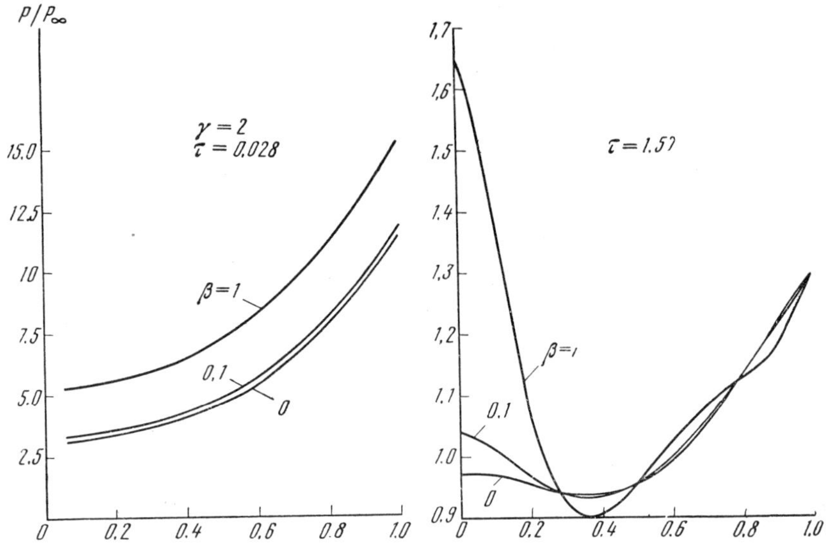

FIGURE 83. Pressure distribution for various parameters β for fixed time, $\gamma = 2$

Computations indicate that the accuracy of calculating the problem in question drops with increasing q. As in problems of gas dynamics, in order to compute the parameters of blast waves at large distances from the axis of the explosion (for $q \longrightarrow q_\infty = 1/(1+\beta)$) it is expedient to use the asymptotic laws for the damping of shock waves to improve the accuracy [2], [7]. We shall not go into this in detail here.

2.3. *The self-similar problem for an annular initial field.* If the magnetic field has only the component h_φ, with $h_{\varphi 1} = Br^{-2}$ and $\rho_1 = Ar^{-\omega}$, then the problem of a cylindrical strong explosion in an infinitely conducting gas is self-similar (here we again use the index 1). Indeed, since the dimension of the constant B coincides with the dimension of the energy E_0', the problem has only one additional constant A with dimension independent of E_0'. This, according to Property 2, implies self-similarity. The law of variation of the field $h_{\varphi 1} = Br^{-2}$ shows that the initial field is formed by the flow of a constant linear current for which a concentric cylindrical conductor of very large radius serves as the return circuit. For the system of equations (7.3) this problem was first studied in [5], [8], [9]. We shall consider the method of solving it and the results for the case $\omega = 0$.

We introduce dimensionless variables by the formulas

$$v = \frac{r}{t} V(\lambda), \qquad p = \rho_1 \frac{r^2}{t^2} P(\lambda), \qquad \rho = \rho_1 R(\lambda), \qquad h_\varphi = \rho_1 \frac{r^2}{t^2} G(\lambda),$$

$$r_2 = \left(\frac{E}{\rho_1}\right)^{1/4} t^{1/2}, \quad \lambda = \frac{r}{r_2},$$

where $E = \alpha E_0'$, α is a constant remaining to be determined, E_0' is the finite part of the energy E_0, and $r_2(t)$ gives the law of motion of the shock wave. The corresponding system of four ordinary differential equations for self-similar motions has the form

$$\begin{gathered} R\lambda(V-0{,}5)V' + \lambda G' + \lambda P' = R(V-V^2) - 2(P+G) - 2G, \\ \lambda\Big[(V-\delta)\frac{R'}{R} + V'\Big] = -2V, \\ \lambda\Big[(V-\delta)\frac{P'}{P} + \gamma V'\Big] = 2 - 2(1+\gamma)V, \\ \lambda\Big[(V-\delta)\frac{G'}{G} + 2V'\Big] = 2 - 4V. \end{gathered} \tag{7.16}$$

The boundary conditions of the problem follow from the necessity of satisfying the conservation laws on the front of the shock wave (7.4) and the symmetry condition $v(0, t) = 0$ on the axis of the explosion. The system of differential equations has three first algebraic integrals [5], [8]: the adiabatic, energy, and frozen-in integrals (cf. §6, Chapter 1):

$$p = R^{\gamma-1}(V-0.5)^{-1}\lambda^{-4}\kappa_1, \qquad G = (V-\delta)^{-2}\lambda^{-4}\kappa_2,$$
$$(p+G)V + (V-0.5)\Big(0.5RV^2 + \frac{1}{\gamma-1}P + G\Big) = \lambda^{-4}\kappa_4.$$

The arbitrary constants contained in the asymptotic expression for the integrals are found from the boundary conditions $\kappa_2 = 0.25G_1$ and $\kappa_4 = -0.5G_1$ on the front of the shock wave. By means of these integrals the problem is reduced to the investigation of a single ordinary differential equation which can be obtained from (7.16) and written as follows:

$$\frac{d\ln y}{dV} = \frac{2(\gamma V-1)\Big[\psi_1(y_1V) + \frac{\gamma-2}{\gamma} - y(V-0{,}5)^2\Big] + (\gamma-1)\psi_2}{(V-0.5)\psi_2}, \tag{7. 17}$$

where

$$\psi_1 = (V-0.5)V^{-2}\Big[\frac{1}{\gamma}\Big(2V - \frac{1}{\gamma-1}\Big) - 0.5yV^2\Big],$$

$$\psi_2 = V(V-1)(V-0.5)y + \frac{1}{\gamma} - 2V, \quad y = \frac{R}{\gamma P}.$$

The basic difficulty of the problem consists in integrating (7.17), since from the unknown function $y(V)$ all the unknown functions can be found by first integrals and quadratures. If we denote by the index 1 the quantities characterizing the unperturbed state of the gas and by the index 2 the quantities directly behind the front of the shock wave, then, using the conditions on the shock wave, we find the dependence of y_2 on V_2 and γ:

$$y_2^{-1} = \gamma(\gamma-1)\Big[0.5V_2^2 + \Big(1 + \frac{0.5}{V_2-0.5} + 2V_2\Big)G_1\Big],$$

$$G_1 = \frac{0.5V_2\left[(0.5 - V_2)\frac{1}{\gamma - 1} - 0.5V_2\right]}{\frac{1}{\gamma - 1}(V_2 - 0.5) + \frac{\gamma - 2}{4(\gamma - 1)}\frac{1}{V_2 - 0.5} + V_2 + 0.5}.$$

From the condition $D_2 \geqslant H^2_{\varphi 1}/4\pi\rho_1$ and the equation for $y_2(V_2)$ it follows that

$$\frac{(\gamma + 1)^2}{0.5\gamma(\gamma - 1)} < y_2 < \infty, \quad \frac{1}{\gamma + 1} > V_2 > 0 \quad \left(G_1 < \frac{1}{8}\right).$$

We point out that in the self-similar gas-dynamical PEP the relation between V_2 and y_2 for fixed γ was represented in the V, y plane as a point, while for the problem being considered the presence of the magnetic field leads to the dependence of $y_2(V_2)$ given by relations on the shock wave. A qualitative analysis was made of the behavior of the integral curves for (7.17). Investigation of (7.17) indicates that to the center of symmetry of the flow (the axis of the explosion) there corresponds the point $y = 0$, $V = 0$. Qualitative analysis of the behavior of the solutions of (7.17) also showed that to the solution of the strong explosion problem in the V, y plane there correspond integral curves beginning at points of the curve $y_2(V_2)$ and passing to the singular point (0, 0) tangent to the axis of the abscissa. (The presence of a set of curves passing to the point (0, 0) along the axis of the abscissa can be proved on the basis of Perron's criterion [10].) For small values of V, (7.17) has the asymptotic solution [5]

$$y = c_1 V^{-2(\gamma-2)/(\gamma-1)}[V - 0.5]^{-(3-2\gamma)/(\gamma-1)}[1 - 2\gamma V]^{-\gamma/(\gamma-1)} \exp\left(\frac{-2}{(\gamma - 1)V}\right), \quad (7.18)$$

where c_1 is an arbitrary constant. Examples of integral curves $y(V)$ found numerically are shown in Figure 84. The system of self-similar equations of the problem was integrated numerically for a wide range of values of G_1 and a number of values of γ. The computation near the center of the explosion was carried out on the basis of special asymptotic formulas following from (7.18). The corresponding computations for the values $\gamma = 1.4$ and $\gamma = 5/3$ were made subsequently in [11]. Some results of the computations for $\gamma = 5/3$ and $\gamma = 2$ are shown in Figure 85 for the value $V_2 = 0.3$. We mention the following differences in the solution found from the corresponding solution of the gas-dynamical problem considered in Chapter 2.

1) The intensity of the shock wave is less than in the gas-dynamical case.

2) $H_\varphi \neq 0$ near the center $H_\varphi \sim 1/r$, i.e. a constant concentrated current flows along the axis of symmetry.

3) The temperature in the flow region decreases when the effect of the magnetic field is considered.

4) The formula for the integral energy balance has a more complicated appearance.

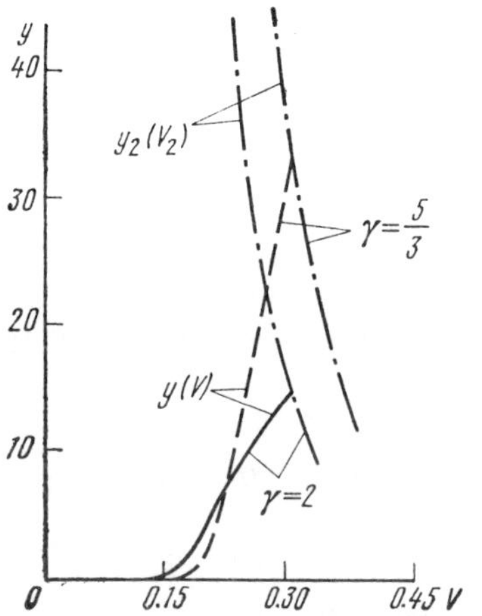

FIGURE 84. Examples of integral curves in the y, V plane

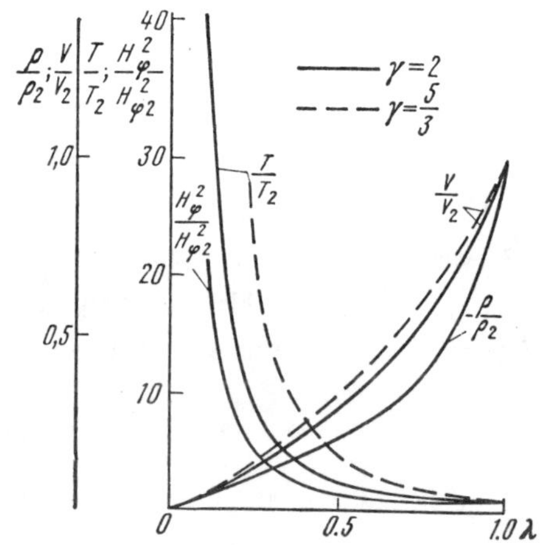

FIGURE 85. Distribution of relative densities, velocities, temperatures, and magnetic fields for strong cylindrical explosion in an annular field

We shall discuss the integral law of energy conservation in more detail. Analysis of the solution of the problem shows that the total energy of the moving gas remains infinite if the initial (infinite energy) is subtracted from it, and the corresponding integral diverges as $\lambda \longrightarrow 0$ like $\ln|\ln \lambda|$. This points up the fact that in order to realize the solution in question we had to allow the release of infinite energy on the axis $r = 0$ at the initial moment of time. A possible means of defining this energy is given in [11].

We take the energy balance in the form

$$E_0 = 2\pi \int\limits_0^{r_2} \left(\frac{\rho v^2}{2} + \frac{p}{\gamma - 1} + \frac{H^2}{8\pi} - \frac{H_1^2}{8\pi}\right) r dr - A_e.$$

Here A_e is the supplementary (infinite) part of the energy evolved at $t = 0$. The quantity A_e can be defined as the electromagnetic energy introduced at the initial moment. For it we can write [11]

$$A_e = \frac{I_0}{c} \Delta\Phi_e,$$

where $\Delta\Phi_e$ is the difference in the flux of the induction field computed per unit length at the initial moment and the moment t and I_0 is the concentric current flowing along the axis $r = 0$.

Considering the dependence of $\Delta\Phi_e$ on H_φ and passing to dimensionless formulas, we find

$$\frac{\alpha}{2\pi}=\int_0^1\left[\frac{1}{2}RV^2+\frac{P}{\gamma-1}+G-G_1\left(\frac{2V}{0.5-V}+1\right)\lambda^{-4}\right]\lambda^3 d\lambda. \qquad (7.19)$$

It can be shown that the quantity α defined by (7.19) will be finite. Specific values of it for various γ and G_1 (or V_2) were obtained numerically. Thus for $V_2 = 0.3$ we have $\alpha = 0.07$ $(\gamma = 5/3)$ and $\alpha = 0.05$ $(\gamma = 2)$.

For $\gamma = 2$ the problem is easily generalized to the case of a screw field with H_{z1} = const. For this it suffices to take

$$v=v_\varphi, \qquad p=p_\varphi-\frac{H_z^2}{8\pi}, \qquad \rho=\rho_\varphi, \qquad H=H_\varphi, \qquad H_z=H_{z1}\frac{\rho}{\rho_1},$$

where the quantities with the index φ correspond to the problem considered with $H_z = 0$ and $\gamma = 2$. This follows from the enumerated properties of the differential equations of MHD and the boundary conditions under the hypothesis that $p_2 + (H_{z2}^2/8\pi) \gg p_1 + (H_{z1}^2/8\pi)$.

2.4. *The self-similar problem for the equations of magnetohydrodynamics of a rarefied plasma.* The problem of strong cylindrical explosion in a quiescent medium considered in the preceding section can also be investigated for the MHD equations of a rarefied plasma (the Chew-Goldberger-Low equations) presented in §3 of Chapter 1. In the region outside the zone of the perturbed motion we neglect the components $p_{\perp 1}$ and $p_{\parallel 1}$ of the stress tensor. For simplicity we assume the initial density to be constant, $B_{\varphi 1} \sim 1/r$ and $B_{z1} = 0$. From the system (1.81)–(1.85) we have for the case of cylindrical symmetry

$$\rho\frac{dv}{dt}=-\frac{\partial}{\partial r}(p_\perp+h)+\frac{p_\parallel-p_\perp}{r}-\frac{2h}{r},$$

$$\frac{d\rho}{dt}=-\rho\left(\frac{\partial v}{\partial r}+\frac{v}{r}\right), \qquad \frac{d}{dt}\frac{h}{r^2\rho^2}=0, \qquad h=\frac{H_\varphi^2}{8\pi},$$

$$\frac{d}{dt}\frac{p_\perp}{h^{1/2}\rho}=0, \qquad \frac{d}{dt}\frac{p_\parallel h}{\rho^3}=0. \qquad (7.20)$$

If it is supposed that the flow is accompanied by the appearance of a shock wave, the following conditions are to be satisfied on the shock wave:

$$\rho_2(v_2-D)=-\rho_1 D, \qquad \rho_2 v_2(v_2-D)+p_{\perp 2}+h_2=h_1,$$

$$(v_2-D)\left(p_{\perp 2}+0.5p_{\parallel 2}+\frac{1}{2}\rho_2 v_2^2+h_2\right)+(p_{\perp 2}+h_2)v_2=-Dh_1,$$

$$h_2(v_2-D)^2=h_1 D^2, \qquad p_{\perp 2}=0.$$

These conditions follow from (1.119)–(1.123). This problem is self-similar. In the dimensionless variables $\lambda, V, P_\parallel, P_\perp, R$ and G we obtain from (7.20) a self-similar system of five equations:

$$R\lambda(V-0.5)V'+\lambda P'_{\perp}+\lambda G'=R(V-V^2)-3P_{\perp}-4G+P_{\parallel},$$

$$\lambda\left[(V-\delta)\frac{R'}{R}+V'\right]=-2V,\ \lambda\left[(V-\delta)\frac{P'_{\perp}}{P_{\perp}}+2V'\right]=2-5V, \qquad (7.21)$$

$$\lambda\left[(V-\delta)\frac{P'_{\perp}}{P_{\parallel}}+V\right]=2-6V,\ \lambda\left[(V-\delta)\frac{G'}{G}+2V'\right]=2-4V.$$

The system (7.21) has four algebraic integrals (cf. §4, Chapter 1): the energy integral (1.154), the frozen-in integral (1.149), which we write in the form

$$G=(V-0.5)^{-2}\lambda^{-4}\kappa_2, \qquad (7.22)$$

and the integrals of the adiabatic invariants (1.151) and (1.152), which take the form

$$P_{\perp}=G^{1/2}R^{1/2}(V-0.5)^{-1/2}\lambda^{-2}\kappa_3, \qquad (7.23)$$

$$P_{\parallel}=G^{-1}R(V-\delta)^{-2}\lambda^{-8}\kappa_4. \qquad (7.24)$$

The presence of these integrals enables us to reduce the system (7.21) for any constants $\kappa_2, \kappa_3, \kappa_4, \kappa_7$ to a single differential equation in the variables R and V of the form

$$\frac{dR}{dV}=f(R,\ V,\ \kappa_7,\kappa_1,\kappa_3,\kappa_4).$$

However, it is more expedient to reduce the system (7.21) to a single equation of another type in the V, λ plane. From the boundary condition $P_{\perp 2}=0$ and the integral (7.23) it follows that $\kappa_3=0$, i.e. $p_{\perp}=0$ everywhere in the flow. The conditions on the shock wave in dimensionless variables assume the form

$$R_2(V_2-0.5)=-0.5,\qquad R_2(V_2-0.5)V_2+G_2=G_1,$$

$$G_2V_2+(V_2-0.5)\left(\frac{1}{2}R_2V_2^2+0.5P_{\parallel}+G_2\right)=-0.5G_1, \qquad (7.25)$$

$$G_2(V_2-0.5)^2=G_1(0.5)^2,$$

where $G_1=B/\alpha\epsilon_0'$ and ϵ_0' is the specific energy.

From conditions (7.25) we find that $\kappa_7=-0.5G_1$ and $\kappa_2=(0.5)^2G_1$; the constant κ_4 is found from (7.24) and (7.25). Using the algebraic integrals, we obtain from (7.21)

$$\lambda\frac{dV}{d\lambda}=\frac{V^2\left(V-V^2+\frac{\kappa_4}{\kappa_2}\lambda^{-4}\right)}{V^2(V-0.5)+0.5\left(V^2+\frac{\kappa_4}{\kappa_2}\lambda^{-4}\right)}. \qquad (7.26)$$

Once (7.26) has been integrated, the dependence $R(\lambda)$ is found from the continuity equation, and $G(\lambda)$ and $P_{\parallel}(\lambda)$ are found from (7.22) and (7.24). Near the center of symmetry, i.e. for small λ, neglecting terms of order λ^4V^2 and higher on the right side of (7.26), we find the asymptotic formula $V=-c/\ln\lambda$ ($c=$ const). Corresponding asymptotic formulas can be obtained for the other functions as well.

The energy parameter α contained in the relation $r_2(\alpha\epsilon_0'/\rho_1)^{1/4}t^{1/2}$ is found from the integral law of energy conservation

$$\alpha = 2\pi\int_0^1 (0.5RV^2 + 0.5P_\parallel + G - G_1\lambda^{-4})\lambda^3 d\lambda - A_0,$$

where, as before, A_0 is the (dimensionless) work on the field of currents sustaining the current I_0 at the center of explosion (cf. the preceding subsection). We mention that if instead of the continuity condition for $p_\perp$ we had taken the condition $[p_\parallel] = 0$, we would have arrived at the conclusion that $p_\parallel = 0$ everywhere, and the problem reduces to the corresponding problem for the MHD equations with $\gamma = 2$.

If initially $p_{\parallel 1}$ and $p_{\perp 1}$ are nonzero, the problem ceases to be self-similar. Approximate consideration of the counterpressures $p_{\parallel 1}$ and $p_{\perp 1}$ can be carried out by the method of linearization considered in Chapter 3 for the equations of gas dynamics.

§3. Plane explosion with an inclined magnetic field

Suppose that in an unbounded, ideally conducting gas there is an instantaneous release of energy over a plane, i.e. a plane explosion occurs. We denote the energy released per unit area by E_0. We assume that in the initial state the gas was at rest; the pressure p_∞ and initial density ρ_∞ are constant. The initial magnetic field H_∞ is constant in magnitude and directed at some angle α_0 to the plane of the explosion. It may be assumed with no loss of generality that the field has components H_x and H_y, where the y axis is parallel to the plane of the explosion and the x axis is perpendicular to it. From the symmetry of the problem it follows that it is one-dimensional, and all the initial characteristics of the flow depend only on the time t and the coordinate x. The plane shock wave which arises propagates in the direction of the x axis. Behind the front of the shock wave a nonstationary flow arises; to describe it we shall use the equations for a nonviscous perfect gas which does not conduct heat. For the flow in question we have the system of equations

$$\frac{\partial\rho v_x}{\partial t} = -\frac{\partial}{\partial x}(\rho v_*^2 + p^*), \qquad \frac{\partial\rho v_y}{\partial t} = -\frac{\partial}{\partial x}\left(\rho v_y v_x - \frac{H_x}{4\pi}H_y\right),$$

$$\frac{\partial\rho}{\partial t} = -\frac{\partial}{\partial x}(\rho v_*), \qquad \frac{\partial p^{1/\gamma}}{\partial t} = -\frac{\partial}{\partial x}(v_x p^{1/\gamma}), \tag{7.27}$$

$$\frac{\partial H_y}{\partial t} = -\frac{\partial}{\partial x}(v_x H_y - H_x v_y), \quad H_x = H_{x\infty} = \text{const}.$$

Here v_x and v_y are the velocity components of the gas, and $p^* = p + H_y^2/8\pi$.

From (7.27) and the thermodynamic relations the energy equation can be obtained in the form

$$\frac{\partial \varepsilon^*}{\partial t} = -\frac{\partial}{\partial x}\left[v_x(\varepsilon^* + p^*) - \frac{1}{4\pi} H_x v_y H_y\right], \quad \varepsilon^* = \frac{1}{2}\rho v^2 + \frac{p}{\gamma - 1} + \frac{H_y^2}{8\pi}.$$

The conditions on the shock wave (1.107)–(1.111) take the form

$$(v_{xn} - D) H_{yn} - H_x v_{yn} = -DH_{y\infty}, \quad -D\rho_\infty v_{yn} - \frac{1}{4\pi} H_x H_{yn} = -\frac{1}{4\pi} H_x H_{y\infty},$$

$$(v_{xn} - D)\rho_n = -D\rho_\infty, \quad -D\rho_\infty v_{xn} + p_n^* = p_\infty^*, \tag{7.28}$$

$$(v_{xn} - D)\varepsilon_n^* + v_{xn} p^* - \frac{H_x}{4\pi} H_{yn} v_{yn} = -D\left(\frac{p_\infty}{\gamma - 1} + \frac{H_{y\infty}^2}{8\pi}\right), \qquad D = \frac{dx_n}{dt}.$$

Here the index n denotes parameters of the front of the shock wave, and D is the speed of the shock wave. In addition to (7.28), we have a condition at the center of symmetry of the flow:

$$v_x(0, t) = 0. \tag{7.29}$$

To solve the problem it is necessary to find the functions v_x, v_y, p, ρ, H_y and $x_n(t)$ which satisfy (7.27) and the boundary conditions (7.28), (7.29).

For $H_x = 0$ the magnetic field is parallel to the wave front. The corresponding problem here was considered in §2. The solution of the problem for $H_x \neq 0$ is complicated by the presence of a new unknown function v_y and by an increase in the number of differential equations. For its complete solution it is necessary to apply numerical and special approximation methods.

We shall mention certain general properties of the solution of the problem [1]. The system (7.27) and the condition (7.28) are invariant with respect to the transformation $v_y' = -v_y$, $v_x' = -v_x$, $x' = -x$. This implies, in particular, that the function v_y is odd in x, and under the condition of continuity of the solution in a neighborhood of $x = 0$ we have $v_y(0, t) = 0$. Thus at $x = 0$ the velocity of the gas is equal to zero.

We shall consider further the question of the law of damping of the shock wave at large distances from the site of the explosion. As the shock wave propagates away from the center of the explosion it will die out, and the general zone of perturbed motion will increase. Between the center of the explosion and the leading front of the wave there will be a flow of gas accompanied by a complex picture of interaction of MHD compression and rarefaction waves. At large distances the shock wave becomes weak. Since the entropy changes little in passing through a weak shock wave, in some neighborhood of the wave front the flow may be considered isentropic. By analogy with gas dynamics it can be supposed that in some neighborhood behind the front of the MHD shock wave the flow is an isentropic Riemann wave [13]. Applying then, for example, the methods of obtaining the asymptotic damping laws for shock waves considered in [7], [14], we can find an asymptotic law for the change in v_{xn} depending on x_n. In analogy to gas dynamics or the case $H_x = 0$ we have [7]

$$v_{xn} = Cx_n^{-1/2}, \tag{7.30}$$

where C is some constant.

Using the relation (7.28), one can establish the corresponding asymptotic formulas for the other unknown functions as well.

We shall consider several methods of solving the problem. For numerical solution we can use the method of characteristics. By this method it would be possible to give a sufficiently precise description of the system of MHD waves in the flow region at least up until the time that secondary shock waves are formed in the flow. However, this method is rather laborious and unsuitable for calculation on computers. We indicate below some simpler methods of computing.

For $H_{x\infty} < h_{y\infty}$, i.e. when the angle α_0 is small, neglecting terms of order $v_y H_x$ as compared with $H_y v_x$ in (7.27) and (7.28), we find that the problem of determining v_x, p and ρ splits off from the problem of finding v_y. Once the problem of plane explosion for the case $H_x = 0$, $\alpha_0 = 0$ has been solved, we can use the results of its solution to find v_y by integrating the second equation of (7.27). In §2 the method of integral relations was applied to compute H_y, v_x, p and ρ. If the results obtained by this method are used, it is expedient to apply it also to the second equation (7.27) to determine v_y in the case of small α_0.

To treat the problem for arbitrary α_0 either the method of lines or the method of integral relations analogous to those methods developed above is suitable. Here by the method of lines we mean that variant of the method of integral relations in which the integrands containing the unknown functions are approximated by linear functions between the points partitioning the spatial coordinate into strips, i.e. the integrals over space are actually computed by the trapezoidal method. The advantage of using the method of lines is that the system of approximating ordinary differential equations has a simple structure.

The quantities v_n, p_n, ρ_n, H_{yn} and the law of motion of the shock wave $x_n(t)$ are important characteristics of the flow. We shall consider an approximate method of determining these quantities independent of the computation of the entire flow field.

Since at the initial times a finite magnetic field has a weak effect on the gas flow, for small x_n the function $v_n(x_n)$ is close to the gas-dynamical function and has the form

$$V_{xn} = \frac{4}{3}\frac{1}{\gamma+1}\frac{1}{\sqrt{\gamma \alpha R_n}}, \tag{7.31}$$

where $V_{xn} = v_{xn}/a_\infty$, $R_n = x_n p_\infty / E_0$, $a_\infty^2 = \gamma p_\infty / \rho_\infty$ and $\alpha = \alpha(\gamma)$.

The dependence (7.31) is analogous to (7.30), and it is natural to assume (cf. Chapter 4) that $V_{xn}(R_n)$ can be approximated by (7.31) over an arbitrary range of values of x_n. Using (7.28) and (7.31), we can determine all the unknown dimensionless quantities on the front of the wave as a function of R_n or the dimensionless time $\tau = E_0^{-1}\rho_\infty^{-1/2}p_\infty^{3/2}t$. We mention that investigation of this method in application to the problem with $H_x = 0$ (and to the problem of gas dynamics (cf. Chapter 4)) indicated that it has sufficiently high accuracy.

Results of the computations are shown in Figures 86–88. In this problem, in addition to γ, there are two dimensionless independent parameters: $\beta^2 = H_x^2/4\pi\gamma p_\infty$ and $G_\infty^2 = H_{y\infty}^2/4\pi\gamma p_\infty$.

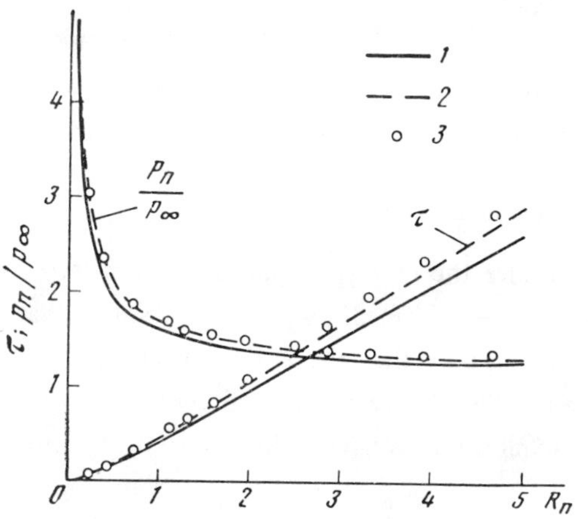

FIGURE 86. Dependence of the dimensionless time τ and pressure on the front of a plane shock wave on the coordinate of the front R_n: 1 corresponds to $\beta = 1$, 2 to $\beta = 0.5$, and 3 to $\beta = 0$.

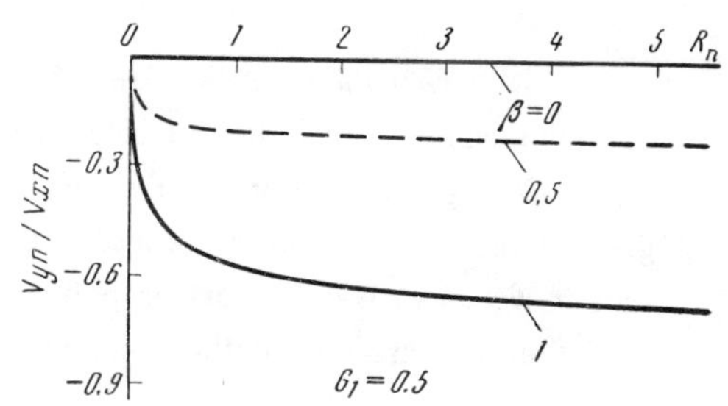

FIGURE 87. Dependence of V_{yn} and G_n on R_n: 1 corresponds to $\beta = 1$, 2 to $\beta = 0.5$, and 3 to $\beta = 0$

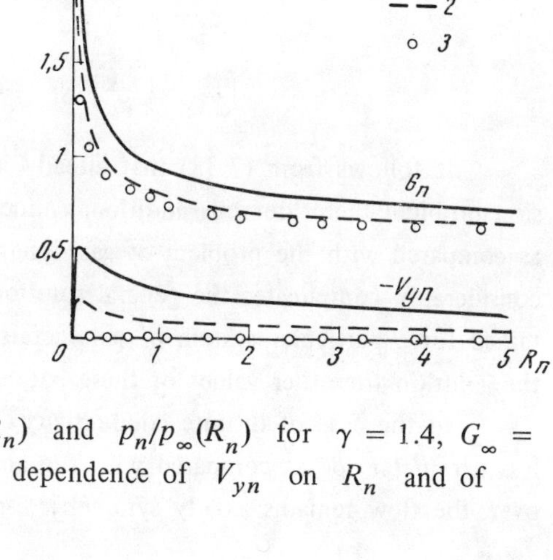

FIGURE 88. Variation of the ratio of the tangential and normal components of the velocity behind the front of the shock wave with increasing R $(G_1 = G_\infty = 0.5)$

Figure 86 shows the graphs of $\tau(R_n)$ and $p_n/p_\infty(R_n)$ for $\gamma = 1.4$, $G_\infty = 0.5$, and various β. Figure 87 shows the dependence of V_{yn} on R_n and of

G_n on R_n for the same values of G_∞ and β, where $V_y = v_y/a_\infty$ and $G^2 = H_y^2/4\pi\gamma p_\infty$. The ratio of the velocity components v_{ny}/v_{nx} as a function of R_n is shown in Figure 88.

From the results given it follows that the component v_{ny} behind the front of the shock wave does not behave monotonically but rather increases in absolute value from zero to a certain maximum at the beginning of the explosion and decreases to zero at a later stage of the explosion. This means that the gas particles near the center of explosion will acquire velocities directed almost along the x axis. For particles located outside some neighborhood of the center the initial velocities will be directed at some angle to the x axis. The gas-dynamical parameters p_n, ρ_n and the velocity component v_{xn} behave qualitatively like the corresponding quantities in the problem of explosion without consideration of a magnetic field.

§4. Spherical point explosion in a constant field

4.1. *General remarks.* We shall consider the case of explosion of a spherical point charge in a quiescent electrically conducting medium. We will assume that $H_1 = |H_1| = \text{const}$. For the time being we suppose that the initial density of the gas and the initial pressure are also constant. We choose a spherical coordinate system r, θ, φ with center at the point of explosion. We have the following system of defining parameters of the problem:

$$r, \quad \theta, \quad t, \quad \gamma, \quad p_1, \quad \rho_1, \quad H_1, \quad \sigma_1, \quad n, \quad m, \quad E_0. \tag{7.32}$$

Since the gas flow will have axial symmetry, the solution of the problem will not depend on the other angular coordinate φ. From the defining parameters (7.32) it is possible to form the following dimensionless combinations:

$$R = \frac{r}{r^0}, \quad \theta, \quad \tau = \frac{t}{t^0}, \quad \beta = \frac{H_1^2}{8\pi p_1},$$

$$\tilde{\sigma}_1 = \rho_1^n p_1^m \sigma_1 / t^0, \quad \gamma, \quad n, \quad m$$

$$\left(r^0 = \left(\frac{p_1}{E_0}\right)^{1/3}, \quad t^0 = r^0 \left(\frac{\rho_1}{p_1}\right)^{1/2}, \quad ct^0 \gg r^0 \right). \tag{7.33}$$

It follows from (7.33) that already in this simplified formulation of the explosion problem there are four additional dimensionless parameters β, $\tilde{\sigma}_1$, n and m as compared with the problem of gas dynamics for a homogeneous medium. This considerably complicates the general solution of the problem, since the solution obtained for a particular system of parameters λ, $\tilde{\sigma}_1$, β, n, m cannot be converted to the solution for other values of these parameters.

In the case of infinite conductivity $\sigma = \infty$ there is the additional dimensionless parameter β as compared with the analogous problem of gas dynamics. Moreover, the flow remains axially symmetric, and the solution will depend on the angle θ.

We proceed to a more detailed treatment of the problem for the case of infinite conductivity of the gas [15], [16]. We take the system of MHD equations in a spherical coordinate system in the form

$$\begin{aligned}
&\frac{\partial}{\partial t}(r^2\rho v_r \sin\theta) = r\sin\theta\,(\pi_{\theta\theta}+\pi_{\varphi\varphi}) - \frac{\partial}{\partial r}(r^2\sin\theta\pi_{rr}) - \frac{\partial}{\partial\theta}(r\sin\theta\pi_{r\theta}),\\
&\pi_{rr} = \rho v_r^2 + p^* - \frac{H_r^2}{4\pi}, \qquad \pi_{\theta\theta} = \rho v_\theta^2 + p^* - \frac{H_\theta^2}{4\pi},\\
&\pi_{\varphi\varphi} = p^* = p + \frac{H^2}{8\pi}, \qquad \pi_{r\theta} = \rho v_r v_\theta - \frac{1}{4\pi}H_r H_\theta = \pi_{\theta r},\\
&\sin\theta\frac{\partial r^3\rho v_\theta}{\partial t} = -\pi_{\varphi\varphi} r^2\cos\theta + \frac{\partial}{\partial\theta}(r^2\sin\theta\pi_{\theta\theta}) + \frac{\partial}{\partial r}(r^3\sin\theta\pi_{\theta r}),\\
&\frac{\partial r^2\rho\sin\theta}{\partial t} = -\frac{\partial}{\partial r}(r^2\rho v_r\sin\theta) - \frac{\partial}{\partial\theta}(r\rho v_\theta\sin\theta), \qquad\qquad (7.34)\\
&\frac{\partial}{\partial t}(r\sin\theta H_r) = \frac{\partial}{\partial\theta}[\sin\theta\,(v_r H_\theta - v_\theta H_r)],\\
&\frac{\partial r H_\theta}{\partial t} = -\frac{\partial}{\partial r}[r\,(v_r H_\theta - v_\theta H_r)],\\
&\frac{\partial}{\partial t}\left[r^2\sin\theta\left(\frac{\rho v^2}{2}+\frac{p}{\gamma-1}+\frac{H^2}{8\pi}\right)\right] = -\frac{\partial}{\partial r}(r^2\sin\theta S_r) - \frac{\partial}{\partial\theta}(r\sin\theta S_\theta),\\
&\frac{\partial}{\partial r}(r^2\sin\theta H_r) + \frac{\partial}{\partial\theta}(rH_\theta\sin\theta) = 0,
\end{aligned}$$

where r is the radius, θ is the latitude reckoned from the direction of the vector H_1 (cf. Figure 89), S_r and S_θ are the components of the energy flux vector, H_r and H_θ are the radial and transverse components of the magnetic field, and v_r and v_θ are the velocity components.

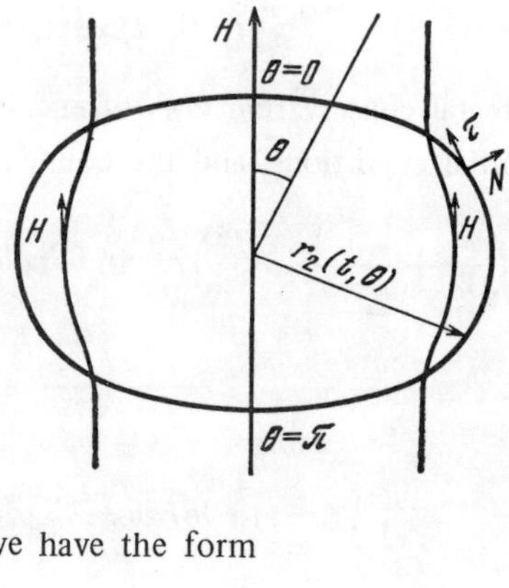

FIGURE 89. Scheme of deformation of the magnetic field and front of the shock wave

The conditions on the shock wave have the form

$$
\begin{aligned}
&[\rho(v_N - D)] = 0, \qquad \left[\rho(v_N - D)v_\tau - \frac{1}{4\pi}H_N H_\tau\right] = 0,\\
&\left[p + \rho(v_N - D)^2 + \frac{1}{8\pi}(H_\tau^2 - H_N^2)\right] = 0,\\
&[S] = \left[\rho(v_N - D)\left(\frac{v^2}{2} + \frac{\gamma p}{\rho(\gamma - 1)}\right) + \frac{1}{4\pi}\{(v_N - D)H^2 - H_N(\boldsymbol{vH})\}\right] = 0,\\
&[H_N \boldsymbol{v}_\tau - \boldsymbol{H}_\tau (v_N - D)] = 0, \qquad [H_N] = 0.
\end{aligned}
\tag{7.35}
$$

Here H_τ is the tangential component of the magnetic field, H_N is the normal component, the square brackets denote the difference of the quantities on different sides of the surface of discontinuity, and D is the speed of the shock wave. We shall denote quantities on the wave front by the index 2.

If $r_2(\theta, t)$ denotes the length of the radius vector to the shock wave, we have

$$
D = W\frac{\partial r_2}{\partial t}, \qquad W = \left[1 + \left(\frac{1}{r_2}\frac{\partial r_2}{\partial \theta}\right)^2\right]^{-1/2}.
$$

In the case being considered H_{N2} and $H_{\tau 2}$ are related to H_{r2} and $H_{\theta 2}$ as follows:

$$
H_{N2} = H_{\theta 2}\sin\psi + H_{\tau 2}\cos\psi, \qquad H_{\tau 2} = H_{\theta 2}\cos\psi - H_{\tau 2}\sin\psi,
$$
$$
\sin\psi = -\frac{W}{r_2}\frac{\partial r_2}{\partial \theta}, \qquad \cos\psi = W.
$$

Analogous relations hold also for the components of the velocity $\boldsymbol{v}$.

From the requirement of symmetry of the motion we obtain a condition for v_θ:

$$
v_\theta(r, 0, t) = v_\theta(r, \pi, t) = 0. \tag{7.36}
$$

The integral conservation laws of energy and mass follow from the system (7.34), the initial conditions, and the boundary conditions:

$$
E_0 = 2\pi\int_0^\pi\int_0^{r_2}\left(\frac{\rho v^2}{2} + \frac{p}{\gamma - 1} + \frac{H^2}{8\pi}\right)r^2\sin\theta\, dr\, d\theta
- \frac{2\pi}{3}\left(\frac{p_1}{\gamma - 1} + \frac{H_1^2}{8\pi}\right)\int_0^\pi r_2^3(\theta, t)\sin\theta\, d\theta, \tag{7.37}
$$

$$
\int_0^\pi\int_0^{r_2}\rho r^2\sin\theta\, dr\, d\theta = \frac{1}{3}\rho_1\int_0^\pi r_2^3(\theta, t)\sin\theta\, d\theta. \tag{7.38}
$$

Solution of the problem reduces to integrating the system (7.34) with conditions (7.35) and (7.36). Its solution for a broad range of times can be given by numerical methods and by applying the asymptotic laws for the damping of shock waves at large distances. The following methods are sketched for the numerical solution of this problem.

1. The MHD equations are taken in the divergence form (7.34). All the unknown functions are represented by trigonometric polynomials of degree n taking into account their symmetry. The coefficients of these polynomials will be certain unknown functions of r and t. Choosing s rays in the octant $0 \leqslant \theta \leqslant \pi/2$, we require that the equations of (7.34) be satisfied on these rays.

These conditions give the necessary number of additional approximating equations involving partial derivatives with respect to t and r. The approximating system obtained is solved by one of the known numerical methods (the method of finite differences, the method of characteristics, the method of integral relations) with the boundary conditions appropriate to the problem. We remark that analogous methods were developed for certain problems of gas dynamics.

2. We successively apply the method of integral relations. Introducing interpolation polynomials in r and θ, we arrive at an approximating system of ordinary differential equations in time which is integrated numerically on a computer.

3. We use finite-difference methods with the introduction of an artificial viscosity or difference schemes with smearing-out of the shock waves analogous to those widely used in ordinary gas dynamics.

In computing the problem by one of these methods the initial data can be assigned by using the results of solving the problem for small t which we discuss in more detail below.

4.2. *Determination of the magnetic fields in the initial stage of the flow.* For sufficiently small t the shock wave will be rather strong, and for moderate magnetic fields in the initial stage of the explosion one may ignore the effect of the field on the motion of the gas and determine only the perturbation of the magnetic field by the shock wave. For a rough estimate of the limits of applicability of this approach we make use of the fact that the magnetic field will have a weak effect on the motion if the total magnetic energy contained in the perturbed gas is small as compared with the energy of the explosion.

If r_* is the admissible radius of the shock wave, then for equality of the energies the condition

$$E_0 = 4\pi \int_0^{r_*} \frac{H_1^2}{8\pi} r^2 dr \tag{7.39}$$

must be satisfied, or

$$E_0 = \frac{H_1^2}{6} r_*^3. \tag{7.40}$$

This implies the condition for small effect of the field

$$r < r_*, \qquad r_* = \left(\frac{6E_0}{H_1^2}\right)^{1/3}. \tag{7.41}$$

By analogous arguments we obtain an estimate for neglecting the counterpressure p_1:

$$r < r_{**} = \left(\frac{3(\gamma-1)}{4\pi}\right)^{1/3} r^0. \tag{7.42}$$

If $r > r_*$ and $r > r_{**}$, it is necessary to consider both the counterpressure and the effect of the field H_1 on the motion of the gas. For $r_{**} < r < r_*$ the counterpressure must be considered, but the effect of the field on the motion may be ignored. For $r < r_*$, $r < r_{**}$ neither the counterpressure nor the effect of the field on the motion need be considered. We shall assume that both inequalities (7.41) and (7.42) are satisfied.

For a strong wave in the conditions on the shock wave the quantity p_1 may be neglected as compared with $\rho_2 v_2^2$, and H_1 may be neglected in the conditions following from the law of conservation of momentum and energy. For v_2, p_2 and ρ_2 we have the usual gas-dynamical conditions, and the jump in the field is determined from the relation $\{H_\tau(v_2 - D)\} = 0$. Since in the case being considered v does not depend on θ and H_1 and is a known function from the solution of the gas-dynamical problem, it follows that the induction equations of the system (7.34) are integrable, and the dependence of the components of the magnetic field on r, θ and t is defined by

$$\begin{aligned} H_r &= H_1 \cos\theta \left[\frac{\gamma+1}{\gamma-1}\left(1 - \frac{5}{2}V\right)\right]^{\beta_3} \left[\frac{\gamma+1}{\gamma-1}\left(\frac{5}{2}\gamma V - 1\right)\right]^{\beta_4} \\ &\qquad \times \left[\frac{5(\gamma+1)}{7-\gamma}\left(1 - \frac{3\gamma-1}{2}V\right)\right]^{\beta_5}, \\ H_\theta &= -\frac{\gamma+1}{\gamma-1} H_1 \sin\theta \left[\frac{\gamma+1}{\gamma-1}\left(1 + \frac{5}{2}V\right)\right]^{\beta_3 - 1} \left[\frac{\gamma+1}{\gamma-1}\left(\frac{5}{2}\gamma V - 1\right)\right]^{\beta_4} \\ &\qquad \times \left[\frac{5(\gamma+1)}{7-\gamma}\left(1 - \frac{3\gamma-1}{2}V\right)\right]^{\beta_5}, \\ \lambda &= \frac{r}{r_{20}} = \left[\frac{5(\gamma+1)}{4}V\right]^{2/5} \left[\frac{\gamma+1}{\gamma-1}\left(\frac{5}{2}\gamma V - 1\right)\right]^{\beta_2} \left[\frac{5(\gamma+1)}{7-\gamma}\left(1 + \frac{3\gamma-1}{2}V\right)\right]^{\beta_1}, \\ \beta_1 &= \beta_2 + \frac{\gamma+1}{3\gamma-1} - \frac{2}{5}, \quad \beta_2 = \frac{\gamma-1}{2\gamma-1}, \quad \beta_3 = \frac{2\gamma}{3(\gamma-2)}, \quad \beta_4 = \frac{2}{2\gamma+1}, \\ &\qquad \beta_5 = -\frac{2}{3}\frac{5}{(\gamma-2)}\beta_1, \end{aligned} \tag{7.43}$$

where

$$r_{20} = \left(\frac{E_0}{\alpha\rho_1}\right)^{1/5} t^{2/5}, \qquad \alpha = \text{const}, \qquad \frac{2}{5\gamma} \leqslant V \leqslant \frac{4}{5(\gamma+1)}. \tag{7.44}$$

Formulas (7.43) are directly applicable for computation of the distribution of H_r and H_θ under the condition $\gamma \neq 2$. The case $\gamma = 2$ is singular (cf. Chapter 2) and must be considered separately.

Since by assumption the field does not affect the motion, the form of the solution for the unknown functions v_r, ρ and p coincides with the gas-dynamical solution (2.5)–(2.10).

The solution (7.43) written out above was first obtained in [15]. Generalization of the solution (7.43) to the case of explosion with variable initial density $\rho_1 = Ar^{-\omega}$ has been carried out by E. V. Rjazanov [16]. He found, in particular, that for $\omega = (7-\gamma)/(\gamma+1)$ the solution for the field has the simple form

$$H_r = H_1 \lambda^{4/(\gamma-1)} \cos\theta,$$
$$H_\theta = -\frac{\gamma+1}{\gamma-1} H_1 \lambda^{4/(\gamma-1)} \sin\theta. \qquad (7.45)$$

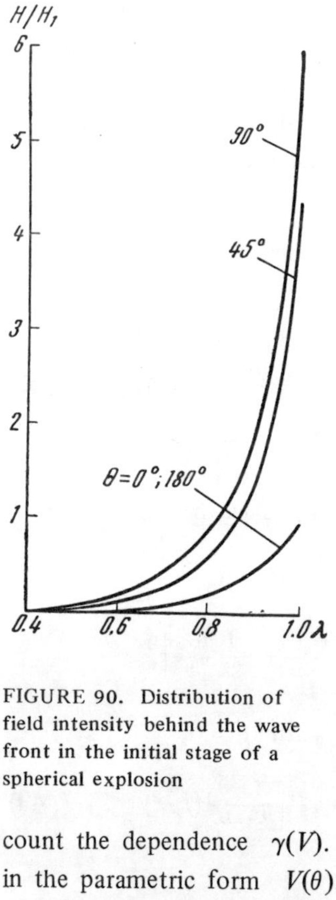

FIGURE 90. Distribution of field intensity behind the wave front in the initial stage of a spherical explosion

It follows from (7.41) and (7.43) that the j_r and j_θ are equal to zero, and j_φ is proportional to $\sin\theta$. The dependence of H/H_1 on $\lambda = r/r_{20}$ for $\theta = 0, 45, 90, 180°$ with $\gamma = 1.4$ is shown in Figure 90.

Using (7.43) and the differential equations for the force lines, it is possible to determine the deformation of the force lines by the flow of gas. For the case in question the differential equation of the force lines has the form

$$\frac{dr}{H_r} = \frac{r\,d\theta}{H_\theta}.$$

Using the formulas for H_r and H_θ from (7.43), we find that

$$\frac{d\lambda}{\lambda\left(1-\frac{5}{2}V\right)} = -\cot\theta. \qquad (7.46)$$

Equation (7.46) can be integrated, taking into account the dependence $\gamma(V)$. As a result we obtain the equation of the force lines in the parametric form $V(\theta)$. Since this expression is rather unwieldy, we consider a simple means of solving (7.46) approximately.

It follows from (7.44) that the parameter V is weakly varying in the region of γ from 0 to 1. Setting $V = V_*$ in (7.46), we find that

$$r = r_{20} C (\sin\theta)^{-a}, \qquad (7.47)$$

where C is a constant of integration and $a = 1 - 5/2V_* > 0$. Equation (7.47) gives the approximate form of the force lines of the magnetic field behind the front of the shock wave. A qualitative picture of the variation of the force lines is shown in Figure 89.

4.3. *Account of the effect of the magnetic field on the motion of the gas.* It is possible to apply the method of linearization in order to consider the effect of the field on the flow in question. It was noted in [16] that the effect of the magnetic field can be found by linearization using the small parameter $\mu = (\alpha/E_0)^{1/2} r_{20}^{3/2} H_1$, where r_{20} is the radius of the shock wave in the corresponding problem with no magnetic field and $\alpha = \text{const}$. Preliminary investigations of the problem by this method have been carried out by E. V. Rjazanov [16].

We shall consider this idea of linearization in detail, applying, in our view, a more suitable system of dimensionless variables and taking for the linearization parameter the quantity $q = H_1^2/4\pi D_0^2 \rho_{10}$, where D_0 is the speed of the shock wave in a strong explosion with no field and $\rho_{10} = ar_{20}^{-\omega}$ ($\omega = \text{const}$). We shall assume that the counterpressure is equal to zero ($p_1 = 0$) and that the initial density is variable: $\rho_1 = ar^{-\omega}$. We remark that counterpressure can be considered by using the principle of superposition of linear corrections.

We introduce the following dimensionless variables:

$$
\begin{gathered}
v_r = D_0 f_r(\lambda, q, \theta), \qquad v_\theta = D_0 f_\theta(\lambda, q, \theta) \qquad \rho = \rho_{10} g(\lambda, q, \theta), \\
p = \rho_{10} D_0^2 P(\lambda, q, \theta), \qquad H_r = H_1 G_r(\lambda, q, \theta), \qquad H_\theta = H_1 G_\theta(\lambda, q, \theta), \\
\lambda = \frac{r}{r_{20}}, \qquad r_{20} = \left(\frac{E_0}{\alpha a}\right)^{\delta/2} t^\delta, \qquad \delta = \frac{2}{5-\omega}.
\end{gathered} \tag{7.48}
$$

After transformations the system of MHD equations (7.34) takes the form

$$
\begin{gathered}
g L_{\lambda\theta}(f_r) - \frac{g f_\theta^2}{\lambda} = -\frac{\partial P}{\partial \lambda} + q\left[\frac{G_\theta}{\lambda}\frac{\partial G_r}{\partial \theta} - \frac{1}{2\lambda^2}\frac{\partial}{\partial \lambda}(G_\theta^2 \lambda^2)\right], \\
g L_{\lambda\theta}(f_\theta) + g\frac{f_r f_\theta}{\lambda} = -\frac{1}{\lambda}\frac{\partial P}{\partial \theta} + q\left[G_r \frac{\partial G_\theta}{\partial \lambda} + \frac{G_r G_\theta}{\lambda} - \frac{1}{2\lambda}\frac{\partial G_r^2}{\partial \theta}\right], \\
-\omega g + L_\lambda(g) + \frac{1}{\lambda^2}\frac{\partial}{\partial \lambda}(g f_r \lambda^2) + \frac{1}{\lambda \sin\theta}\frac{\partial}{\partial \theta}(g f_\theta \sin\theta) = 0, \\
\left(2 - \omega - \frac{2}{\delta}\right) P + L_\lambda(P) + f_r \frac{\partial P}{\partial \lambda} + \frac{f_\theta}{\lambda}\frac{\partial P}{\partial \theta} + \gamma P\left[\frac{1}{\lambda^2}\frac{\partial}{\partial \lambda}(f_r \lambda^2)\right. \\
\left. + \frac{1}{\lambda \sin\theta}\frac{\partial}{\partial \theta}(f_\theta \sin\theta)\right] = 0, \\
L_\lambda(G_r) - \frac{1}{\lambda \sin\theta}\frac{\partial}{\partial \theta}[\sin\theta\,(f_r G_\theta - G_r f_\theta)] = 0, \\
L_\lambda(G_\theta) + \frac{1}{\lambda}\frac{\partial}{\partial \lambda}[\lambda (f_r G_\theta - G_r f_\theta)] = 0.
\end{gathered} \tag{7.49}
$$

Here $L_{\lambda\theta}$ and L_λ are the operators

$$
L_{\lambda\theta} = \kappa + L_\lambda + f_r \frac{\partial}{\partial \lambda} + \frac{f_\theta}{\lambda}\frac{\partial}{\partial \theta},
$$

$$L_\lambda = -\lambda \frac{\partial}{\partial \lambda} + q \frac{b}{\delta} \frac{\partial}{\partial q}, \qquad \kappa = 1 - \frac{1}{\delta}, \qquad b = 2 + \delta(\omega - 2).$$

In addition, from the equation div $H = 0$ we have

$$\frac{\partial}{\partial \lambda}(G_r \lambda^2) + \frac{\lambda}{\sin\theta} \frac{\partial}{\partial \theta}(G_\theta \sin\theta) = 0. \tag{7.50}$$

We further introduce the notation

$$R = \frac{r_2(t, \theta)}{r_{20}}, \qquad U = \frac{D}{D_0}. \tag{7.51}$$

The conditions on the shock waves (7.35) in the dimensionless variables (7.48), (7.51) are written as follows:

$$\begin{gathered} P_2 - U f_{N2} R^{-\omega} + \frac{q}{2}(G_{\tau 2}^2 - G_{\tau 1}^2) = 0, \\ R^{-\omega} U f_{\tau 2} + q(G_{N1} G_{\tau 2} - G_{N1} G_{\tau 1}) = 0, \\ g_2(f_{N2} - U) = -U R^{-\omega}, \quad G_{N1} f_{\tau 2} - G_{\tau 2}(f_{N2} - U) = G_{\tau 1} U, \\ \frac{P_2}{\gamma - 1} + \frac{P_2}{2}(1 - g_2 R^\omega) + \frac{q}{4}(1 - g_2 R^\omega)(G_{\tau 2} - G_{\tau 1})^2 = 0, \quad G_{N2} = G_{N1}. \end{gathered} \tag{7.52}$$

We remark that the next to the last condition in (7.52) is the shock adiabatic. We use it in place of the energy equation. In this approximation the form of the shock wave will be different from a sphere (cf. Figure 89). By means of the kinematic and geometric relations for the front of the shock wave we find

$$\begin{gathered} U = \frac{\left(R + \frac{b}{\delta} q R_q'\right) R}{\Delta}, \qquad f_{N2} = f_{r2} \frac{R}{\Delta} - f_{\theta 2} \frac{R_\theta'}{\Delta}, \\ f_{\tau 2} = f_{r2} \frac{R_\theta'}{\Delta} + f_{\theta 2} \frac{R}{\Delta}, \qquad G_{Ni} = G_{ri} \frac{R}{\Delta} - G_{\theta i} \frac{R_\theta'}{\Delta}, \\ G_{\tau i} = G_{ri} \frac{R_\theta'}{\Delta} + G_{\theta i} \frac{R}{\Delta}, \qquad \Delta = \sqrt{R^2 + (R_\theta')^2}, \\ R_q' = \frac{\partial R}{\partial q}, \qquad R_\theta' = \frac{\partial R}{\partial \theta}, \qquad i = 1, 2. \end{gathered} \tag{7.53}$$

We represent any one of the unknown functions $F(\lambda, q, \theta)$ in the form $F(\lambda, q, \theta) = F_0(\lambda) + qF_1(\lambda, \theta) + o(q)$ and let $f_{r0}(\lambda) = f_0(\lambda)$.

Linearizing the MHD system (7.49), (7.50) with respect to the parameter q, we obtain equations for the functions F_1:

$$\begin{gathered} g_0\left[(k + \kappa) f_{r1} - \lambda \frac{\partial f_{r1}}{\partial \lambda} + k f_{r1} \frac{\partial f_0}{\partial \lambda} + f_0 \frac{\partial f_{r1}}{\partial \lambda}\right] + g_1\left(\kappa f_0 - \lambda \frac{d f_0}{d \lambda} + f_0 \frac{\partial f_0}{\partial \lambda}\right) \\ = -\frac{\partial P_1}{\partial \lambda} + \frac{G_{\theta 0}}{\lambda} \frac{\partial G_{r0}}{\partial \theta} - \frac{1}{2\lambda^2} \frac{\partial}{\partial \lambda}(G_{\theta 0}^2 \lambda^2) \qquad (k = b/\delta), \end{gathered} \tag{7.54}$$

$$\begin{gathered} g_0\left[\kappa f_{\theta 1} - \lambda \frac{\partial f_{\theta 1}}{\partial \lambda} + k f_{\theta 1} + f_0 \frac{\partial f_{\theta 1}}{\partial \lambda} + \frac{f_0 f_{\theta 1}}{\lambda}\right] \\ = -\frac{1}{\lambda} \frac{\partial P_1}{\partial \theta} + G_{r0} \frac{\partial G_{\theta 0}}{\partial \lambda} + \frac{G_{r0} G_{\theta 0}}{\lambda} - \frac{1}{2\lambda} \frac{\partial G_{\theta 0}^2}{\partial \theta}, \end{gathered} \tag{7.55}$$

$$-\omega g_1 - \lambda \frac{\partial g_1}{\partial \lambda} + kg_1 + \frac{1}{\lambda^2} \frac{\partial}{\partial \lambda} \lambda^2 [g_0 f_{r1} + g_1 f_0]$$

$$+ \frac{1}{\lambda \sin \theta} \frac{\partial}{\partial \theta} (g_0 f_{\theta 1} \sin \theta) = 0, \tag{7.56}$$

$$-\lambda \frac{\partial P_1}{\partial \lambda} + kP_1 + f_0 \frac{\partial P_1}{\partial \lambda} + f_{r1} \frac{\partial P_0}{\partial \lambda} + \gamma P_0 \Big[\frac{1}{\lambda^2} \frac{\partial}{\partial \lambda} (f_{r1} \lambda^2)$$

$$+ \frac{1}{\lambda \sin \theta} \frac{\partial}{\partial \theta} (f_{\theta 1} \sin \theta) \Big] + \gamma P_1 \frac{1}{\lambda^2} \frac{d}{d\lambda} (\lambda^2 f_0) = 0, \tag{7.57}$$

$$-\lambda \frac{\partial G_{r1}}{\partial \lambda} + kG_{r1} = \frac{1}{\lambda \sin \theta} \frac{\theta}{\partial \theta} [\sin \theta \, (f_0 G_{\theta 1} + f_{r1} G_{\theta 0} - G_{r0} f_{\theta 1})], \tag{7.58}$$

$$-\lambda \frac{\partial G_{\theta 1}}{\partial \lambda} + kG_{\theta 1} = -\frac{1}{\lambda} \frac{\partial}{\partial \lambda} [\lambda \, (f_0 G_{\theta 1} - f_{r1} G_{\theta 0} - G_{r0} f_{\theta 1})], \tag{7.59}$$

$$\frac{\partial}{\partial \lambda} (G_{r1} \lambda^2) + \frac{\lambda}{\sin \theta} \frac{\partial}{\partial \theta} (G_{\theta 1} \sin \theta) = 0. \tag{7.60}$$

The last equation has been obtained from (7.50) and can be used in place of one of the equations (7.58) or (7.59) (in this regard see the remark in §3 of Chapter 1).

This MHD system has a remarkable property; namely, the equations defining the gas-dynamical functions f_{r1}, $f_{\theta 1}$, P_1 and g_1 separate from the equations for G_{r1} and $G_{\theta 1}$. Now for the functions with index 0 the solution of the gas-dynamical system is known and is presented in Chapter 2, while for G_{r0} and $G_{\theta 0}$ we have the equations

$$\frac{\partial G_{r0}}{\partial \lambda} = \frac{f_0}{\lambda^3} \frac{\partial}{\partial \lambda} (G_{r0} \lambda^2), \qquad \frac{\partial G_{\theta 0}}{\partial \lambda} = \frac{1}{\lambda^2} \frac{\partial}{\partial \lambda} (\lambda f_0 G_{\theta 0}). \tag{7.61}$$

The solution of (7.61) was found in the preceding section (cf. (7.43)), where we studied the problem for weak fields. One can see formally that the functions (7.43) satisfy (7.61).

If the gas-dynamical functions f_{r1} and $f_{\theta 1}$ are known, equations (7.58)–(7.60) provide the possibility of determining the corrections to the field G_{r0}, $G_{\theta 0}$. Thus the basic problem consists in determining the gas-dynamical corrections f_{r1}, $f_{\theta 1}$, g_1 and P_1 from the system of inhomogeneous linear equations (7.54)–(7.57).

For the coordinate of the shock wave and its velocity U we take

$$R = 1 + qR_1, \qquad U = 1 + qU_1, \tag{7.62}$$

then, by (7.53), between U_1 and R_1 there is the relation

$$U_1 = (1 + k) R_1. \tag{7.63}$$

Since the shock wave differs from a sphere, in the linear approximation for any function $F(\lambda, q, \theta)$ we have

$$F_2 = F_0(1) + qF_1(1, \theta) + \left(\frac{\partial F}{\partial \lambda}\right)_{\substack{q=0\\ \lambda=1}} R_1 q \quad (F_2 = F|_{r=r_2}). \tag{7.64}$$

Linearization of the conditions on the shock wave, taking account of (7.62)–(7.64), gives the following relations:

$$G_{r02} = G_{r01} = \cos\theta, \qquad G_{\theta 02}(f_{02} - 1) = -G_{\theta 1}, \qquad -\frac{\gamma+1}{\gamma-1}\sin\theta = G_{\theta 02}, \tag{7.65}$$

$$P_{12} - f_{r12} + [(P_0')_2 - (f_0')_2 - \omega f_{02}] R_1 - u_1 f_{02} + \frac{1}{2}(G_{\theta 02}^2 - \sin^2\theta) = 0,$$

$$\left((F_0')_2 = \frac{\partial F}{\partial \lambda}\bigg|_{q=0,\ \lambda=1}\right),$$

$$f_{02}\frac{\partial R_1}{\partial \theta} + f_{\theta 12} + \cos\theta\,[G_{\theta 02} + \sin\theta] = 0,$$

$$g_{02} f_{r12} + g_{12}(f_{02} - 1) + \{(f_0')_2 g_{20} + [(g_0')_2 + \omega g_{02}](f_{02} - 1) - \omega\} R_1$$
$$+ (1 - g_{02}) U_1 = 0, \tag{7.66}$$

$$\cos\theta\left[f_{02}\frac{\partial R_1}{\partial \theta} + f_{\theta 12}\right] - G_{\theta 02}[f_{r12} + (f_0')_2 R_1 - U_1]$$

$$-\left[G_{r02}\frac{\partial R_1}{\partial \theta} + G_{\theta 12} + (G_{\theta 0}')_2 R_1\right](f_{02} - 1) = -\sin\theta\, U_1 + \cos\theta\frac{\partial R_1}{\partial \theta},$$

$$\left[-\frac{1}{2}P_{02}g_{12} - \frac{1}{2}(g_0')_2 P_{02} - \omega\frac{P_{02}}{2}g_{02}\right]R_1 + \frac{1-g_0}{4}(G_{\theta 02} + \sin\theta)^2 = 0,$$

$$G_{r12} + (G_{r0}')_2 R_1 - G_{\theta 02}\frac{\partial R_1}{\partial \theta} = \sin\theta\frac{\partial R_1}{\partial \theta}.$$

Linearization of the integral laws of conservation of energy and mass gives

$$\int_0^\pi \left[\int_0^1 \left(\frac{g_1 f_0^2 + 2f_1 f_0 g_0}{2} + \frac{P_1}{\gamma - 1} + \frac{1}{2}G_0^2\right)\lambda^2 d\lambda + \right.$$
$$\left. + R_1\left(\frac{f_{02}^2 g_{02}^2}{2} + \frac{P_{02}}{\gamma - 1}\right)\right]\sin\theta\, d\theta = \frac{1}{3}, \tag{7.67}$$
$$\int_0^\pi \left[\int_0^1 g_1\lambda^2 d\lambda + R_1(g_{20} - 1)\right]\sin\theta\, d\theta = 0.$$

The problem consists in solving the system of linear partial differential equations (7.54)–(7.57) with boundary conditions (7.66) on the shock wave (for $\lambda = 1$). In addition, we have the symmetry condition $f_{\theta 1}(\lambda, \pi) = f_{\theta 1}(\lambda, 0) = 0$. By means of this condition (and the integral relations) we can find the shape of the shock wave in the approximation considered.

Analysis of the equations and boundary conditions indicates that the solution of the problem should be sought in the form

$$
\begin{aligned}
&g_1 = M_0(\lambda) + \sin^2\theta M_1(\lambda), \quad P_1 = \pi_0(\lambda) + \sin^2\theta \pi_1(\lambda),\\
&f_{r1} = \varphi_{r0}(\lambda) + \varphi_{r1}(\lambda)\sin^2\theta, \qquad (7.68)\\
&f_{\theta 1} = F_{\theta 1}(\lambda)\sin 2\theta, \quad R_1 = R_{10} + R_{11}\sin^2\theta.
\end{aligned}
$$

We mention also that in correspondence with (7.43) we have

$$
G_{r0} = \psi_r(\lambda)\cos\theta, \quad G_\theta = -\psi_\theta(\lambda)\sin\theta. \qquad (7.69)
$$

If the solution (7.68), taking account of (7.69), is substituted into (7.54)–(7.57), then the variables separate, and for the seven unknown functions in (7.68) we obtain systems of ordinary differential equations and corresponding boundary conditions for $\lambda = 1$. The solution of these linear systems can be obtained numerically using the methods developed in Chapter 3 in considering one-dimensional problems. In solving the linear systems the integral laws (7.67) are to be used both for control of the accuracy of the computations and for determining the constants R_{10} and R_{11}. We note that for the singular solution (2.28) with $\omega = \omega_1$ the linear systems are systems with constant coefficients. In this case (as also for one-dimensional problems) the problem of finding the solution reduces to a purely algebraic one in the sense that it consists in finding the eigenvalues and arbitrary constants entering in the solution.

Thus our analysis leads us to the conclusion that the solution of the problem in question for small q has the form

$$
\begin{aligned}
&p = ar_{20}^{-\omega} D_0^2 [P_0(\lambda) + q(\pi_0(\lambda) + \pi_1(\lambda)\sin^2\theta) + o(q)],\\
&\rho = ar_{20}^{-\omega}[g_0(\lambda) + q(M_0(\lambda) + M_1(\lambda)\sin^2\theta) + o(q)],\\
&v_r = D_0[f_0(\lambda) + q(\varphi_{r0}(\lambda) + \varphi_{r1}(\lambda)\sin^2\theta) + o(q)],\\
&v_\theta = D_0[q\varphi_{\theta 1}(\lambda)\sin 2\theta + o(q)],\\
&H_r = H_1[\psi_r(\lambda)\cos\theta + qG_{r1}(\lambda, \theta) + o(q)],\\
&H_\theta = H_1[-\psi_\theta(\lambda)\sin\theta + qG_{\theta 1}(\lambda, \theta) + o(q)],\\
&r_2 = r_{20}(t)[1 + q(R_{10} + R_{11}\sin^2\theta) + o(q)],\\
&q = \frac{H_1^2}{4\pi D_0^2 \rho_{10}}.
\end{aligned}
$$

We remark that in our solution there is an analogy with the case of density which varies according to height (with no magnetic field) when the density distribution in z is symmetric with respect to $z = 0$ (cf. [17]). We mention also that the case of motion of a piston in an infinitely conducting gas with a magnetic field (both the self-similar and the non-self-similar linearized solution) has been considered in the work of American authors (cf., for example, [18]).

To compute the parameters of the wave front at large distances from the point of the explosion one can use the asymptotic damping laws for weak waves which we studied in [7]. In that paper asymptotic formulas are given which make it possible to construct an approximate theory of the shape of a shock wave based on a method analogous to that presented in Chapters 4 and 5 and in §3 of the present chapter. We shall not go into this in detail here.

§5. On the perturbation of an arbitrary weak magnetic field

The problem of the perturbation of a constant weak magnetic field H_1 during the propagation of a strong shock wave from a spherical point explosion in an infinitely conducting quiescent gas can be generalized to the case of any weak field H_1 both for spherical and for planar and cylindrical explosions [19]. We shall also assume the shock wave to be sufficiently strong and that $\rho_2 v_2^2 > H_1^2/8\pi$, $r < r_*$, where r_* is computed on the basis of some mean value for the field H_1. In first approximation we neglect the effect of the field H_1 on the perturbed motion of the gas. We shall follow our work [19].

In the case of a spherical shock wave $(\nu = 3)$ we consider the problem in spherical coordinates r, θ, φ, in the case of a cylindrical wave $(\nu = 2)$ we take cylindrical coordinates r, φ, z, and in the case of a plane wave $(\nu = 1)$ we choose Cartesian coordinates r, y, z.

To determine the perturbed field H one can use the induction equation (1.76). However, here it is convenient to proceed by a simpler path.

If we consider the continuity equation and go over to Lagrange variables (for the Lagrange coordinate we take the initial coordinate of the particle), the induction equation is transformed to the form (1.78) (for more details see §3, Chapter 1):

$$\frac{\boldsymbol{H}}{\rho} = \left(\frac{\boldsymbol{H}_1}{\rho_1}\,\mathrm{grad}_0\right)\boldsymbol{r};$$

here we have taken $H_0 = H_1$ and $\rho_0 = \rho_1$; and r is the radius vector of the particle. Since the gas-dynamical variables can be assumed to depend only on the Lagrange coordinate r_0 and t, from (1.78) we find on considering the gas-dynamical equation of continuity the following formulas for the components of the magnetic field in the zone of the perturbed motion [19]:

$$H_1 = H_{11}\left(\frac{r_0}{r}\right)^{\nu-1}, \quad H_2 = H_{21}\frac{\rho}{\rho_0}\left(\frac{r}{r_0}\right)^{\omega_*}, \quad H_3 = H_{31}\frac{\rho}{\rho_0}\left(\frac{r}{r_0}\right)^{\frac{1}{2}(\nu-2)(\nu-1)} \tag{7.70}$$

Here $\omega_* = 1$ for $\nu = 2, 3$, $\omega_* = 0$ for $\nu = 1$, and the H_i are the components of the magnetic field vector respectively in spherical, cylindrical, and Cartesian coordinate systems. It follows from (7.70) that the corresponding conditions on the shock wave for the field are satisfied.

In the sequel we shall assume for simplicity that the initial density ρ_0 is constant. In this case for r/r_0 we have the dependence (cf. Chapter 2)

$$\left(\frac{r}{r_0}\right)^{\nu}=\frac{p}{p_2}\left(\frac{\rho}{\rho_2}\right)^{-\gamma}\left(\frac{r}{r_2}\right)^{\nu}.$$

Here r_2 is the coordinate of the shock wave, p_2 is the pressure behind the wave front, and the dependence of p/p_2, ρ/ρ_2 and ρ/ρ_0 on r/r_2 is known from the analytic solution of the problem of a strong explosion (cf. Chapter 2).

We shall consider the special case of a constant initial magnetic field H_1. Suppose that in the spherical case the initial field is directed along the z axis, in the cylindrical case it makes an angle α with the z axis, and in the planar case it has the components H_{r1}, H_{y1} and H_{z1}. For the spherical and cylindrical cases we have

$$H_{11}=H_1\cos\theta,\qquad H_{21}=H_1\sin\theta,\qquad H_{31}=0\qquad(\nu=3);$$
$$H_{11}=H_1\sin\alpha,\qquad H_{21}=H_1\sin\alpha\sin\varphi,\qquad H_{31}=H_1\sin\alpha\cos\varphi\qquad(\nu=2).$$

In the case being considered we obtain from (7.70)

$$\begin{aligned}
&H_r=H_1\cos\theta\left(\frac{r_0}{r}\right)^2, && H_\theta=-H_1\sin\theta\frac{\rho r}{\rho_1 r_0}, \quad H_\varphi=0 \qquad (\nu=3),\\
&H_r=H_1\sin\alpha\sin\varphi\,\frac{r}{r_0}, && H_\varphi=H_1\sin\alpha\cos\varphi\,\frac{\rho r}{\rho_1 r_0},\\
& && H_z=H_1\cos\alpha\,\frac{\rho}{\rho_1}\qquad(\nu=2),\\
&H_r=H_{r1}, && H_y=H_{y1}\frac{\rho}{\rho_1},\quad H_z=H_{z1}\frac{\rho}{\rho_1}\ (\nu=1).
\end{aligned}\tag{7.71}$$

As we noted above, the particular case of a constant initial magnetic field was first considered in [15]. In this work the method of direct integration of the induction equation was applied to determine the components of the magnetic field. For $\nu=3$ and $H_1=\text{const}$ one can obtain from (7.71) the solution (7.43)–(7.44). The case of a cylindrical explosion when the field has only a single component H_z was considered in §2. Questions of the interaction of a plane shock wave with a weak magnetic field not related to problems in the theory of explosions have been studied in [20]. We mention also that in the case $\nu=3$ the perturbation of the field can be found for explosion in an incompressible fluid. In this case in (7.70) one must take $\rho=\rho_1$. From the equation of continuity in the Lagrange variables we have

$$\frac{r_0}{r}=\Omega^{1/3}\quad\left(\Omega=1-\frac{l^3}{r^3}\right).$$

Here l is the radius of the cavity; its dependence on t is known (cf. §7 of Chapter 1). In this case from (7.70) we find that $H_r = H_{r1}\Omega^{3/2}$, $H_\theta = H_{\theta 1}\Omega^{-1/3}$ and $H_\varphi = H_{\varphi 1}\Omega^{-1/3}$. Unfortunately, this solution becomes meaningless near the boundary of the cavity, since H_θ and H_φ become infinite at $r = l$.

§6. Propagation of shock waves in a medium with finite electrical conductivity

6.1. *Deformation of a weak field.* The assumption of infinite conductivity of the medium can fiequently not be justified. Thus, for the motion of air behind the front of a strong shock wave (M = 20) with characteristic dimensions about 1 m the magnetic Reynolds numbers have order unity. We shall present some well-known values of the conductivities for air behind the front of a shock wave, for air with the addition of cesium, for the ionosphere at a height of 250 km, for sea water, and for mercury:

Medium	Air M=20	Air M=12	Air +0.01% Cs M=12	Ionosphere (350 m)	Water (sea)	Mercury
σ (sec^{-1})	$2\cdot 10^{12}$	10^{11}	$2\cdot 10^{13}$	10^{11}	$2\cdot 10^{10}$	10^{16}

For air the values of σ behind the front of the shock wave are given; M is the Mach number. When the characteristic dimensions of the phenomenon are large, for example in explosions in the ionosphere, the numbers R_m may be greater than 10, and here the application of the approximation of infinite conductivity can be completely justified.

Assuming that the value of σ is finite, we take the case of weak magnetic fields. Suppose that the initial magnetic field is homogeneous. We shall consider only the case of cylindrical explosion which we studied in 1960 and first published in [9]. Analogous investigations were later carried out by Sakurai [21].

We take the equations of motion in the form

$$-\rho\frac{dv}{dt}=\frac{\partial p^*}{\partial r}+\frac{2(\nu-1)h_\varphi}{r},\quad -\frac{1}{\rho}\frac{d\rho}{dt}=\frac{\partial v}{\partial r}+\frac{\nu-1}{r}v, \tag{7.72}$$

$$-\frac{1}{2}\frac{dh_z}{dt}=h_z\left(\frac{\partial v}{\partial r}+\frac{\nu-1}{r}v\right)-r^{1-\nu}h^{1/2}\frac{d}{dr}\left(\nu_m r^{\nu-1}\frac{dh_z^{1/2}}{dr}\right), \tag{7.73}$$

$$-\frac{1}{2}\frac{dh_\varphi}{dt}=h_\varphi\frac{\partial v}{\partial r}-h_\varphi^{1/2}\frac{\partial}{\partial r}\left[\nu_m r^{-1}\frac{\partial}{\partial r}\left(rh_\varphi^{1/2}\right)\right], \tag{7.74}$$

$$-\frac{dp}{dt}=\gamma p\left(\frac{\partial v}{\partial r}+\frac{(\nu-1)v}{r}\right)$$
$$-2(\gamma-1)\nu_m\left\{\frac{1}{r^2}\left[\frac{\partial}{\partial r}\left(rh_\varphi^{1/2}\right)\right]^2+\left(\frac{\partial h_z^{1/2}}{\partial z}\right)^2\right\}, \tag{7.75}$$

$$h=h_z+(\nu-1)h_\varphi,\quad p^*=p+h,\quad h_\varphi=\frac{H_\varphi^2}{8\pi},\qquad h_z=\frac{H_z^2}{8\pi}.$$

Let $h_\varphi = 0$ and $h_z \neq 0$. We introduce the dimensionless variables

$$\frac{v}{D}=f, \quad G=\frac{H}{H_\infty}, \quad \frac{\rho}{\rho_\infty}=g, \quad \lambda=\frac{r}{r_2}, \quad P=\frac{p}{\rho_\infty D^2},$$

$$q=\frac{a_*^2}{D^2}, \quad a_*^2=\frac{\gamma}{\rho_\infty}\left(p_\infty+\frac{H_*^2}{8\pi}\right).$$

If we seek a solution of (7.72)–(7.73) in the form

$$f=f_0(\lambda)+qf_1(\lambda)+o(q), \qquad g=g_0(\lambda)+qg_1(\lambda)+o(q),$$
$$P=P_0(\lambda)+qP_1(\lambda)+o(q), \qquad G=G_0(\lambda)+qG_1(\lambda)+o(q),$$

then on repeating the arguments of the preceding section we find that in first approximation the quantities f_0, g_0 and P_0 will satisfy the equations and relations for a strong explosion in a gas, while for G_0 we find from (7.73) that

$$(f_0-\lambda)\frac{dG_0}{d\lambda}+G_0\left(\frac{df_0}{d\lambda}+\frac{f_0}{\lambda}\right)-\frac{k_0}{\lambda}\frac{\partial}{\partial\lambda}\left(\lambda\frac{dG_0}{d\lambda}\right)=0,$$
$$k_0=\frac{c^2}{4\pi\sigma}\left(\frac{\alpha\rho_\infty}{E_0}\right)=\text{const.} \tag{7.76}$$

From the conditions on the shock wave ($\lambda = 1$) for the case of finite conductivity (cf. Chapter 1) we obtain (the gas is at rest in front of the wave)

$$G_{02}=G_{01}, \qquad f_{02}G_{02}-k_{02}\left(\frac{\partial G}{\partial\lambda}\right)_{02}=-k_{01}\left(\frac{\partial G}{\partial\lambda}\right)_{01}.$$

The gas-dynamical functions will satisfy the usual conditions on a strong wave. We assume that $k_{02} \neq k_{01}$ in general. Flow arises by virtue of the continuity of the magnetic field in advance of the shock wave (region 1). In the first approximation, however, the velocity f_0 in advance of the shock wave is equal to zero; therefore the induction equation (7.76) in region 1 takes the simple form

$$\frac{k_{01}}{\lambda}\frac{\partial}{\partial\lambda}\left(\lambda\frac{\partial G_0}{\partial\lambda}\right)+\lambda\frac{\partial G_0}{\partial\lambda}=0. \tag{7.77}$$

In region 2 (behind the front of the shock wave) equation (7.76) with $k_0 = k_{02}$ holds. Equation (7.77) is integrable, and its solution taking into account the condition $G \longrightarrow 1$ for $\lambda \longrightarrow \infty$ has the form

$$G_{01}=1-A_1\int_\lambda^\infty e^{-x^2/2k_{01}}\frac{dx}{x},$$

where A_1 is an arbitrary constant.

For region 2 from (7.76) we obtain the system of equations

$$G_0'=\Omega_0/\lambda, \qquad \Omega_0'=\frac{1}{K_{02}}\left[(f_0-\lambda)\Omega_0+\lambda G_0\left(f_0'+\frac{f_0}{\lambda}\right)\right], \tag{7.78}$$

where $f_0(\lambda)$ is a function which is known from the solution of the gas-dynamical problem. Investigation of the system (7.78) shows that it admits a bounded solution near $\lambda = 0$, and moreover this solution has the expansion

$$G_0 = c_1 [1 + a_1\lambda^2 + O(\lambda^4)],$$

where c_1 is an arbitrary constant and $a_1 = 1/\gamma k_{02}$.

If we introduce a new variable $\widetilde{G}_0 = G_0/c_1$, then at the center of symmetry we have the condition

$$\tilde{G}(0) = 1, \qquad \tilde{\Omega}(0) = 0. \tag{7.79}$$

The system of equations for $\widetilde{G}$ and $\widetilde{\Omega}$ in region 2 was integrated numerically from the center to the shock wave. Further, from the conditions on the shock wave the constants c_1 and A_1 were determined for the function G, and hence the function G_0 was specified in regions 1 and 2. The integral in the formula for G_{01} was found from tables.

The solution we have obtained enables us in first approximation to determine the structure of the magnetic field during passage of the shock wave. Computations were carried out for various values of the parameters k_{0i} and γ. Results of the computations for $\gamma = 5/3$ are shown in Figure 91. Our investigations indicate that with decrease of the parameters k_{01} ($k_{02} \geqslant k_{01}$) the maximum of H/H_∞ increases and tends to the ideal value $(\gamma + 1)/(\gamma - 1)$ as $k_{0i} \to 0$, while the distribution of H/H_∞ tends to $H/H_\infty = \rho/\rho_\infty$ (this follows from the frozen-in integral for infinite conductivity).

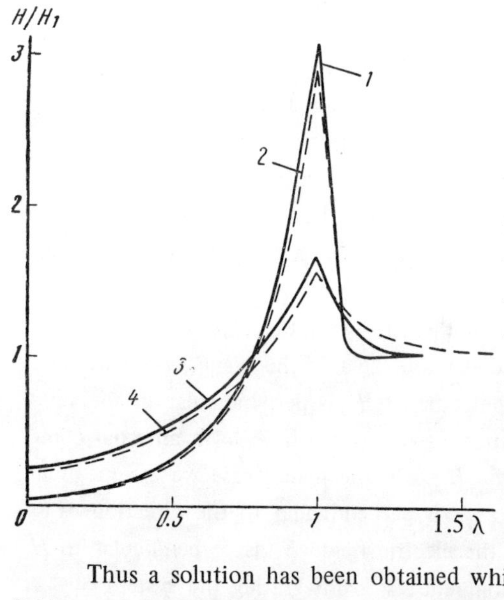

FIGURE 91. Spatial distribution of the magnetic field for various values of k_{0i} ($\gamma = 5/3$, $\nu = 2$)
1 — $k_{0i} = k_{02} = 0{,}01$; 2 — 0,02; 3 — 0,1; 4 — $k_{01} = 0{,}1$, $k_{02} = 0{,}2$

Thus a solution has been obtained which characterizes the structure of the field variation. Finding higher approximations will make it possible to determine the interaction of the motion and the field and to determine more precisely the structure of the field variation during passage of a shock wave in an electrically conducting gas. Analogous investigations could be carried out for planar and spherical explosions, where in the case of explosion of a spherical charge it is necessary to determine two components of the field, H_r and H_θ, which will depend both on λ and on the angle θ.

6.2. *Flows for small and finite numbers* R_m. In the theory of shock waves in tubes and in experiments with discharges along conductors the necessity may arise of studying the propagation of plane and cylindrical shock waves for small magnetic numbers R_m. These questions arise also in the theory of explosive MHD generators. The problem of cylindrical and plane explosions can also be utilized in studying questions of hypersonic flow around slender blunt bodies in the presence of a magnetic field.

The results presented below were for the most part published in 1962 [22]. Flows for small R_m were later studied by the American scientist P. Lykoudis [23].

We shall assume that the conductivity of the gas σ is finite and that

$$\sigma = \sigma_1 \rho^n p^m, \tag{7.80}$$

where σ_1, n and m are constants, p is the pressure, and ρ is the density of the gas.

Suppose that the magnetic Reynolds numbers are small:

$$R_m = \frac{ul}{\nu_m} < 1 \quad \left(\nu_m = \frac{c^2}{4\pi\sigma}\right). \tag{7.81}$$

Here u is a characteristic velocity, ν_m is the magnetic viscosity, and l is a characteristic length which is taken as either the height of the shock tube or the average radius of the shock wave. For small R_m it is possible to neglect the reverse effect of the gas flow on the magnitude of the electric and magnetic fields.

For large numbers R_m one must consider the effect of the gas flow on the variation of the magnetic field. We shall assume that the gas is perfect with a constant ratio of specific heat capacities.

We denote by ϵ_0 the magnitude of the energy of the explosion computed per unit length in the cylindrical case and per unit area in the plane case. We also assume that the stress vector of the magnetic field H is perpendicular to the velocity vector of the flow v. In the cylindrical case H can have both an axial component H_z and an azimuthal component H_φ. In the planar case we assume that the magnetic field is directed along the z axis perpendicular to the direction of motion of the gas. The stress vector of the electric field E is perpendicular to H.

Solution of the problem in question reduces to integrating the system of MHD equations which we take in the form

$$\rho \frac{d\boldsymbol{V}}{dt} + \nabla p = \boldsymbol{f}, \quad \frac{d\rho}{dt} + \rho \operatorname{div} \boldsymbol{v} = 0, \quad \boldsymbol{f} = \frac{1}{c} j \times \boldsymbol{H},$$

$$\frac{dp}{dt} + \gamma p \operatorname{div} \boldsymbol{v} = (\gamma - 1)\frac{j^2}{\sigma}, \qquad \boldsymbol{j} = \sigma\left(\boldsymbol{E} + \frac{1}{c}\boldsymbol{v} \times H\right), \tag{7.82}$$

$$\operatorname{curl} \boldsymbol{H} = \frac{4\pi}{c}\boldsymbol{j}, \quad \operatorname{curl} \boldsymbol{E} = -\frac{1}{c}\frac{d\boldsymbol{H}}{dt}, \quad \operatorname{div} \boldsymbol{H} = 0. \tag{7.83}$$

It is to be noted that the vector v is directed perpendicular to E and H, and all the unknown functions depend only on time and the single coordinate r of distance from the axis of symmetry or from the plane of the explosion. From the system (7.82)–(7.83) we have for the given case the energy equation

$$\frac{d}{dt}\left\{r^{\nu-1}\left(\frac{\rho v^2}{2}+\frac{p}{\gamma-1}\right)\right\}+\frac{d}{dr}\left\{r^{\nu-1}v\left(\frac{\rho v^2}{2}+\frac{\gamma p}{\gamma-1}\right)\right\}=r^{\nu-1}\boldsymbol{jE}. \tag{7.84}$$

On explosion a shock wave begins to propagate through the gas. The conditions on the shock wave have the form

$$\begin{gathered}[\rho(v-D)]=0, \qquad [\rho v(v-D)+p]=0,\\ \left[(v-D)\left(\frac{\rho v^2}{2}+\frac{p}{\gamma-1}\right)+pv\right]=0,\end{gathered} \tag{7.85}$$

$$[\boldsymbol{H}]=0, \qquad [\boldsymbol{E}_\tau]=0, \tag{7.86}$$

where the square brackets denote the difference in the values of the quantities on different sides of the discontinuity, D is the speed of the shock wave, and E_τ is the tangential component of the electric field. If we assume that the shock wave is strong, we can neglect the magnitude of the initial pressure of the gas as compared with the pressure behind the front of the shock wave.

We shall consider the problem of strong cylindrical or plane explosions in a gas under the assumption that the R_m are small, the initial density of the gas ρ_1 is constant, there is no electric field initially, and the initial magnetic field depends on the coordinate as follows:

$$H^2=\kappa_1 r^{\omega} \qquad (\kappa_1=\text{const}), \tag{7.87}$$

$$\omega=0, \quad \text{if} \quad H=H_z, \quad \omega=-2, \quad \text{if} \quad \nu=2,\ H=H_\varphi.$$

The expressions for j^2 and f_r, the projection of the force f on the direction of r, will be

$$j^2=\frac{\sigma^2}{c^2}\left\{H_z^2+(\nu-1)H_\varphi^2\right\}v^2, \qquad f_2=-\frac{\sigma}{c^2}\left\{H_z^2+(\nu-1)H_\varphi^2\right\}v. \tag{7.88}$$

In this case the system of equations of motion of the gas obtained from (7.82) contains a dimensional constant

$$\kappa=\kappa_1\sigma_1 c^{-2}, \qquad [\kappa]=M^{1-m-n}L^{m+3n-3-\omega}T^{2m-1}.$$

The solution of the problem will depend on the defining parameters r, t, ϵ_0, ρ_1 and κ. From dimension theory it follows that the problem of explosion will be self-similar if the condition $\nu(2m-1)-2(\omega+1)=0$ is satisfied.

We introduce the dimensionless variables V, R, P, and λ by the formulas

$$v = \frac{r}{t} V(\lambda), \qquad \rho = \rho_1 R(\lambda), \qquad p = \rho_1 \frac{r^2}{t^2} P(\lambda), \qquad \lambda = \frac{r}{r_2}, \tag{7.89}$$

where r_2 is the radius of the shock wave.

Because of the self-similarity we have for r_2

$$r_2 = \left(\frac{\varepsilon}{\rho_1}\right)^{1/\nu+2} t^{2/\nu+2}, \qquad \varepsilon_0 = \alpha\varepsilon, \tag{7.90}$$

where ϵ is a fixed constant having the dimension of ϵ_0 and α is a quantity which is calculated in the manner explained below.

Considering (7.80) and (7.87)–(7.90), from (7.82) we obtain the following differential equations for the self-similar functions V, R and P:

$$\begin{gathered} \lambda\{(V-\delta)RV' + P\} = R(V - V^2) - 2P - k\lambda^a V R^n P^m, \\ \lambda\{(V-\delta)R' + RV'\} = -\nu RV, \\ \lambda\{(V-\delta)P' + \gamma PV'\} = 2P - (2+\nu\gamma)PV + (\gamma-1)k\lambda^a V^2 R^n P^m. \end{gathered} \tag{7.91}$$

Here

$$\delta = \frac{2}{\nu+2}, \qquad a = (\nu+2)(m-0.5), \qquad k = \kappa\rho_1^{\,m+n-1}\left(\frac{\xi}{\rho_1}\right)^{m-0.5}.$$

In the case considered $(E = 0)$ it follows from (7.84) that the system (7.91) has the energy integral

$$\lambda^{\nu+2}\left\{(V-\delta)\left(\frac{RV^2}{2} + \frac{P}{\gamma-1} + PV\right)\right\} = c_1. \tag{7.92}$$

We denote by the index 2 the quantities behind the front of the shock wave. Since $D = dr_2/dt$ and the gas in front of the shock wave is at rest, on passing to dimensionless variables we find from (7.85)

$$\begin{gathered} R_2(V_2-\delta) + \delta = 0, \qquad R_2V_2(V_2-\delta) + P_2 = 0, \\ (V_2-\delta)\left(\frac{R_2V_2^2}{2} + \frac{P_2}{\gamma-1}\right) + P_2V_2 = 0. \end{gathered} \tag{7.93}$$

Thus the unknown functions $V(\lambda)$, $R(\lambda)$ and $P(\lambda)$ must satisfy (7.93) on the shock wave (for $\lambda = 1$). Moreover, the condition of constant total energy in the region occupied by the moving gas must be satisfied, i.e.

$$2\{(\nu-1)\pi - (\nu-2)\}\int_0^{r_2}\left(\frac{\rho v^2}{2} + \frac{p}{\gamma-1}\right) r^{\nu-1}dr = \varepsilon_0.$$

Passing to dimensionless variables, we find

$$\alpha(\nu, \gamma, m, n, k) = 2\{(\nu-1)\pi - (\nu-2)\}\int_0^1\left(\frac{RV^2}{2} + \frac{P}{\gamma-1}\right)\lambda^{\nu+1}d\lambda. \tag{7.94}$$

From the last equation of (7.93) it follows that the constant c_1 in the energy integral (7.92) is equal to zero. We introduce the new variables

$$y = \lambda^{\beta} R, \qquad z = \lambda^{\beta} P, \qquad \beta = \frac{(\nu+2)(m-0.5)}{m+n-1} \qquad (m+n-1 \neq 0).$$

The system (7.91) in the new variables assumes the form

$$\begin{aligned} &\lambda \{(V-\delta) y V' + z'\} = y(V - V^2) + (\beta - 2) z - kVy^n z^m, \\ &\lambda \{(V-\delta) y' + yV'\} = \{(\beta - \nu)V - \beta\delta\} y, \\ &\lambda \{(V-\delta) z' + \gamma z V'\} = (\beta - 2 - \nu\gamma) V z + (2 - \beta\delta) z + (\gamma - 1) k V^2 y^n z^m. \end{aligned} \tag{7.95}$$

The energy integral (7.92) can be written as follows:

$$z = y \frac{(\gamma-1)(\delta - V) V^2}{2(\gamma V - \delta)}. \tag{7.96}$$

Using (7.96), the solution of (7.95) can always be reduced to integrating an equation of first order. Thus, eliminating z' from the first and third equations of (7.95) and introducing the new variable $\mu = \ln \lambda$, we find

$$\frac{d\mu}{dV} = \chi(y, V) = \frac{yz - (\delta - V)^2 y}{(2 - 2\delta - \gamma\nu V) z + (1 - V)(\delta - V) V y + k(\gamma V - \delta) V y^n z^m}. \tag{7.97}$$

Using (7.97), we obtain from the second equation of (7.95)

$$\frac{dy}{dV} = \frac{y}{\delta - V} \{1 + ((\nu - \beta) V + \beta\delta) \chi(y, V)\}. \tag{7.98}$$

The function $z(y, V)$ is defined by (7.96). If (7.98) has been integrated, i.e. the dependence $y(V)$ has been found, then from (7.96) it is possible to determine $z(V)$ and from (7.97) to obtain $\mu(V)$ by quadrature; the complete solution of the problem can thus be found.

It was assumed above that $m + n - 1 \neq 0$. It can be shown that also in the case $m + n - 1 = 0$ solution of (7.91) can be reduced to integrating one differential equation of first order of the form $d\lambda/aV = F(\lambda, V)$ and a single quadrature.

Analysis of the solution of (7.98) and the system (7.91) shows that for $n = m = 0$ ($\sigma = \text{const}$) near the center the following asymptotic formulas hold:

$$\begin{aligned} &P = \left[c_1 - \frac{\delta}{\gamma} \ln (k\lambda)\right] \lambda^{-2}, \qquad R = c_2 \lambda^{2/\gamma - 1}, \\ &y = \lambda^2 R, \qquad V = \frac{\delta}{\gamma} + \frac{y(\gamma - 1)}{(\gamma - 1) c_1 - 4k\gamma^2 \ln y}. \end{aligned} \tag{7.99}$$

For $m = 1$, $n = 1$, and $\nu = 2$ the solution has a strongly nonintegrable singularity for the function P and loses its meaning. The case $n = m = 0$ was investigated numerically in detail. For the numerical solution the system (7.91) was used directly. It was solved for the derivatives and integrated on a computer; the Cauchy problem was solved with the initial data at $\lambda = 1$. In a neighborhood of the center of the explosion the computation was carried out on the basis of asymptotic formulas of the type (7.99). The energy integral was used for general control of computational accuracy. In the process of computing (7.91) the constant α was determined by integration. The maximum relative error in determining P from (7.91) and (7.96) did not exceed 0.01%. Some results of the computation for $\nu = 1$ and $\nu = 2$ are shown in Figures 92 and 93. In the cylindrical case with $\gamma = 1.4$ it

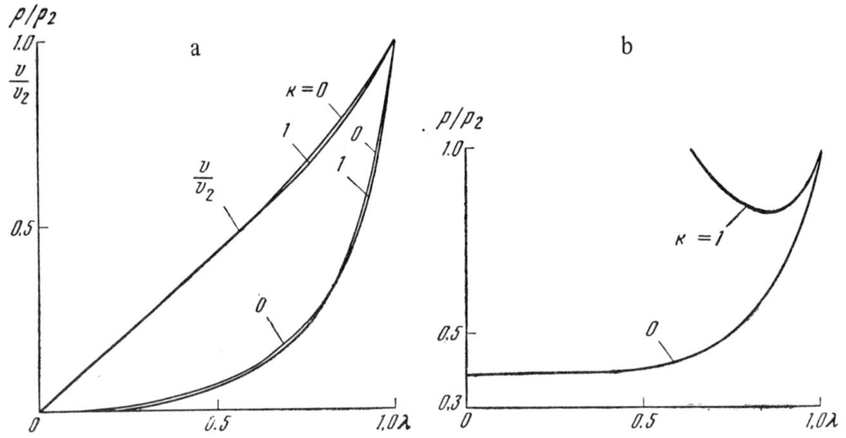

FIGURE 92. Distribution of velocities, densities (a) and pressures (b) behind a plane wave for small values of R_m ($\gamma = 1.4, \nu = 1, n = 0, m = 0$)

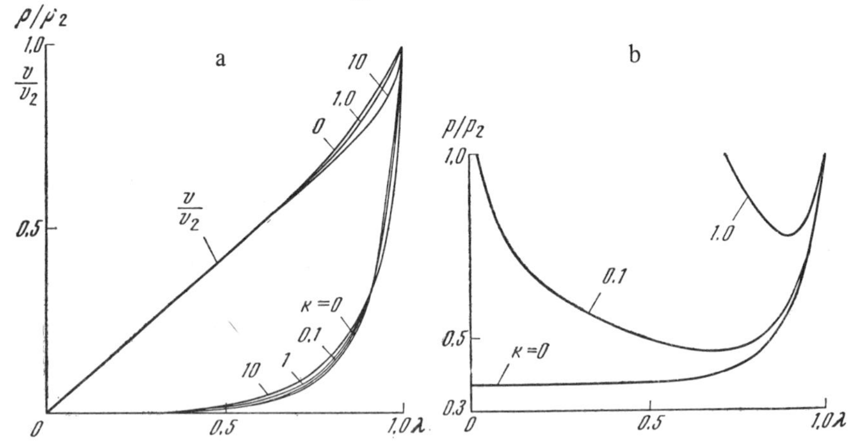

FIGURE 93. Distribution of velocities, densities (a) and pressures (b) behind a cylindrical wave for small values of R_m and various k

was found that $\alpha = 0.998$ ($k = 0.01$), $\alpha = 1.13$ ($k = 0.1$), $\alpha = 1.56$ ($k = 1$) and $\alpha = 15.2$ ($k = 10$), where

$$k = \frac{\kappa_1 \sigma_1}{c^2 \rho_1} \left(\frac{\varepsilon_0}{\alpha \rho_1} \right)^{-1/2}$$

is a parameter characterizing the interaction between the gas and the magnetic field.

It follows from (7.88) that the electromagnetic force f and the velocity v have opposite directions. These lead to a retardation of the velocity of the flow as compared with the gas velocity in an explosion where there is no magnetic field. Joule dissipation also occurs in the flow, resulting in a supply of heat to the particles.

A current flows through the gas whose density is given by the first relation of (7.88). If the current is led off from the system, the system can be considered a nonstationary MHD generator. We mention that certain systems using the energy of an explosive for generating electric current have been considered in [24]. We also remark that in [25] it is shown that the results of the theory correspond satisfactorily to experiments on cylindrical explosions in a magnetic field for small numbers R_m.

The solutions considered can also be found for the case of anisotropic conductivity (consideration of Hall currents).

In the case of finite numbers R_m, to solve the strong explosion problem one should consider the variation in the magnetic field and make use of the full system of MHD equations (7.82), (7.83).

We take the case of cylindrical symmetry under the condition $H_z = 0$, $m = n = 0$, i.e. the problem of a strong cylindrical point explosion when the conductivity of the gas is constant in the region of continuity of the gas-dynamical parameters, but the initial magnetic field is variable with $H^2 = H_\varphi^2 = \kappa_1 r^{-2}$. In this case one can take the system of equations (7.72), (7.74), (7.75) and note the relations

$$j_z = (c1/4\pi r)\ [d/dr/(rH_\varphi)],\ E_z = (j_z/\sigma) - (v/c)H_\varphi.$$

If dimensionless variables H^0, E^0 and $\mathcal{J}$ are introduced by the formulas

$$h_\varphi = \rho_1 \frac{r^2}{t^2} H^0, \quad E_z = \frac{\sqrt{8\pi\varepsilon}}{ct} E^0, \quad j_z = \frac{c}{t} \sqrt{\frac{\rho_1}{2\pi}}\, \mathcal{J}$$

and formulas 97.89) are used, this system can be written as follows:

$$\lambda\left\{\left(V - \frac{1}{2}\right) RV' + P' + H^{0\prime}\right\} + RV(V-1) + 2(P + H^0) = 0,$$

$$\lambda\left\{\left(V - \frac{1}{2}\right) R^1 + RV^1\right\} + 2RV = 0; \qquad (7.100)$$

$$\lambda\left\{\left(V - \frac{1}{2}\right) P^1 + \gamma PV^1\right\} - 2P + 2(\gamma+1)PV - \frac{2(\gamma-1)}{\lambda^4} A \left\{\frac{d}{d\lambda}\left(\lambda^2 \sqrt{H^0}\right)\right\}^2 = 0,$$

$$\lambda\left\{\left(V - \frac{1}{2}\right) H^{0\prime} + 2H^0 V^1\right\} - 2H^0 + 4H^0 V -$$

$$-\frac{2}{\lambda}\sqrt{H^0}\,A\frac{d}{d\lambda}\left\{\frac{1}{\lambda}\frac{d}{d\lambda}\left(\lambda^2\sqrt{H^0}\right)\right\}=0.$$

For E^0 and $\mathcal{J}$ we have

$$E^0=\frac{A}{\lambda}\frac{d}{d\lambda}\left(\lambda^2\sqrt{H^0}\right)-\lambda^2 V\sqrt{H^0},\quad \mathcal{J}=\frac{1}{\lambda}\frac{d}{d\lambda}\left(\lambda^2\sqrt{H^0}\right)\left(A=\nu_m\left(\frac{\rho_1}{\varepsilon}\right)^{1/2}\right). \tag{7.101}$$

Here A is a dimensionless constant parameter. We remark that a system of self-similar equations to (7.100) was indicated in [27]. The system (7.100) has the following integrals:

$$\frac{RV^2}{2}+\frac{P}{\gamma-1}+H^0-2V\left(\frac{RV^2}{2}+\frac{\gamma}{\gamma-1}P+2H^0\right)+$$
$$+\frac{4A}{\lambda^3}\sqrt{H^0}\frac{d}{d\lambda}(\lambda^2\sqrt{H^0})=c_2, \tag{7.102}$$
$$\lambda^2(V-0.5)\sqrt{H^{0\prime}}-A\frac{1}{\lambda}\frac{d}{d\lambda}(\lambda^2\sqrt{H^0})=c_3.$$

The last integral was pointed out by Greenspan [26] in studying the analogous problem for motions with a conductivity jump. The constants c_2 and c_3 are found from the boundary conditions. Using this integral, one can reduce the order of the system (7.100).

We shall consider the case in which the motion of the gas is accompanied by the appearance of a shock wave. From (7.85) and (7.86) we find the boundary conditions on the front of the shock wave (for $\lambda = 1$) for the dimensionless functions V, R, P and H^0:

$$[R(V-\delta)]=0,\quad [RV(V-\delta)+P]=0,$$
$$\left[(V-\delta)\left(\frac{RV^2}{2}+\frac{P}{\gamma-1}\right)+PV\right]=0, \tag{7.103}$$
$$[A(4H^0+H^{0\prime})-2VH^0]=0,\quad [H^0]=0.$$

In addition, the conditions $V_\infty = 0$, $R_\infty = 1$, $P_\infty = 0$ and $H^0_\infty = 0$ for $\lambda = \infty$ and the condition that the velocity be zero on the axis of symmetry must be satisfied. As before, we denote by the index 1 the quantities in front of the shock wave (region 1) and by the index 2 those behind the wave (region 2). The radius of the shock wave is determined by (7.90). From the integral law of energy conservation we find the formula for α.

For the complete solution of the problem it is necessary to integrate the system (7.100) in regions 1 and 2 taking into account the energy integral of (7.102). The boundary conditions (7.103), the conditions at infinity, and the condition at the center of symmetry must hereby be satisfied, and the constant α must be computed. The current density and the electric field are found from (7.101).

We mention [28] where the problem of cylindrical dispersion of a gas for finite R_m was considered.

§7. On the excitation of electromagnetic waves by strong shock waves with a conductivity jump

In the propagation of strong shock waves through a gas the properties of the gas are abruptly changed after passage of the front of the shock wave: the pressure, temperature, density, and electrical conductivity are strongly increased. The presence of interaction between a strong shock wave and the electromagnetic field is related to the abrupt increase in the electrical conductivity of the gas behind the front of the jump. For large nuclear explosions the increase in electrical conductivity may be caused by thermal ionization of the gas and other factors. A considerable increase in the electrical conductivity of the gas is also noted for blast waves arising from the detonation of chemical explosives [29].

If a shock wave propagates in space where there are magnetic and electric fields, then the change in properties of the medium on passing through the front of the shock wave causes perturbations of the magnetic and electric fields which propagate in the form of electromagnetic waves. Questions related to the radiation of electromagnetic waves by shock waves have been investigated in a number of works [1], [15], [30]–[32].

Below we shall consider problems of the radiation of electromagnetic waves by spherical and plane shock waves as they propagate in weak magnetic and electric fields. It will be assumed that the mechanism of exciting electromagnetic waves is related to the appearance of nonstationary currents and a conductivity jump on passage of the shock wave through the gas. We shall follow our papers [15], [30].

Suppose that a strong shock wave propagates with velocity $D(t)$ in a gas occupying an infinitely large volume; here t is the time. Since the electric and magnetic fields in advance of the front of the shock wave are assumed to be weak, we shall neglect their effect on the motion of the gas behind the front of the wave. This means that the parameters of the wave front will be the same as when they are obtained from solving the gas-dynamical problem. We shall assume that the gas-dynamical problem has been completely solved and hence that the velocity $D(t)$ is known.

We consider the initial and boundary conditions. Suppose that at the initial instant of time $t = 0$ the stress vectors of the magnetic field H and electric field E are constant. We assume that the conductivity of the gas is equal to zero before the front of the shock wave and equal to infinity behind it. Because of the absence of magnetic charges on the front of the shock wave, for the normal component of the magnetic field H_n we have the condition of continuity (we take the magnetic permeability equal to one). It follows from Maxwell's equations that the tangential component of the electric field vector E_τ is also continuous.

We mention that the conditions of continuity of H_n and E_τ coincide with the analogous conditions on shock waves propagating in a medium with infinite conductivity. We shall assume [1], [32] that the tangential component of the magnetic field H_τ is continuous. In accordance with the results of [1], [32] we suppose that the coefficient of magnetic viscosity ν_m in the shock layer is larger than the other dissipative coefficients. Denoting by the index 1 the quantities before the front of the shock wave and by the index 2 quantities directly behind it, we can write

$$\boldsymbol{H}_1 = \boldsymbol{H}_2, \qquad \boldsymbol{E}_{\tau 1} = \boldsymbol{E}_{\tau 2}. \tag{7.104}$$

Because of the infinite conductivity of the gas behind the front of the shock wave, in the system of coordinates coupled with the wave there is the following relation between H and E:

$$\boldsymbol{E}_2 = -\frac{1}{c}\boldsymbol{v}_2 \times \boldsymbol{H}_2 \tag{7.105}$$

(c is the speed of light).

Considering (7.104), in the stationary coordinate system we have

$$\boldsymbol{E}_{\tau 1} = -\frac{1}{c}[(\boldsymbol{v}_2 - \boldsymbol{D}) \times \boldsymbol{H}_1]_\tau . \tag{7.106}$$

The boundary conditions (7.104) and (7.106) on the front of the shock wave must be considered in solving the problems in question on the radiation of electromagnetic waves. Since the solution of the gas-dynamical problem is assumed known, the quantities v_2 and D in (7.106) are considered given.

The progagation of electromagnetic waves in a medium with zero conductivity and the values $\mu = \epsilon = 1$ is described by the system of Maxwell's equations

$$\frac{1}{c}\frac{dE}{dt} = \operatorname{curl} \boldsymbol{H}, \qquad \frac{1}{c}\frac{\partial \boldsymbol{H}}{\partial t} = -\operatorname{curl} \boldsymbol{E}, \qquad \operatorname{div} \boldsymbol{H} = 0, \qquad \operatorname{div} \boldsymbol{E} = 0. \tag{7.107}$$

The system (7.107) with boundary conditions (7.106) and the given initial conditions enables us to determine the laws of propagation of the electromagnetic waves.

Condition (7.104) and the MHD equations are to be used to find the electric and magnetic fields and the current distribution in the region of motion behind the front of the shock wave.

7.1. *Spherical shock waves.* Let the initial magnetic field be H_0 and the initial electric field E_0. We introduce a spherical coordinate system r, θ, φ, where the angle θ is reckoned from the direction of the vector H_0. If it is noted that v and D are directed along a radius, the boundary conditions (7.105) and (7.106) can be written componentwise as follows:

$$H_{r1} = H_{r2}, \qquad H_{\theta 1} = H_{\theta 2}, \qquad H_{\varphi 1} = H_{\varphi 2}, \tag{7.108}$$

$$E_{\theta 1}=\frac{1}{c}(v_2-D)\,H_{\varphi 1}, \qquad E_{\varphi 1}=-\frac{1}{c}(v_2-D)\,H_{\theta 1}. \tag{7.109}$$

Since for strong shock waves in a perfect gas with ratio of heat capacities γ we have $v_2 = 2D/(\gamma+1)$, conditions (7.108) and (7.109) can be rewritten as

$$E_{\theta 1}=-\frac{a}{c}DH_{\varphi 1}, \quad E_{\varphi 1}=\frac{a}{c}DH_{\theta 1} \quad \left(a=\frac{\gamma-1}{\gamma+1}\right). \tag{7.110}$$

We shall consider examples of solving problems on the propagation of electromagnetic waves. Suppose that $E_0 = 0$, $H_0 \neq 0$. We assume that the law of motion of the shock wave is given by the relations

$$\begin{gathered} D=\frac{dr_2}{dt}=\frac{c\varphi(\xi_2, r_2)}{a\psi(\xi_2, r_2)}, \qquad \xi_2=ct-r_2, \quad r_2(0)=0, \\ \varphi(\xi_2, r_2)=\xi_2\sum_{k=0}^{m} g_k\xi_2^k\,[\xi_2+(k+2)\,r_2], \qquad 0\leqslant m<\infty, \\ \psi(\xi_2 r_2)=r_2^2-\xi_2\sum_{k=0}^{m} g_k\xi_2^k\,[\xi_2+(k+2)\,r_2+\xi_2^2 r_2^{-1}(k+3)^{-1}]. \end{gathered} \tag{7.111}$$

Here the g_k are constants.

In this case the solution of the problem is given by the formulas

$$\begin{gathered} E_r=E_\theta=0, \qquad H_\varphi=0, \\ E_\varphi=H_\theta \sin\theta\frac{\xi}{r^2}\sum_{k=0}^{m} g_k\xi^k\,[\xi+(k+2)\,r], \\ H_r=-\frac{2H_0}{r^3}\xi^2\cos\theta\sum_{k=0}^{m}\frac{g_k}{k+3}\xi^k\,[\xi+(k+3)\,r]+H_0\cos\theta, \\ H_\theta=-\frac{H_0}{r}\xi\sin\theta\sum_{k=0}^{m} g_k\xi^k\left[k+2+\frac{\xi}{r}+\frac{\xi^2}{(k+3)\,r^2}\right]-H_0\sin\theta. \end{gathered} \tag{7.112}$$

We note that the value $\xi = 0$ corresponds to the front of the electromagnetic wave. It follows from (7.112) that for $\xi = 0$ the components of the vectors E and H assume their initial values.

Suppose that $D = D_0 = \text{const}$, $H_0 \neq 0$, $E_0 \neq 0$, and assume for simplicity that the vectors H_0 and E_0 are mutually perpendicular. A strong shock wave with constant velocity can be caused, for example, by a spherical piston expanding into the gas from some point (taken as the origin) with constant speed. In this case the problem is self-similar, and its exact solution can be found by introducing the self-similar independent variable $\gamma = r/D_0 t$, separating variables in the system (7.107), and solving the system of linear ordinary differential equations, with subsequent choice of the arbitrary constants from the boundary conditions. The solution has the form

$$\begin{aligned} H_r &= H_0 f_1(\lambda)\cos\theta, & H_\theta &= H_0 f_2(\lambda)\sin\theta + E_0 f_3(\lambda)\cos\varphi, \\ H_\varphi &= -E_0 f_3(\lambda)\sin\varphi\cos\varphi, & E_r &= E_0 f_4(\lambda)\sin\varphi\sin\theta, \\ E_\theta &= E_0 f_5(\lambda)\sin\varphi\cos\theta, & E_\varphi &= H_0 f_3(\lambda)\sin\theta + f_5(\lambda)\cos\varphi, \end{aligned} \tag{7.113}$$

where the $f_i(\gamma)$ $(i = 1, \cdots, 5)$ are known functions; they are given in [30].

We mention that the solution (7.113) for the case $E_0 = 0$ can be obtained from (7.112) with $m = 0$.

The solution of Maxwell's equations given by (7.112) can be used for finding approximately the parameters of electromagnetic waves for laws $D(t)$ which differ from (7.111) and model the dependence for an explosion. To this end one must insert in (7.112) the constants g_k found from the condition of approximation of the given dependence $D(t)$ or $D(r_2)$ by means of (7.111). In particular, this solution can be used when $D(t)$ and $r_2(t)$ are given in tables. It must here be noted that if the initial position of the electromagnetic wave is characterized by the quantity $r = r_0$, then in (7.112) the variable ξ must be taken in the form $\xi = r - r_0 - ct$.

7.2. *The case of plane waves.* We use a Cartesian coordinate system, taking the axis Oy parallel to the vector E and the axis Oz parallel to the vector H; in the unperturbed region we assume E and H to be constant, equal to E_{y0} and H_{z0}. Condition (7.106) then takes the form

$$E_{y1} = -\frac{1}{c}(v_2 - D) H_{z1}. \tag{7.114}$$

From (7.107) we have

$$\frac{\partial E_y}{\partial x} = -\frac{1}{c}\frac{\partial H_z}{\partial t}, \quad -\frac{\partial H_z}{\partial x} = \frac{1}{c}\frac{\partial E_y}{\partial t}. \tag{7.115}$$

It follows that E_y and H_z satisfy wave equations; moreover,

$$E_y + H_z = \Phi(\xi), \quad E_y - H_z = F(\eta), \tag{7.116}$$

where $\xi = x - ct$, $\eta = x + ct$, and $\Phi(\xi)$ and $F(\eta)$ are arbitrary functions.

We shall consider concrete problems.

A. *Propagation of a plane shock wave.* Suppose that at the initial instant $t = 0$ a strong shock wave begins to propagate through the gas, and that the pressure on its front varies according to the law

$$p_2 = p_0\left(\frac{x_0}{x_2}\right)^\beta, \tag{7.117}$$

where p_0 and β are positive constants, x_0 is the initial position of the shock wave, and $x_2(t)$ is the coordinate of the shock wave. The case $\beta = 0$ in (7.117) was considered earlier in [1]. The gas-dynamical conditions on the shock wave have the form

$$v_2 = \frac{2}{\gamma+1} D, \qquad D^2 = \frac{\gamma+1}{2} \frac{p_2}{\rho_1} \quad \left(D = \frac{dx_2}{dt}\right). \tag{7.118}$$

From (7.117) and (7.118) we obtain the law of motion of the shock wave:

$$x_2 = \left[x_0 + (0.5\beta + 1)\left(\frac{p_0}{2} \frac{\gamma+1}{\rho_1}\right)^{1/2} t\right]^{1/1+0,5\beta}. \tag{7.119}$$

The electromagnetic wave travelling in front of the shock wave in the positive direction of the axis Ox changes the initial field E_{y0}, H_{z0}, while

$$E_{y1} - H_{z1} = E_{y0} - H_{z0}. \tag{7.120}$$

On the front of the shock wave E_{y1} and H_{z1} are connected by (7.106), which, considering (7.118), takes the form

$$E_{y1} = \frac{D}{c} \frac{\gamma-1}{\gamma+1} H_{z1}. \tag{7.121}$$

Using (7.116) and (7.119)–(7.121), we find

$$H_{z1} = \frac{E_{y0} - H_{z0}}{\left(\frac{p_0}{\rho_1} \frac{\gamma+1}{2}\right)^{1/2} \left(\frac{\gamma-1}{\gamma+1}\right) \left[\frac{x_0}{f(\xi)}\right]^{\beta/2} \frac{1}{c} - 1}, \quad E_{y1} = H_{z1} + E_{y0} - H_{z0},$$

where $f(\xi)$ is found from the relation

$$\xi = f + c\left(\frac{2}{\gamma+1}\right)^{1/2} \left[\left(\frac{f}{x_0}\right)^{1+0,5\beta} - 1\right] \frac{1}{1+0,5\beta}.$$

Assuming the solution of the gas-dynamical problem to be completely known and using the frozen-in condition $H_z = \psi(s)\rho$ for the motion of the gas behind the front of the shock wave, where $\psi(s)$ is an arbitrary function of the Lagrange coordinate s, we can also find the dependence $H_z(x, t)$ or $H_z(s, t)$ in the region of gas flow. This dependence has the form

$$H_z = \frac{\gamma-1}{\gamma+1} \frac{E_{y0} - H_{z0}}{\rho_1} \frac{\rho(s, t)}{\left(\frac{p_0}{\rho_1} \frac{\gamma+1}{2}\right)^{1/2} \left(\frac{\gamma-1}{\gamma+1}\right) \left(\frac{s}{x_0}\right)^{\beta/2} \frac{1}{c} - 1}.$$

B. *Strong explosion over a plane.* At $t = 0$ there occurs an explosion of a charge in the form of a plane. The solution of the gas-dynamical problem is known, and the law of motion of the shock wave is

$$x_2 = \left(\frac{\tilde{E}}{\rho_1}\right)^{1/3} t^{2/3},$$

where $\widetilde{E}$ is a constant related to the energy of explosion (cf. Chapter 2).

We shall distinguish two cases of solving the problem.

a) $E_{y0} \equiv 0$; $H_{z0} = 0$ for $0 \leqslant x < x_0$ and $H_{r0} \neq 0$ for $x > x_0$, where $x_0 \geqslant \widetilde{E}/\rho_1 c^2$.

b) At time $t_0 > \widetilde{E}/\rho_1 c^3$ a plane electromagnetic wave is incident on the shock wave arising from the explosion.

Case a). Repeating arguments analogous to those of problem A, we find that for $x < x_0$

$$H_{z1} = 0, \qquad E_{y1} = 0;$$

and for $x \geqslant x_0$

$$H_{z1} = 1 - \frac{H_{z0}}{1 - \frac{2}{3c}\left(\frac{\widetilde{E}}{\rho_1}\right)^{1/3} [\tau(\xi)]^{1/3} \left(\frac{\gamma - 1}{\gamma + 1}\right)}, \qquad E_{y1} = H_{z1} - H_{z0},$$

where $\tau(\xi)$ is determined by

$$\tau = \frac{1}{c}\left[\left(\frac{\widetilde{E}}{\rho_1}\right)^{1/3} \tau^{2/3} - \xi\right] \qquad (\tau \geqslant t_0).$$

Case b). The solution of the problem on the reflection of the electromagnetic wave from the surface of the shock wave can be found by using the preceding results. It must only be noted that the quantities E_{y0} and H_{z0} will not be arbitrary in this case but are rather connected by the relation $E_{y0} = H_{z0}$. This follows from the fact that before collision of the shock and electromagnetic waves there were no electric and magnetic fields before the front of the shock wave.

In conclusion we mention that the results obtained in this section can be used to determine the parameters of the gas-dynamical shock wave if the parameters of the radiated electromagnetic wave are known.

BIBLIOGRAPHY

1. A. G. Kulikovskiĭ and G. A. Ljubimov, *Gas-ionizing magnetohydrodynamic shock waves,* Dokl. Akad. Nauk SSSR **129** (1959), 52–55 = Soviet Phys. Dokl. **4** (1959/60), 1185–1188. MR **22** #6333.

2. V. P. Korobeĭnikov, *The calculation of one-dimensional flows for cylindrical and plane explosions in an ideally conducting gas with the back pressure and the magnetic field taken into account,* Dokl. Akad. Nauk SSSR **165** (1965), 1019–1022 = Soviet Phys. Dokl. **10** (1965/66), 1131–1134.

3. J. D. Murray, *Strong cylindrical shock waves in magnetogasdynamics,* Mathematika **8** (1961), 99–120.

4. V. P. Korobeinikov and E. V. Rjazanov, *Some solutions of the equations of one-dimensional magnetohydrodynamics and their application to problems of shock wave propagation,* Prikl. Mat. Meh. **24** (1960), 111–120 = J. Appl. Math. Mech. **24** (1960), 144–158. MR **22** #8970.

5. V. P. Korobeĭnikov, *On one-dimensional motions of a gas in a magnetic field accompanied by shock waves,* PMTF Ž. Prikl. Meh. i Tehn. Fiz **1960**, no. 2, 47–53. (Russian)

6. A. Sakurai, *Interaction of cylindrical shock wave with applied magnetic field parallel to the axis,* J. Phys. Soc. Japan **17** (1962), 1663.

7. V. P. Korobeĭnikov, *The decay of weak magnetohydrodynamic shock waves,* Magnitnaja Gidrodinamika **1967**, no. 2, 25–30 = Magnetohydrodynamics **3** (1967), no. 2, 15–18.

8. ———, *Unidimensional automodel models of a conducting gas in a magnetic field*, Dokl. Akad. Nauk SSSR **121** (1958), 613–615 = Soviet Phys. Dokl. **3** (1958/59), 739–742. MR **21** #4005.

9. ———, *On cylindrical explosion and rectilinear discharge in an electrically conducting medium taking into account the magnetic field*, Voprosy Magnitnoĭ Gidrodynamiki i Dinamiki Plazmy. II, Izdat. Akad. Nauk. Latv. SSR, Riga, 1962, pp. 197–206. (Russian)

10. E. Kamke, *Differentialgeleichungen. Lösungsmethoden und Losungen. Teil I: Gewohnliche Differentialgleichungen*, 3rd ed., Math. und ihre Anwendung in Phys. und Tech., Band 18, Geest & Portig, Leipzig, 1944; Russian transl., Fizmatgiz, Moscow, 1961. MR **9**, 33.

11. C. Greifinger and J. D. Cole, *Similarity solution for cylindrical magnetohydrodynamic blast waves*, Phys. Fluids 5 (1962), 1597–1607. MR **26** #2134.

12. V. P. Korobeĭnikov, *Plane explosion in an electrically conducting gas with an oblique magnetic field*, Dokl. Akad. Nauk SSSR **171** (1966), 545–548 = Soviet Phys. Dokl. **11** (1966/67), 929–931.

13. A. G. Kulikovskiĭ and G. A. Ljubimov, *Magnetohydrodynamics*, Fizmatgiz, Moscow, 1962; English transl., Adiwes Internat. Ser. Phys., Addison-Wesley, Reading, Mass., 1965.

14. L. D. Landau and E. M. Lifšic, *Mechanics of continuous media*, 2nd ed., GITTL, Moscow, 1953; English transl., Part I: *Fluid mechanics*; Part II: *Theory of elasticity*, Course of Theoretical Physics, vols. 6, 7, Pergamon Press, New York; Addison-Wesley, Reading, Mass., 1959. MR **16**, 417; **21** #6839.

15. V. P. Korobeĭnikov and V. P. Karlikov, *On the interaction of strong explosion waves with an electromagnetic field*, Dokl. Akad. Nauk SSSR **133** (1960), 764–767 = Soviet Phys. Dokl. **5** (1960/61), 679–683. MR **25** #4743.

16. V. P. Korobeĭnikov and E. V. Rjazanov, *On the propagation of magnetohydrodynamic shock waves arising during explosions*, Voprosy Magnitnoĭ Gidrodynamiki. IV, Izdat. Akad. Nauk Latv. SSR, Riga, 1964, pp. 33–41. (Russian)

17. V. P. Karlikov, *The linearized problem of the propagation of a strong explosion in an inhomogeneous atmosphere*, Candidate's Dissertation, Moscow State Univ., Moscow, 1958. (Russian)

18. N. Neso, J. B. Fanucci and L. J. Kijewski, *Nonuniform expansion of a piston into an ionized medium with a weak magnetic field*, Phys. Fluids 6 (1963), 1241–1249.

19. V. P. Korobeĭnikov, *On the interaction of shock waves in an ideally conducting gas with weak magnetic fields*, PMTF Ž. Prikl. Meh. i Tehn. Fiz. **1964**, no. 4, 113–114. (Russian)

20. J. M. Burgers, *Penetration of a shock wave into a magnetic field*, Magnetohydrodynamics (A Symposium), edited by Rolf K. M. Landshoff, Stanford Univ. Press, Stanford, Calif., 1957, pp. 36–53; Russian transl., Atomizdat, Moscow, 1958.

21. A. Sakurai, *Blast wave theory*, Basic Developments in Fluid Dynamics, edited by M. Holt, Vol. 1, Academic Press, New York, 1965, pp. 309–375.

22. V. P. Korobeĭnikov and E. V. Rjazanov, *On the effect of a magnetic field on the propagation of plane and cylindrical shock waves*, PMTF Ž. Prikl. Meh. i Tehn. Fiz. **1962**, no. 4, 47–51. (Russian)

23. P. S. Lykoudis, *Magnetofluidmechanic blast waves in a medium with finite electrical conductivity*, Phys. Fluids 7 (1964), 1372–1380.

24. J. G. Linhart, *Plasma physics*, 2nd ed., Ser. in Phys., North-Holland, Amsterdam; Interscience, New York, 1960. MR **23** #B354.

25. G. C. Vlases and D. L. Jones, *Experimental study of cylindrical magnetohydrodynamic blast waves*, Phys. Fluids 11 (1968), 987–992.

26. H. P. Greenspan, *Similarity solution for a cylindrical shock-magnetic field interaction*, Phys. Fluids 5 (1962), 255–259. MR **27** #1095.

27. F. A. Baum, S. A. Kaplan and K. P. Stanjukovič, *Introduction to cosmic gas dynamics*, Fizmatgiz, Moscow, 1958. (Russian) MR **21** #1210.

28. P. P. Volosevič and V. S. Sokolov, *The self-similarity problem on the dispersion of an electrically conducting gas in a medium with a given axial magnetic field*, Magnitnaja Gidrodinamika **1967**, no. 1, 43–46 = Magnetohydrodynamics **3** (1967), no. 1, 26–28.

29. A. A. Briš, M. S. Tarasov and V. A. Cukerman, *Electric conductivity of the explosion of condensed explosives*, Ž. Èksper. Teoret. Fiz. 37 (1959), 1544–1550 = Soviet Phys. JETP **10** (1960), 1095–1100.

30. V. P. Karlikov and V. P. Korobeĭnikov, *Disturbances in an electromagnetic field caused by shock waves under the influence of a sudden increase in conductivity*, Prikl. Mat. Meh. 25 (1961), 554–556 = J. Appl. Math. Mech. 25 (1961), 819–823. MR 25 #895.

31. R. Gallet, *Propagation and production of electromagnetic waves in a plasma*, Il Nuovo Cimento (10) 13 (1959), Supplement, 234–256.

32. A. A. Barmin and A. G. Kulikovskiĭ, *On shock waves of ionized gases in an electromagnetic field*, Dokl. Akad. Nauk SSSR 178(1968), 53–58 = Soviet Phys. Dokl. 13 (1968/69), 4–6.

CHAPTER 8

ON THE PROPAGATION OF PERTURBATIONS FROM SOLAR FLARES

§1. Some data on the basic parameters of solar flares and the interplanetary medium

Questions of the occurrence of chromospheric solar flares and the propagation of the perturbations thereby arising through the interplanetary medium have been considered in a number of works. Below we shall study only questions of the propagation of perturbations caused by a flare through the gas of the solar corona and the interplanetary medium.

It is known (cf., for example, [1]–[4]) that solar chromospheric flares originate and develop in a relatively small volume (the area of a flare occupies about 0.1% of the area of the solar disk, and the height of the flare layer is of the order 10^9 cm). The time of development of processes at the seat of the flare is as a rule somewhere between 3×10^2 and 4×10^3 sec.

A considerable energy is released during the development of a flare in the chromosphere. The energy E_0^* is different for different flares and varies in order of magnitude between 10^{29} and 10^{34} erg. An appreciable part of this energy is dissipated into the surrounding space as light and Roentgen radiation and also as fast particles (solar cosmic rays). However, the main part of the energy is consumed in motion of the gas in the solar corona and in the interplanetary medium, i.e. in the creation of so-called corpuscular streams. This part of the energy we denote by E_0. In the sequel we shall be interested only in the energy E_0 and will consider processes of motion of the medium outside the seat of the flare, where the details of the process of inception and development of the flare are not of essential importance.

The propagation of perturbations from a solar flare takes place in the solar corona and in the interplanetary medium, with the leading front of the peturbation reaching distances of the order of an astronomical unit (a.u.), as a rule, within 20 to 50 hours after the start of the flare. As is known, magnetic storms occur on earth during these time intervals. Some particular observed data are given in §6.

We shall consider the basic characteristics of the quiet-day interplanetary medium, i.e. the period before a powerful flare begins.

A stream of particles, called the solar wind [1], continuously leaves the sun. On the earth's orbit the particle density is of the order 1–100 cm^{-3}. The velocity v_1 of particles of the solar wind does not decrease with distance from the sun. Thus, according to measurements based on the radiolocation of the solar corona [4], it was found that the velocity of the solar wind at a height of 3.5×10^{10} cm above the photosphere at the instant of measurement was approximately equal to 1.6×10^6 cm/sec, while at the earth's orbit this velocity exceeded 3×10^7 cm/sec. Data from many rocket measurements indicate that generally at distances of 1 a.u. the velocity v_1 has magnitude 330–700 km/sec. The density of the interplanetary plasma drops with increasing distance from the sun [1], [6].

We introduce a spherical polar system of coordinates r, θ, φ with center at the sun for which the axis of rotation of the sun is the polar axis. With respect to this coordinate system the density ρ_1 of the gas, the velocity v_1 of the solar wind, the pressure p_1 of the gas, the magnetic field H_1, and other parameters of the initial interplanetary medium depend in general not only on the radial coordinate r but on the angles θ and φ as well and also on the time t. This situation and also the presence of a large number of unknown functions and auxiliary parameters leads to substantial mathematical difficulties in the theoretical solution of the gas motion caused by a solar flare. For what follows we shall make certain simplifying assumptions. First of all, we will assume that the velocity of the quiet-day solar wind v_1 is directed along a radius. Secondly, on the basis of results of work [1], [4]–[8] devoted to theoretical and experimental questions of determining the parameters of the interplanetary gas, we can conclude that for the initial density ρ_1 and the velocity v_1 of the quiet-day solar wind within some solid angle $\kappa < 2\pi$ with origin at the sun and in a region between the sun and the earth's orbit the following approximate relations can be adopted:

$$\rho_1 = Ar^{-\omega}, \qquad v_1 = Br^{\omega-2}. \tag{8.1}$$

Here A, B and ω are quantities which we assume to be constant, not depending on the angles θ or φ or on the time t. Concrete values for A, B and ω can be found, for example, on the basis of data from rocket measurements and radiolocation measurements (cf. [1], [4], [5], [8]). The following theoretical models may be invoked to describe the motion of the gas.

1) Kinetic description of processes in a magnetized, high-temperature, three-component plasma consisting of electrons, ions, and neutral particles.

2) Use of the magnetohydrodynamic equations for a two- or three-component plasma.

3) Application of hydrodynamical approximations of a plasma with anisotropic "pressure" [9], [10] (cf. Chapter 1).

4) The hydrodynamic model. In this model the hydrodynamic approximation is used to describe motion of the medium, and the equations of ordinary gas dynamics or single-fluid magnetohydrodynamics are applied.

In view of the extreme complexity of the phenomenon in question we shall limit ourselves to studying only the hydrodynamic model.

§2. Use of dimensional analysis and the simplest similarity laws

We shall make use of the single-fluid hydrodynamic approximation to describe the motion of the medium, which we assume to be a perfect gas. If we neglect the time of energy release, the effect of the initial pressure of the gas p_1, electromagnetic forces, and the viscosity and thermal conductivity, then the system of basic characteristic parameters for the motion of the gas has the form

$$r,\ \theta,\ \varphi,\ t,\ E_0,\ A,\ B,\ \omega,\ \gamma,\ g_\odot,\ R_\odot,\ l,\ \Omega, \tag{8.2}$$

where γ is the effective adiabatic exponent, Ω is the mean angular speed of rotation of the sun, $g_\odot$ is the acceleration due to the force of gravity on the surface of the sun, $R_\odot$ is the solar radius, and l is a characteristic linear dimension of the seat of the flare.

The quantities A, B and ω may assume different values depending on the time of year and on the state of the interplanetary medium, while the energy E_0 is different for different flares. It therefore makes sense to clarify those basic dimensionless parameters which characterize the phenomenon under study and to consider questions of converting functions characterizing the motion of the gas when the parameters of the medium and the energy E_0 change [25].

Since gravitation and the proper rotation of the sun have a weak effect on the propagation of perturbations during flares, the parameters Ω and $g_\odot$ may be neglected for rough estimates of the characteristics of the moving gas. From the dimensional parameters E_0, A and B it is possible to form the following quantities with dimensions of length and time: the kinetic characteristic length $r^* = (E_0/AB^2)^{1/(\omega-1)}$ and the kinetic characteristic time $t^* = (r^*)^{3-\omega}/B$. Then for any dimensionless characteristic of the flow (for example, the density $g = \rho/\rho_1$) on the basis of the π theorem of dimension theory we can write

$$g = g\left(\frac{r}{r^*},\ \frac{t}{t^*},\ \theta,\ \varphi,\ \gamma,\ \omega,\ \alpha_1,\ \alpha_2\right), \tag{8.3}$$

where $\alpha_1 = R_0/r^*$, $\alpha_2 = l/r^*$ and $\alpha_2 \ll \alpha_1$. It follows from (8.3) that for fixed γ, α_1 and α_2 the dimensionless functions of the form (8.3) will describe a class of flows for different parameters E_0, A and B. For characteristic dimensions of the flow region much greater than $R_\odot$ one may neglect the effect of the finite radius of the sun on the motion of the gas in some neighborhood of the leading front of the perturbations, i.e. the effect of α_1 and hence also α_2 may be neglected.

If inside some solid angle it is assumed that the velocity of the solar wind is directed along a radius and the condition of energy release corresponds to spherical symmetry of the flow, then the gas flow inside the solid angle in question may be assumed spherically symmetric.

We introduce the notation $x = r/r^*$ and $\tau = t/t^*$. For the spherically symmetric flow model functions of the form (8.3) will depend only on the two variable parameters x and τ, i.e.

$$g = g(x, \tau, \gamma, \omega). \tag{8.4}$$

For a time of arrival of the leading front of a perturabtion of shock-wave type at the point $r = r_a$ we have $t_a = t^* \tau(x_a, \gamma, \omega)$ and $x_a = r_a / r^*$, where the dependence $\tau(x_a, \lambda, \omega)$ is determined in solving the problem.

Naturally, in considering the process of propagation of perturbations during flares globally the dependence of the initial density and velocity components of the solar wind on the angles θ and φ and possibly also on the time t must be considered. In this case the desired solution of the hydrodynamic problem will depend on a large number of variables, and the functions entering in this solution will have the form (8.3).

In the sequel we shall consider only the spherically symmetric model of motion of the medium. If the motion of the quiet-day solar wind is assumed isothermic and the question is posed of accounting for the initial pressure p_1, then for the one-dimensional model we have

$$p_1 = C r^{-\omega}. \tag{8.5}$$

In this case we have a new dimensional constant C which enables us to form a new characteristic length $r^0 = (E_0/C)^{1/3-\omega}$. In formulas of the form (8.4) there is now still another dimensionless parameter

$$\alpha_3 = r^0/r^* = E_0^{\frac{1}{(3-\omega)} - \frac{1}{\omega-1}} C^{-1/3-\omega} \times (AB^2)^{1/\omega-1}.$$

Estimates we have made of the total initial energy of the solar wind in the volume enclosed within a certain solid angle (cf. §4) indicate that at distances of the order of an astronomical unit the kinetic energy of the quiet-day solar wind is of greatest importance, and it is approximately an order of magnitude greater than the total initial thermal energy of the gas (and greater than the initial energy of the magnetic field). The parameter α_3 may therefore be neglected for a rough qualitative analysis of the phenomenon in question.

In order to get an ideal of the order of magnitude of the kinetic characteristic length r^*, we refer r^* to some radius r_a and on the basis of the formulas for ρ_1 and v_1 we eliminate AB^2 from the expression for r^*. We then obtain

$$\frac{r^*}{r_a} = \left(\frac{E_0}{k_*}\right)^{1/(\omega-1)}, \qquad k_* = \rho_1(r_a)[v_1(r_a)]^2 r_a^3.$$

For α_3 we can write

$$\alpha_3 = E_0^{1/(3-\omega)} k^{-1/(3-\omega)} k_*^{1/(\omega-1)}, \qquad k = p_1(r_a) r_a^3.$$

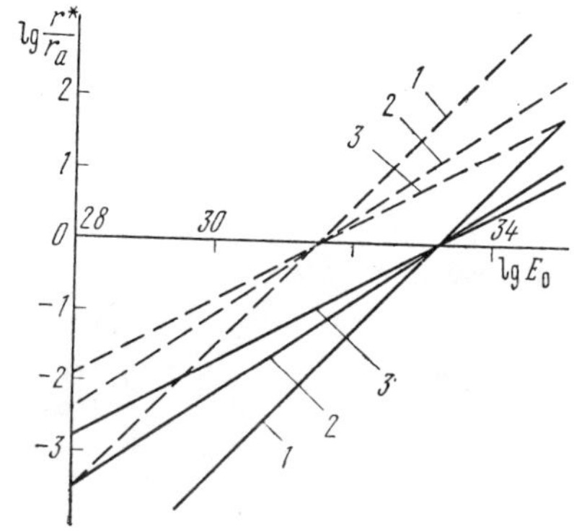

FIGURE 94. Dependence of the characteristic length r^* on the energy of the flare E_0 for two values of k_*

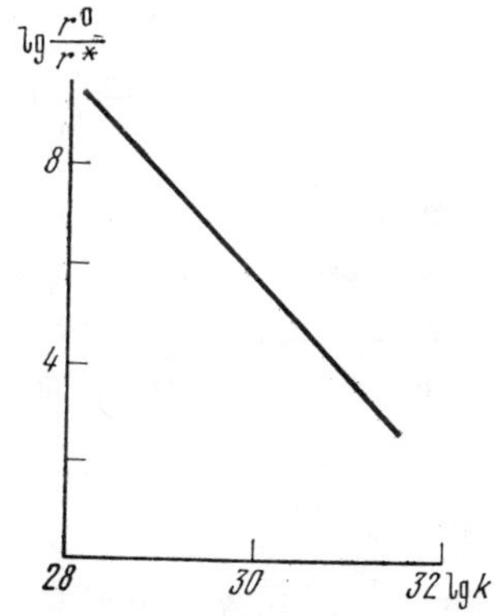

FIGURE 95. Dependence of the ratio of characteristic lengths on the magnitude of the parameter of the mean thermal energy k

Figure 94 shows the dependence of r^*/r_a on E_0 for various ω in the case where $r_a = 1.5 \times 10^{13}$ cm (the astronomical unit), $k_* = (1.5)^3 \times 10^{31}$ g $\cdot$ cm^2/sec^2 (dashed lines) and $5 \times (1.5)^3 \times 10^{32}$ g $\cdot$ cm^2/sec^2 (solid lines), and the energy of the flare E_0 varies from 10^{28} to 10^{35} erg. Computations show that for large values of E_0 the quantity r^* may be greater than an astronomical unit. Figure 95 shows curves of $\alpha_3(k)$ for $k_* = (1.5)^3 \times 10^{32}$ g $\cdot$ cm^2/sec^2, $E_0 = 10^{33}$ erg and $\omega = 2.5$ (the range of variation of k given in this graph corresponds to a pressure variation from 10^{-8} to 10^{-11} dyne/cm^2).

The dependence of the unknown functions on the dimensionless parameters can be determined theoretically or experimentally by measurements from rockets during the time of solar flares. In the experimental and theoretical determination of the gas parameters the dependence on the coordinates and time for the density, pressure and velocity of the gas and for the time of arrival of the perturbation at a prescribed point can be found only for some fixed $E = E_{01}, A = A_1$ and $B = B_1$.

We therefore consider the question of converting data obtained to the case of other values of these constants E_{02}, A_2 and B_2. We proceed as follows. We find the dimensionless time τ and dimensionless coordinate x for the state E_{01}, A_1, B_1, and then by the formulas $r = r^*x$ and $t = t^*\tau$ we find the values of the coordinate and time for the state E_{02}, A_2, B_2.

For converting the time we find

$$\tau = \frac{t_{(1)}}{t_1^*}, \qquad t_{(2)} = t_2^*\tau = \frac{t_2^*}{t_1^*} t_{(1)}, \tag{8.6}$$

where

$$t_i^* = (r_i^*)^{3-\omega}/B_i, \qquad r_i^* = \left(\frac{E_{0i}}{A_i B_i^2}\right)^{1/\omega-1}, \qquad i = 1, 2.$$

For converting distances (the Euler or Lagrange coordinate of a gas particle) we have

$$x = \frac{r_{(1)}}{r_1^*}, \qquad r_{(2)} = r_2^* x = \frac{r_2^*}{r_1^*} r_{(1)}. \tag{8.7}$$

Here $t_{(i)}$ and $r_{(i)}$, $i = 1, 2$, are dimensional values of the time and coordinate corresponding to a flare with parameters E_{0i}, A_i, B_i.

Analogous formulas can also be written for converting the velocity, density, and pressure of the gas and other quantities. Thus for the density ρ we have

$$g = \frac{\rho_{(1)}}{\rho_{11}}, \qquad \rho_{(2)} = \rho_{12} g = \frac{\rho_{12}}{\rho_{11}} \rho_{(1)}. \tag{8.8}$$

Formulas (8.6)–(8.8) give (in the chosen flow model) the similarity laws for processes of gas motion caused by a flare. For more complex models of gas flow formulas of the type (8.6)–(8.8) hold only for fixed values of the auxiliary dimensionless constant parameters, for example, $\alpha_j, j = 1, 2, 3$.

An analogous dimensional analysis of the problem can also be performed for the case in which a more complex model of the medium is adopted, for example, the two-component model of a plasma. If the magnetic field has an essential effect on the motion, the flow will not be spherically symmetric. Moreover, in reckoning with strong magnetic fields the system of defining parameters of the problem also contains quantities characterizing the state of the initial magnetic field. Consideration of gravitation also leads to the appearance of new dimensionless variables and further complication of the problem.

The conclusions drawn above are to be considered in theoretical and experimental determination of the parameters of the gas flow during solar flares.

§3. The point explosion problem in an inhomogeneous moving medium

The above-indicated features of the development of a solar flare in the initial stage afford the conclusion that in a number of cases this phenomenon can be modeled by a point explosion in a gas the initial parameters of which are dictated by the state of the solar corona and the interplanetary plasma. We shall consider the formulation of the problem in the simplest case of spherical symmetry. We shall assume that the gas is perfect with constant value γ. We neglect the viscosity and thermal conductivity of the gas. We also neglect the effect of magnetic fields and gravitation on the motion of the gas. We then arrive at the following problem (cf. also §1, Chapter 3) [26].

Suppose that at time $t = 0$ in a gas with initial state $v_1(r), \rho_1(r), p_1(r)$ energy E_0 is released at the point $r = 0$. It is required to determine the ensuing motion of the gas.

If we assume for simplicity that the initial state of the gas is isothermic, then in correspondence with the approximate relations (8.1) indicated above we have the following initial data for v_1, ρ_1 and p_1 (for $r > 0$):

$$v_1 = Br^{\omega-2}, \qquad \rho_1 = Ar^{-\omega}, \qquad p_1 = Cr^{-\omega}. \tag{8.9}$$

To determine the flow of the gas which arises it is necessary to integrate the system of gas dynamics (3.5) for an adiabatic perturbed motion. We denote by $r_2(t)$ the leading front of the perturbed gas, which we assume to be an ordinary shock wave. Then, in addition to the initial conditions (8.9), we have

$$r_2(0) = 0 \tag{8.10}$$

and the condition of finite energy release E_0 at the point $r = 0$.

The boundary conditions on the shock wave for $r = r_2$ have the form (3.6) (cf. Chapter 3). In addition to the conditions on the shock wave we have a condition at the center of symmetry. Thus, if there is no constantly moving mass source at the center, then $v(0, t) = 0$. Here it should again be noted that, strictly speaking, the initial functions $v_1(r), \rho_1(r)$ and $p_1(r)$ themselves must form a solution of the stationary equations of gas dynamics, taking account of gravitational and other forces. Otherwise motion begins in advance of the front of the main perturbation. For a rough qualitative analysis we shall suppose that in the case of approximate perscriptions of v_1, ρ_1 and p_1 the effects of motion of the gas in advance of the front of the shock wave may be neglected. The general picture of the gas flow is shown in Figure 96.

For $v_1 = 0$ and $p_1 = 0$ (or $B = C = 0$) we have the case of the well-studied self-similar problem of a strong explosion in a gas with variable initial density. The exact solution of this problem has been studied in [11]–[13] and in Chapter 2.

We consider certain questions pertaining to the solution of this problem. If $\omega = \omega_1 = (7 - \gamma)/(\gamma + 1)$, then, as we indicated in Chapter 2, the self-similar strong explosion problem has the analytic solution [11]

$$\rho = \frac{\gamma+1}{\gamma-1}\rho_1\frac{r}{r_2}, \qquad p = \frac{2\delta_1^2}{\gamma+1}\rho_1\frac{r_2^2}{t_2}\left(\frac{r}{r_2}\right)^3, \tag{8.11}$$

$$v = \frac{2\delta_1}{\gamma+1}\frac{r}{t}, \qquad \delta_1 = \frac{2}{5-\omega_1}, \tag{8.12}$$

$$r_2 = \left(\frac{E_0}{\alpha A}\right)^{\delta_1/2} t^{\delta_1}, \qquad \alpha = \frac{2\varkappa(\gamma+1)}{3(\gamma-1)(3\gamma-1)^2}. \tag{8.13}$$

If $\gamma = 5/3$, then $\omega_1 = 2$. For other values of γ and $\omega = 2$ the solution has been studied in detail in [13], where the values of α are computed and graphs of the self-similar functions are given.

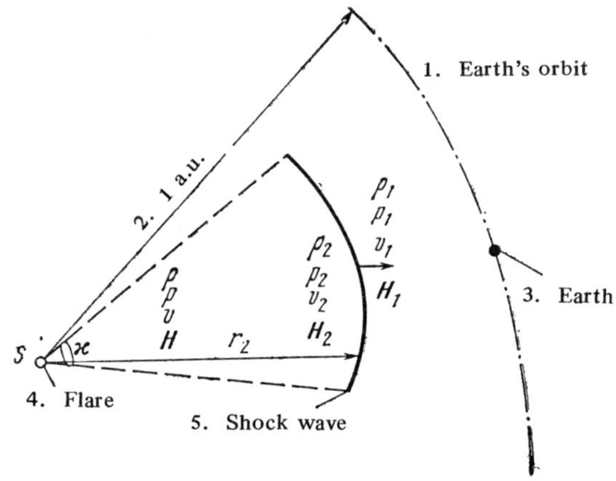

FIGURE 96. Scheme of propagation of the shock wave

For arbitrary γ and $\omega < 3$ we have a formula analogous to (8.13) for the law of motion of the shock wave:

$$r_2 = \left(\frac{E_0}{\alpha A}\right)^{\delta/2} t^{\delta}, \qquad \delta = \frac{2}{5-\omega}, \tag{8.14}$$

where $\alpha(\gamma, \omega)$ is the constant introduced earlier in Chapter 2 and defined from the integral law of energy conservation.

If $B \neq 0$ or $C \neq 0$, then the problem is not self-similar and for its complete solution numerical or approximation methods can be used.

The dimensional analysis presented earlier is altogether applicable to the problem being considered. We mention here only the following. If it is assumed that at the center of the explosion there is a heavy point mass M which interacts with the gas according to Newton's law, then on the right side of the first equation of (3.5) we must add the gravitational acceleration:

$$F_g = -\frac{Mf}{r^2}. \tag{8.15}$$

In this case there appears in the system of defining parameters still another dimensional constant $\zeta = Mf$ and a new characteristic length

$$r_g = \left(\frac{E_0}{\zeta A}\right)^{(2-\omega)/3}. \tag{8.16}$$

If $\omega = 2$, then r_g is a derived parameter, since in this case ζA has the dimension of energy. We remark that for $C = B = 0$ the problem will be self-similar even when the gravitational force is included. Such problems have been investigated in detail by M. L. Lidov [14].

The above dimensional analysis shows that even for the case of the simplest spherically symmetric hydrodynamical flow model in the general case the solution depends in an essential way on the constant parameters γ, ω and α_3. This introduces additional difficulties in solving the problem. In order to clarify the character of the effect of the parameters B and C on the solution it is possible to use the method of linearization about the self-similar solution studied in detail in Chapter 3. Two cases can be considered separately: $B \neq 0$, $C = 0$ and $B = 0$, $C \neq 0$.

The first case corresponds to considering the velocity of the unperturbed medium, and the second case to considering a variable counterpressure. Solution of the problem in these two cases makes it possible to consider (in the linear formulation) both these parameters simultaneously by making use of the principle of superposition of linear solutions. We shall not go into this in detail here, since all the necessary data are given in Chapter 3.

§4. On application of the solutions of the problems to the propagation of perturbations during solar flares

There are three basic restrictions on the applicability of the solutions considered above to propagation of perturbations during solar flares.

The first is the suitability of the hydrodynamic description of the interplanetary medium. As is known, the length of the mean free path of plasma particles computed on the basis of Coulomb scattering is much greater than 1 a.u., and therefore the question arises of the possibilities of using the model of a continuous medium and the theory of shock waves. Investigations recently conducted [1], [15], [16] indicate, however, that in a magnetized plasma such as the interplanetary medium an essential role as a characteristic physical parameter is played by the Larmor radius r_L, and so-called collisionless shock waves may be realized (r_L for the interplanetary plasma is much less than 1 a.u.). On the other hand, it is to be noted that in the initial stage of development of a flare the plasma possesses a rather high density (the density of the chromosphere is 10^{12} cm^{-3}), and application of the hydrodynamical description here is altogether justified. Moreover, starting in 1964 results have been published on the detection of magnetohydrodynamic shock waves and other discontinuities in the interplanetary medium [4], [17], [18]. Here it should also be mentioned that the arrival of the leading front of the basic perturbation at the earth's orbit within approximately 24 hours is difficult to explain without invoking the theory of shock waves.

From the point of view of theoretical investigations Gold advanced the hypothesis more than 10 years ago that geomagnetic disturbances stimulated by a flare were caused by the arrival of the shock wave [19].

The arguments enumerated in favor of the application of the hydrodynamic model afford the hope that the theory of shock waves can be successfully applied in studying questions of solar flares.

The second restriction on the models of phenomena during flares proposed above is that the process of motion of the gas is not one-dimensional, since both the initial characteristics as well as the ensuing motion are essentially three-dimensional. However, the utilization of one-dimensional spherically symmetric solutions can be useful in considering flows inside some solid angle proceeding from the center of the flare or the center of the sun.

The third essential restriction is related to the hypothesis of instantaneous energy release in the flare. Actually, the process of energy release in a number of cases takes a time of the order of 1/20 the time required for the propagation of the shock wave a distance of 1 a.u. Therefore, to describe processes in the one-dimensional formulation it should be assumed that E_0 is a function of time. The simplest cases of such problems have been considered in [1], [11]. The approximate formulation of the problem involving instantaneous energy release may be applied only to estimate flow parameters of the gas in those cases where the time of energy release does not exceed 1/25 the transit time of the wave to some fixed point. If it is assumed that the energy of the flare is evolved over some period of time $E_0 = E_0(t)$, then the flare process can be modelled by the expansion of a piston in a gas, supposing that the velocity of the piston depends on time. It is also possible to study a more general model where after a point explosion with energy E_0 a piston starts to move in the gas with variable velocity, modelling the process of expansion of the gases from the seat of the flare. An analogous problem has already been studied in ordinary gas dynamics [35].

In the models mentioned above the supply of mass to the gas flow has been neglected. In reality, however, mass ejection from the solar chromosphere into the moving stream can have a definite effect on the character of the flow. To reckon with this situation the model of the flare must be complicated still further. The following formulation of the problem can be proposed. Suppose that at the initial instant $t = 0$ at the point $r = 0$ energy E_0 is released, calculated over the solid angle κ. Moreover, at the same point energy $N(t)$ and mass $Q(t)$ are supplied which play a role in establishing the process of motion inside the solid angle κ. If N and Q depend on powers of t, while $E = 0, p_1 = 0$ and $v_1 = 0$, it is possible to distinguish a class of self-similar solutions. For the case of constant initial density the self-similar problem has been investigated in [35]. Qualitative analysis of the self-similar problem shows that the flow here will have a complex character. Contact surfaces and a second shock wave arise in the flow. We remark also that the dependence of the initial parameters ρ_1, p_1 and v_1 on the angles θ and φ can be considered in sectorial approximation using a rough scheme analogous to that proposed in [36].

It is of interest to estimate the initial energy of the quiet-day solar wind per unit solid angle ($\kappa = 1$) at a distance of 1 a.u.

Taking the approximation (8.9), we shall assume that

$$A = a \cdot 10^{-22} r_a^{\omega}, \qquad 0.01 \leqslant a \leqslant 1,$$
$$B = v_1(r_a) r_a^{2-\omega}, \quad v_1(r_a) = 1.5k \cdot 10^7 \text{ cm/sec},$$
$$1.5 \leqslant k \leqslant 4, \qquad C = \beta A \cdot 10^{12}, \qquad 0.1 \leqslant \beta \leqslant 15,$$

$r_a = 1$ a.u., and the constants A, B and C are expressed in grams, centimeters and seconds. Neglecting the solar radius, we have for the initial energy per unit solid angle

$$E_1 = \int_0^{r_a} r^2 \left(\frac{\rho_1 v_1^2}{2} + \frac{p_1}{\gamma - 1} \right) dr. \tag{8.17}$$

Considering (8.9), we find that

$$E_1 \sim A \left[\frac{r_a^{\omega-1}}{\omega - 1} \frac{B^2}{2} + \frac{r_a^{3-\omega}}{(\gamma - 1)(3 - \omega)} \frac{C}{A} \right]. \tag{8.18}$$

In (8.17) and (8.18) the first term corresponds to the initial kinetic energy E_{1k}, and the second term to the thermal energy E_{1T}: other forms of energy are not considered. For $\omega = 2, a = 0.1, b = 5$ and $k = 1.5$ we find $E_{1k} \sim 5 \times 10^{31}$ and $E_{1T} \sim 10^{30}$

Thus the main fraction of the energy is contained in E_{1k}. The thermal and electromagnetic energies are considerably less than the initial kinetic energy. On the other hand, the quantity E_{1k} at distances of the order of 1 a.u. and greater is comparable with the total energy of the flare, and it is desirable to take it into account in considering theoretical models of the gas motion. An example of this was given in §3 of the present chapter and in Chapter 3. For powerful flares which are strongly localized in space and develop rapidly we can apply the results of the theory of strong point explosions. The simplest solution (8.11)–(8.13) in the particular case $\gamma = 5/3$, $\omega_1 = 2$ was used by Parker [1] to describe the gas motion during flares (he did not consider the effect of the solar wind at all).

If the results of the investiagtion of those simplest problems already considered are used, one can attempt a more detailed investigation of questions related to the application of the theory of explosion to solar flares. The following conclusions, which follow from the preceding investigation, may be mentioned regarding applications to solar flares.

1. From the theory of strong explosions it is possible to determine the transit time of the shock wave to the earth's orbit or to another point, once the energy of the flare is known. Indeed, from (8.14) we have

$$t_a = \left(\frac{\alpha A}{E_0}\right)^{1/2} r_a^{(5-\omega)/2}. \tag{8.19}$$

From (8.19) with identical values of ω for different flares we find

$$\frac{t_2}{t_1} = \left(\frac{A_2}{A_1}\frac{E_{01}}{E_{02}}\right)^{1/2}.$$

If $A_2 = A_1$, then

$$\frac{t_2}{t_1} \sim \left(\frac{E_{01}}{E_{02}}\right)^{1/2}. \tag{8.20}$$

Since α and A may be considered known for various γ and ω, on the basis of (8.19) we can determine approximately the time of arrival of the leading front of the perturbation at the earth. (8.20) shows that the ratio of the squares of the transit times of the perturbation to some point is inversely proportional to the ratio of the energies of the flares.

2. It is possible to determine the energy of flares by knowing the time of arrival t_a of the perturbation at a given point:

$$E_0 = \alpha \frac{r_a^{5-\omega}}{t_a^2} A. \tag{8.21}$$

If $t_a = 10^5$ sec and $r_a = 1$ a.u., then for $\omega = 2$ we have $E_0 \sim a\alpha(1.5)^5 \times 10^{33}$ erg. For $\alpha \sim 1$ and $a \sim 0.1$ we have $E_0 \sim 10^{33}$ erg. More precise estimates for specific flares are given in §6.

3. Analogous conclusions and estimates can be made also for the case in which the initial velocity v_1 (cf. also §6) and initial pressure p_1 are considered. We mention here that an analysis of the solutions obtained in Chapter 3 for cases including counterpressure shows that in a certain zone behind the front of the shock wave the density increases with decreasing r, i.e. there is a negative density gradient. This effect is absent if p_1 and ρ_1 are constant.

4. Conclusions of dimension theory and similarity laws may be used to process experimental results of measuring the parameters of shock waves.

5. The effect of the flow on the magnetic field and the effect of gravitation on the gas flow can be considered.

§5. On the variation of the magnetic field during motion of the gas

To describe the magnetic fields we choose the spherical coordinate system introduced in §1. Let Ω be a vector directed along the axis of rotation of the sun, which we take as the polar axis. The components of the magnetic field H at the point r, θ, φ are H_r, H_θ, H_φ, where the component H_r is directed along the sun's radius, H_φ along the vector $\Omega \times r$ and H_θ in the southerly direction (the usual right-handed coordinate system is assumed here). Parker [1] proposed a model of the unperturbed magnetic field in which it was assumed that outside some distance $r = r_0$ the solar wind has a radial direction and constant velocity $v_1 = \text{const}$ (the case $\omega = 2$ in (8.1)). The gas was considered an ideally conducting medium. We shall consider an analogous model for the case of variable velocity $v_1 = Br^{\omega-2}$.

The equation for the magnetic force lines in the r, φ plane has the form

$$\frac{dr}{rd\varphi} = \frac{H_r}{H_\varphi} \sin\theta. \tag{8.22}$$

By analogy with [1], [20], for the components of the initial magnetic field we can write

$$H_{r1} = H_{r0}\left(\frac{r_0}{r}\right)^2, \quad H_{\theta 1} = H_{\theta 0} = 0, \quad H_{\varphi 1} = -H_{\varphi 0}\left(\frac{r_0}{r}\right)^2\left[\frac{\Omega}{v_1}(r - r_0)\sin\theta\right]. \tag{8.23}$$

Here we denote by $H_{r0}, H_{\theta 0}, H_{\varphi 0}$ the components of the field H at some point $(r_0, \theta_0, \varphi_0)$ where a force line originates which passes through the point (r, θ, φ).

Integrating (8.22) with consideration of (8.23), we find the equation for the force lines:

$$\begin{aligned} \varphi &= \varphi_0 + \frac{r_0}{2-\omega}\left[1 - \left(\frac{r_0}{r}\right)^{2-\omega}\right]\frac{\Omega}{v_1} - \frac{r - r_0}{3-\omega}\frac{\Omega}{v_1}, & \omega \neq 2, \\ \varphi &= \varphi_0\left[1 - \frac{r}{r_0} + \ln\frac{r}{r_0}\right]\frac{r_0\Omega}{v_1}, & \omega = 2. \end{aligned} \tag{8.24}$$

Here Ω is the mean angular velocity of rotation of the sun, $\Omega = 2.9 \times 10^{-6}$ rad/sec.

Formulas (8.24) give a family of spirals in the r, φ plane which depend on the parameters $\varphi_0, r_0, \omega, \Omega$ and B.

Relations (8.23) and (8.24) give a more general model (as compared with Parker's model [1]) of the quiet-day interplanetary field. Since the total energy of the magnetic field is considerably less than the flare energy E_0 while the quantity $\rho_2 v_2^2$ is much greater than the initial pressure $H_1^2/8\pi$, in first approximation the effect of the magnetic fields on the motion of the interplanetary plasma need not be considered. Then for spherically symmetry gas flows caused by strong explosion in a medium with variable density $\rho_1 = Ar^{-\omega}$ it is possible to compute the parameters of the magnetic field. In the approximation of single-fluid magnetohydrodynamics we have on the basis of formulas presented in Chapters 2 and 7 (cf. also [11], [12], [22])

$$H_r = H_{r1}\left(\frac{\xi}{r}\right)^2, \qquad H_\theta = H_{\theta 1}\frac{\rho}{\rho_1}\frac{r}{\xi}, \qquad H_\varphi = \frac{\rho}{\rho_1}\frac{r}{\xi}H_{\varphi 1},$$
$$\frac{r}{r_2} = \Phi_1(\omega, \gamma, \mu), \qquad \frac{\xi}{r^2} = \Phi_2(\omega, \gamma, \mu), \qquad \frac{\gamma+1}{2\gamma} \leqslant \mu \leqslant 1. \tag{8.25}$$

Here ξ denotes the Lagrange coordinate of the particle, μ is a parameter, and the functions $\Phi_1(\omega, \gamma, \mu)$ and $\Phi_2(\omega, \gamma, \mu)$ are given by (2.5) and (2.17).

The relations (8.22)–(8.25) enable us to write a differential equation for the force lines in the region of the perturbed motion of the gas:

$$\frac{d\varphi}{dr} = \frac{(r_0 - r)\,\Omega}{v_1(\xi)}\frac{r^2}{\xi^2}\frac{\rho}{\rho_1}.$$

We denote by u the angle between the direction of the radius vector and the magnetic field; for this angle we have

$$\tan u = r\frac{d\varphi}{dr}. \tag{8.26}$$

In the general case of a strong shock wave for the difference between $\tan u$ at any point of the flow and the initial value $\tan u_1$ we have

$$\tan u - \tan u_1 = \frac{H_{\varphi 1}}{H_{r1}\sin\theta}\left[\left(\frac{\xi}{r}\right)^2\frac{\rho_1}{\rho} - 1\right]. \tag{8.27}$$

From (8.27) it is easy to find an expression for the jump in $\tan u$ on passing through the shock wave:

$$\tan u_2 - \tan u_1 = +\frac{2}{\gamma - 1}\frac{H_{\varphi 1}}{H_{r1}\sin\theta}. \tag{8.28}$$

The results presented in this section on determining the configurations of magnetic fields may be useful in processing the results of observations of the interplanetary medium. Computed configurations of the magnetic fields make it possible to solve the problem of the motion of high-energy particles on intersecting a jump of the magnetic fields. The induction electric field behind the front $E = -(1/c)\mathbf{v} \times H$ can accelerate these particles. Calculations of the equations of motion of the particles (cf. §3, Chapter 1) have shown [3], [4] that a twofold acceleration of electrons across the front of a shock wave is possible (with a homogeneous initial magnetic field). We shall not consider this question in detail here.

§6. Comparison of theoretical conclusions with certain observational data

At the present time data have accumulated on the observation of chromospheric solar flares and measurements of the parameters of the shock waves arising during these flares and propagating through interplanetary space. Analysis of optical and radioastronomic measurements, data from observations on the occurrence of magnetic storms on the earth, and measurements from satellites, cosmic rockets, and automatic interplanetary stations of the parameters of shock waves and the intensities of cosmic rays published over the past five years [1], [4], [8], [17], [18], [20], [27]–[33] affords the following general conclusions, some of which have already been noted above:

1) The transit time of the perturbation to the earth's orbit varies from 16 to 70 hours.

2) As the shock wave passes near the earth's orbit the density of the plasma increases 2–3 times.

3) Optical measurements give an estimate of the energy of a flare of the order 10^{30}–10^{32} erg.

4) The averaged velocity of shock waves at the earth's orbit is of the order 500–800 km/sec.

5) The average velocities of shock waves at the earth's orbit is, as a rule, no more than 2 or 3 times greater than the velocity of motion of the quiet-day solar wind.

TABLE 7

Data From Observations of Flares

№№	Flare characteristic	Transit time and distance				Parameters of the medium			Estimate of the energy $E_0 \cdot 10^{-31}$ erg			
		t hr		$r_2 \cdot 10^{-6}$ km					model of strong explosion		wind v_1	
		t_1	t_2	$r_2^{(1)}$	$r_2^{(2)}$	$n_1 \frac{1}{\text{cm}^3}$	$v_1 \frac{\text{km}}{\text{sec}}$	D_{av}	$\gamma = \frac{5}{3}$	$\gamma = 3$	E_0 from $x_2(\tau)$	E_0 from $x_2(\gamma)$
1	4 X 1965 0937 *UT* S24W31, 2+	17.1	26	150	170	6	350	620	201	38.5	165	17
2	18 I 1966 2254 *UT* N20E07, 2*B*	44	55	128	147	9	380	480	28.0	5.42	14.0	9.9
3	7 VII 1966 . 0027 *UT* N34W48, 2*B*	44.5	44.52	152	152,02	4	400	710	21,0	4.00	11.6	14
4	18 IX 1967 . 1918 *UT* N18W84, 2*B*	16	27	116	150	20	450	860	210	40.5	168	55
5	17 IX 1967 . 2314 *UT* N16W61 , 2*B*	28	46	117	150	20	450	510	71,8	14.0	41.0	13.2

The experimental data mentioned above were processed by us in [37], and facts necessary for the sequel are presented in Table 7. This table gives the dates and time of flares (universal time UT), the coordinates on the sun and the point. We shall take the model of a blast wave to describe the process of propagation of the disturbances. In correspondence with the method indicated for determining the energy of the flare on the basis of known values of α and A, and the transit time t_a of the perturbation to a given point r_a, from the formula (8.21) for a strong explosion we have

$$E_0 = \alpha(\gamma) \frac{r_2^3}{t^2} m_p n_{1a} r_a^2 \qquad (\omega = 2), \tag{8.29}$$

where n_{1a} is the density at the earth's orbit or at another point of measurement and m_p is the mass of a proton.

We shall assume that the flare energy goes to motion of the gas inside a solid angle $\kappa = 2\pi$. The effective adiabatic exponent γ can change depending on the state of the interplanetary medium. For a monatomic gas with sufficiently frequent collisions, $\gamma = 5/3$. When effects of ionization and radiation are included this exponent will be close to 1.3. For a rarefied plasma with extremely rare collisions it is meaningful to take $\gamma = 3$. Indeed, as we noted in Chapter 1 in considering the magnetohydrodynamic equations for a rarefied plasma, for $p_{\|} > p_{\perp}$ and with a weak magnetic field it follows from the hydrodynamic equations and the conditions on the shock wave that $\gamma = 3$. We mention that in the case $\pi = 2\pi$ we have the values $\alpha = 1.047$ $(\gamma = 5/3)$ and $\alpha \sim 0.2$ $(\gamma = 3)$. The magnitudes of energies estimated by (8.29) are indicated in Table 7. By reducing (8.16) to dimensionless form, it is possible to verify the similarity law [25] for flares on the basis of the $r_2(t)$ dependence. Here values were taken for E_0 which were computed with $\gamma = 5/3$ using the coordinate of measurement closest to the sun. In those cases where the measurements of the transit time were made by detectors at two points of space the dimensionless parameters $x_2 = r_2/r^*$ and $\tau = t/t^*$ were computed for both points. A check of the position on the graph of points corresponding to $r_2^{(2)}, t_2$ indicated good agreement with the similarity law (Table 7).

For the case of measurement at two points of space the average propagation speeds D_{av} were also determined. The theoretical formula for the speed of the shock wave D without considering the wind has the form

$$\frac{D}{v_1} = \frac{dx_2}{dt} = y^{-1} = \frac{\sqrt{\alpha}}{\delta} x_2^{(\omega-1)/2}. \tag{8.30}$$

When drift by the solar wind v_1 is considered we have, according to the results of §4 of Chapter 3,

$$x_2 = \left(\frac{\delta}{\sqrt{\alpha}}\right)^{2/(\omega-1)} y^{2/(\omega-1)} \exp \frac{2A_{11}y}{\omega - 1}, \tag{8.31}$$

$$\tau = \tau_0 + \int_{x_{20}}^{x_2} y(x)\, dx, \qquad x_{20} = \alpha^{\delta/2} \tau_0^{\delta}. \tag{8.32}$$

Suppose that $\omega = 2$ and $\gamma = 5/3$; then $A_{11} = 0.563$. Flare energies were estimated on the basis of the $x_2(\tau)$ dependence following from (8.31) and (8.32). The corresponding values of E_0 were found to be lower than those computed on the basis of the theory of a strong explosion without considering the wind (cf. Table 7).

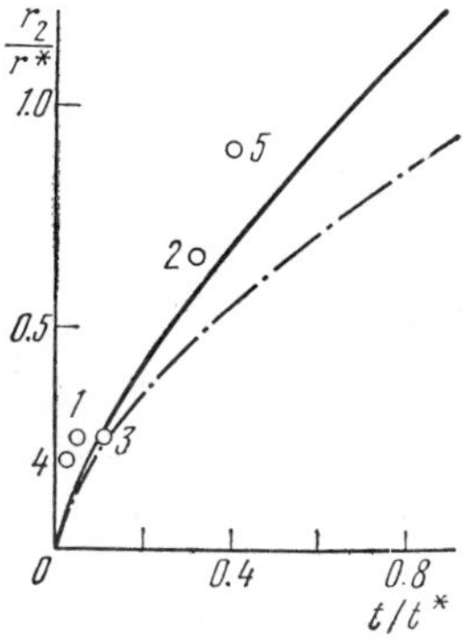

FIGURE 97. Dependence of the coordinate of the shock wave on time. The circles denote data obtained by processing results of measurements.

FIGURE 98. Graph of the dependence of D/v_1 on x_2 for a stong shock wave without considering the wind (a) and with the wind (b). The circles denote data obtained from processing results of measurments.

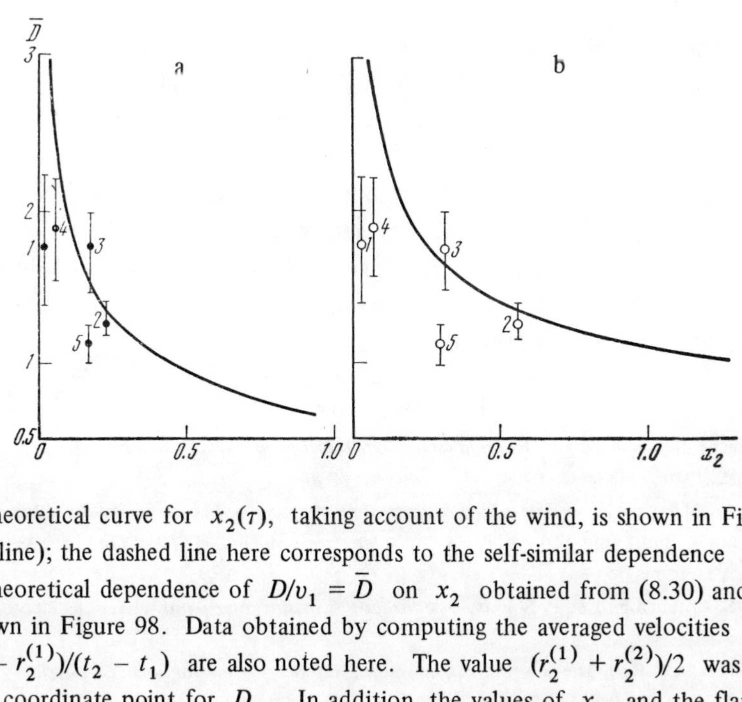

The theoretical curve for $x_2(\tau)$, taking account of the wind, is shown in Figure 97 (solid line); the dashed line here corresponds to the self-similar dependence $x_2(\tau)$. The theoretical dependence of $D/v_1 = \bar{D}$ on x_2 obtained from (8.30) and (8.31) is shown in Figure 98. Data obtained by computing the averaged velocities $D_{av} = (r_2^{(2)} - r_2^{(1)})/(t_2 - t_1)$ are also noted here. The value $(r_2^{(1)} + r_2^{(2)})/2$ was taken as the coordinate point for D_{av}. In addition, the values of x_2 and the flare energy E_0 were determined on the basis of (8.31) and the values of D and v_1; these data were then used to find the corresponding values of τ ($\tau_{av} = t_{av}/t^*$) on the basis of the time $t_{av} = (t_1 + t_2)/2$. These data on the connection between x_2 and τ are noted by circles in Figure 97. The vertical lines in Figure 98 correspond to the expected errors based on results of determining D_{av} [30], [32]. The numbers of the points correspond to the flare numbers indicated in Table 7. The circles denote data obtained by processing results of measurements.

Another possible method of determining E_0 consists in measuring the ratio of densities ρ_2/ρ_1, computing y on the basis of them from the conditions on the shock wave, and determining E_0 from the $x_2(y)$ dependence according to (8.31). Naturally, we could also consider the counterpressure.

The results presented indicate satisfactory agreement between the observed data and the theoretical values.

We mention also that estimates of the effective γ from the conditions on the shock waves, the values of ρ_2/ρ_1 given in [32], and the quantities D_{av} indicate [34] that the values of γ lie between 2 and 3, while the smaller n_1 and the greater D_{av}, the closer γ is to the value 3.

The comparison presented here of data from the theory of explosion and from observations enable one to judge the possibility of applying this theory for a qualitative and rough quantititive analysis of the process of propagation of disturbances from strong solar flares.

This chapter contains results the author obtained during the years 1961–1969. Preliminary results on the application of the theory of point explosions to the phenomenon of solar flares were communicated by the auther in his lectures to students of the Mechanics-Mathematics Faculty of Moscow State University (1961/62 school year) and in a report at the Conference on Magnetohydrodynamics (Riga, 1963) [23]. The main results of §§1–5 were reported by the author at the Third All-Union Congress on Theoretical and Applied Mechanics (Moscow, 1968) [24] and were published in the author's papers [25], [26], [38]. The results of §6 of this chapter were obtained jointly with Ju. M. Nikolaev [37].(1)

BIBLIOGRAPHY

1. Eugene Newman Parker, *Interplanatary dynamical processes*, Interscience, New York, 1963; Russian transl., "Mir", Moscow, 1966.

2. D. Ja. Martynov, *A course of general astronomy*, "Nauka", Moscow, 1966. (Russian)

3. Henry J. Smith and Elske v. P. Smith, *Solar flares*, Macmillan, New York, 1963; Russian transl., "Mir", Moscow, 1966.

4. D. S. Colburn and C. P. Sonett, *Discontinuities in the solar wind*, Space Sci. Rev. 5 (1966), 439–506.

5. A. Maxwell, R. J. Defouw and P. Cummings, *Radio evidence for solar corpuscular emission*, Planetary and Space Sci. 12 (1964), 435–449.

6. Y. C. Whang, C. K. Liu and C. C. Chang, *A viscous model of the solar wind*, Astrophys. J. 145 (1966), 255–269.

7. G. S. Bisnovatyĭ-Kogan, *Flow of an ideal gas in a spherically symmetric gravitational field with allowance for radiant heat conduction and radiation pressure*, Prikl. Mat. Meh. 31 (1967), 762–769 = J. Appl. Math. Mech. 31 (1967), 776–783.

(1) *Added in proof.* A number of our results are contained in the recently published work [39].

8. V. V. Vitkevič and V. I. Vlasov, *Radio astronomical investigations of motions and dimensions of the small-scale irregularities in the interplanetary plasma*, Dokl. Akad. Nauk SSSR 181 (1968), 572–574 = Soviet Phys. Dokl. 13 (1968/69), 624–626.

9. G. F. Chew, M. L. Goldberger and F. E. Low, *The Bolzmann equation and the one-fluid hydromagnetic equations in the absence of particle collisions*, Proc. Roy. Soc. London Ser. A 236 (1956), 112–118. MR 18, 92.

10. R. K. Jaggi, *Wave motion in a plasma with anisotropic pressure*, Phys. Fluids 5 (1962), 949–962.

11. L. I. Sedov, *Similarity and dimensional methods in mechanics*, 4th ed., GITTL, Moscow, 1957; English transl., Academic Press, New York, 1959. MR 19, 898; 21 #6840.

12. V. P. Korobeĭnikov, N. S. Mel'nikova and E. V. Rjazanov, *Theory of point explosions*, Fizmatgiz, Moscow, 1961. (Russian) MR 29 #5494.

13. H. Mirels and J. F. Mullen, *Aerodynamic blast simulation in hypersonic tunnels*, AIAA J. 3 (1965), 2103–2108.

14. M. A. Lidov, *Automodel motion of a gas with spherical symmetry in the field of a gravitating center*, Astronom. Ž. 34 (1957), 603–608 = Soviet Astronomy AJ 1 (1957/58), 588–593. MR 19, 1143.

15. R. Z. Sagdeev, *Collective processes and shock waves in a rarefied plasma*, Voprosy Teorii Plazmy, vyp. 4, Atomizdat, Moscow, 1964, pp. 20–80. (Russian)

16. Ju. A. Berezin, R. H. Kurtmullaev and Ju. E. Nesterihin, *Collisionless shock waves in a rarefied plasma*, Fiz. Gorenija i Vzryva 2 (1966), no. 1, 3–28 = Combustion, Explosion and Shock Waves 2 (1966), no. 1, 1–14.

17. J. T. Gosling et el., *Measurements of the interplanetary solar wind during the large geomagnetic storm of April* 17–18, 1965, J. Geophys. Res. 72 (1967), 1813–1821.

18. ———, *Satellite observations of interplanetary shock waves*, J. Geophys. Res. 73 (1968), 43–50.

19. T. Gold, in *Discussion on shock waves and rarefied gas dynamics*, Gas Dynamics of Cosmic Clouds (Sympos, Cambridge, 1953), I.A.U. Sympos. Ser., no. 2, North-Holland, Amsterdam; Interscience, New York, 1955, pp. 103–104.

20. Paul J. Coleman, Jr., *Variations in the interplanetary magnetic field*: *Mariner* 2. 1: *Observed properties*, J. Geophys. Res. 71 (1966), 5509–5531.

21. A. G. Kulikovskiĭ and G. A. Ljubimov, *Magnetohydrodynamics*, Fizmatgiz, Moscow, 1962; English transl., Adiwes Internat. Ser. Phys., Addison-Wesley, Reading, Mass., 1965.

22. V. P. Korobeĭnikov, *On the interaction of shock waves in an ideally conducting gas with weak magnetic fields*, PMTF Ž. Prikl. Meh. i Tehn. Fiz. 1964, no. 4, 113–114. (Russian)

23. V. P. Korobeĭnikov and E. V. Rjazanov, *On the propagation of magnetohydrodynamic shock waves arising during explosions*, Voprosy Magnitnoĭ Gidrodynamiki. IV, Izdat. Akad. Nauk Latv. SSR, Riga, 1964, pp. 33–41. (Russian)

24. V. P. Korobeĭnikov, *Application of the theory of explosion to questions of shock waves during solar flares*, III All-Union Congr. Theoret. Appl. Mech., Abstracts of Reports, "Nauka", Moscow, 1968, p. 171. (Russian)

25. ———, *Application of dimensional analysis to problems of interplanetary gas motion associated with solar flares*, Dokl. Akad. Nauk SSSR 185 (1969), 1254–1257 = Soviet Phys. Dokl. 14 (1969/70), 308–311.

26. ———, *On the gas flow due to solar flares*, Solar Phys. 7 (1969), 463–469.

27. Robert J. Mackin, Jr., and Marcia Neugebauer (editors), *The solar wind* (Proc. Conf., Pasadena, Calif., 1964), Pergamon Press, New York, 1966; Russian transl,, "Mir", Moscow, 1968.

28. K. I. Gringauz, V. V. Bezrukih and L. S. Musatov, *Solar wind observations with the Venus 3 space probe*, Kosmičeskie Issled. 5 (1967), 251–258 = Cosmic Res. 5 (1967), 216–222.

29. S. N. Vernov et al., *Investigation of cosmic rays during flights of automatic stations*, Proc. Fifth All-Union Winter School on Cosmic Physics, Apatity, 1968. (Russian)

30. G. P. Ljubimov, *Retardation of shock waves from solar flares in cosmic space*, Astronom. Cirkuljar, no. 488 (1968) (Russian)

31. S. J. Bame et al., *Solar wind and magnetosheath observations during the January 13–14, 1967, geomagnetic storm*, J. Geophys. Res. 73 (1968), 5761–5768.

32. Murray Dryer and Donald L. Jones, *Energy deposition in the solar wind by flare-generated waves*, J. Geophys. Res. 73 (1968), 4875–4881.

33. V. G. Gorbackiĭ, *Cosmic explosions*, "Nauka", Moscow, 1967. (Russian)

34. G. G. Černyĭ, *Gas flow with large hypersonic velocity*, Fizmatgiz, Moscow, 1959; English transl., *Introduction to hypersonic flow*, Academic Press, New York, 1961.

35. S. S. Grigorjan, G. V. Marčenko and Ju. L. Jakimov, *On nonstationary gas motions in shock tubes of variable cross section*, PMTF Ž. Prikl. Meh. i Teh. Fiz. 1961, no. 4, 109–113. (Russian).

36. Dallas D. Laumbach and Ronald F. Probstein, *A point explosion in a cold exponential atmosphere*, J. Fluid Mech. 35 (1969), 53–75.

37. V. P. Korobeĭnikov and Ju. M. Nikolaev, *Propagation of solar wind disturbances caused by chromospheric flares*, Kosmičeskie Issled. 7 (1969), 891–894 = Cosmic Res. 7 (1969), 798–801.

38. V. P. Korobeĭnikov, *On the motion of interplanetary gas caused by a solar flare*, Problems of Appl. Math. and Mech. (Dedicated to Academician Anatoli Alekseevič Dorodnicyn on his Sixtieth Birthday), "Nauka", Moscow, 1971, pp. 211–221. (Russian)

39. V. P. Korobeĭnikov and Ju. M. Nikolaev, *Shock waves and magnetic field configuration in interplanetary space*, Cosmic Electrodynamics 3 (1972), 3–24; English transl., ibid, 25–44.

APPENDIX

1. *Generalized Cotes coefficients.* We shall consider the numerical values of the generalized Cotes coefficients A_{lk} introduced in §3 of Chapter 4.

A. Suppose that we use through interpolation by polynomials of order n. Then by (4.49) we obtain the following values of the A_{lk}:

$A_{00} = A_{01} = ½$ for $n = 1$;

$A_{00} = A_{02} = 1/6, A_{10} = -1/24, A_{11} = 1/3$ and $A_{12} = 5/24$ for $n = 2$.

For $n = 3$ and 4 and $k = 0, 1, 2, 3, 4$ the values of A_{lk} are indicated in Table 8.

TABLE 8

Cotes Coefficients

l	$n=3;\ 72\,A_{lk}$				$n=4;\ 2880\,A_{lk}$				
	0	1	2	3	0	1	2	3	4
0	9	27	27	9	224	1024	384	1024	224
1	0	8	32	8	−27	378	648	918	243
2	1	−5	19	9	− 8	32	192	992	232
3	—	—	—	—	−19	106	−264	646	251

B. For piecewise interpolation by polynomials of degree less than n we shall consider cases of piecewise linear interpolation for any n and the case of interpolation by two matching polynomials of fourth order for $n = 9$.

For the case of piecewise linear interpolation evaluation of the integrals (4.48) gives the values $A_{lk} = 0$ for $l > k, A_{ll} = ½n = A_{ln}$ and $A_{lk} = 1/n\ (k \neq l)$. We remark that in the case of piecewise linear approximation it is expedient to resolve the approximating system in such a manner that it does not contain the determinant Δ and its minors. This is always possible (the corresponding system of approximating equations is often called the system of the method of lines), as follows from the fact that the matrix corresponding to Δ is triangular.

In interpolation with two matching polynomials of fourth order, direct evaluations of the integrals J_{sl} for k from 1 to 8 give the values shown in Table 9. Here the numerical factor $\delta^* = 2.2880$.

TABLE 9

Reduced Cotes Coefficients

A_{lk}	0	1	2	3	4	5	6	7	8
$90\,A_{0k}$	7/2	16	6	16	7	16	6	16	3,5
$\delta^* A_{1k}$	—27	378	648	918	467	1024	384	1024	224
$\delta^* A_{2k}$	— 8	32	192	992	456	1024	384	1024	224
$\delta^* A_{3k}$	—19	106	—264	646	475	1024	384	1024	224
$\sigma^* A_{4k}$	0	0	0	0	224	1024	384	1024	224
$\sigma^* A_{5k}$	0	0	0	0	—27	378	648	918	243
$\delta^* A_{6k}$	0	0	0	0	— 8	32	192	992	232
$\delta^* A_{7k}$	0	0	0	0	—19	106	—264	646	251

Values of pressures and

q	τ	R^*						
		0.1490		0.1929		0.2382		0.2856
		P	U	P	U	P	U	P
0.10	0.0209	11.500	2.250					
0.15	0.0340	3.488	0.2096	7.611	1.720			
0.20	0.0500	2.160	0.0445	2.919	0.2689	5.667	1.333	
0.25	0.0690	1.613	0.0118	1.849	0.0716	2.573	0.2384	4.500
0.30	0.0905	1.314	0.0038	1.419	0.0241	1.699	0.0802	2.364
0.35	0.1167	1.142	0.0015	1.188	0.0090	1.304	0.0306	1.585
0.40	0.1483	1.025	0.0006	1.047	0.0033	1.096	0.0107	1.234
0.45	0.1864	0.9342	0.0002	0.9538	0.0012	0.9773	0.0036	1.046
0.50	0.2343	0.8792	0	0.8887	0.0003	0.9063	0.0009	0.9405
0.55	0.2964	0.8518	0	0.8520	0	0.8635	0.0001	0.8849
0.60	0.3773	0.8492	0	0.8441	0	0.8474	0	0.8596
0.65	0.4856	0.8706	0	0.8669	0	0.8656	0.0001	0.8674
0.70	0.6377	0.9067	0	0.9076	0	0.9077	0	0.9067
0.75	0.8622	0.9441	0	0.9439	0	0.9434	0	0.9426
0.80	1.211	0.9725	0	0.9718	0	0.9709	0	0.9700
0.85	1.863	0.9895	0	0.9909	0	0.9926	0	0.9944
0.90	3.167	0.9963	0	0.9961	0	0.9958	0	0.9954
0.95	6.400	0.9991	0	0.9995	0	1.000	0	1.001

q	τ	R^*					
		0.6008		0.7020		0.8282	
		P	U	P	U	P	U
0.50	0.2343	2.167	0.2976				
0.55	0.2964	1.468	0.0742	1.9545	0.2250		
0.60	0.3773	1.125	0.0125	1.349	0.0470	1.778	0.1667
0.65	0.4856	0.9569	0.0003	1.065	0.0046	1.247	0.0271
0.70	0.6377	0.9059	0.0005	0.9298	0.0001	0.9933	0.0006
0.75	0.8622	0.9262	0.0005	0.9206	0.0007	0.9214	0.0007
0.80	1.211	0.9646	0.0001	0.9620	0.0001	0.9561	0.0003
0.85	1.863	1.001	0	0.9995	0	0.9945	0
0.90	3.167	0.9926	0	0.9918	0	0.9913	0
0.95	6.400	1.007	0	1.010	0	1.014	0

2. *Table of pressure and velocity head at fixed points of space.* The values of $P = p/p_\infty$ and the function $U = gf^2$ proportional to the velocity head were found by the method considered in §3 of Chapter 4 for $\nu = 2$ and $\gamma = 1.4$. It should be mentioned that a detailed analysis of the cylindrical case with $\gamma = 1.4$ and $\gamma = 5/3$ was recently completed by N. A. Arhangel'skiĭ (see Chapter 4, reference [61]).

TABLE 10

velocity head

	0.3342		0.3885		0.4495		0.5186	
U	*P*	*U*	*P*	*U*	*P*	*U*	*P*	*U*
1.042								
0.2787	3.722	0.8167						
0.0880	2.129	0.2361	3.167	0.6402				
0.0301	1.470	0.0722	1.935	0.1928	2.750	0.5000		
0.0096	1.169	0.0233	1.382	0.0570	1.772	0.1520	2.426	0.3878
0.0024	1.001	0.0059	1.110	0.0153	1.294	0.0405	1.612	0.1103
0.0003	0.9157	0.0008	0.9640	0.0024	1.052	0.0079	1.294	0.0246
0	0.8772	0	0.8997	0.0001	0.9335	0.0005	0.9988	0.0027
0.0002	0.8724	0.0003	0.8807	0.0002	0.8929	0.0001	0.9134	0
0.0001	0.9045	0.0002	0.9014	0.0004	0.8986	0.0005	0.8986	0.0006
0	0.9413	0.0001	0.9392	0.0001	0.9361	0.0002	0.9318	0.0004
0	0.9690	0	0.9680	0	0.9670	0.0001	0.9660	0.0001
0	0.9963	0	0.9982	0	1.000	0	1.001	0
0	0.9950	0	0.9945	0	0.9940	0	0.9933	0
0	1.003	0	1.002	0	1.004	0	1.005	0

0.9901		1.209		1.521		1.989	
P	*U*	*P*	*U*	*P*	*U*	*P*	*U*
1.623	0.1200						
1.153	0.0125	1.500	0.0833				
0.9500	0.0001	1.068	0.0035	0.389	0.0548		
0.9447	0.0005	0.9365	0.0008	0.9994	0.0002	1.292	0.0333
0.9865	0	0.9790	0.0001	0.9725	0.0002	0.9506	0.0007
0.9913	0	0.9925	0	0.9949	0	0.9951	0
1.029	0	1.0245	0	1.020	0	1.035	0